Condensed Isogeometric Analysis for Plate and Shell Structures

Condensed Isogeometric Analysis for Plate and Shell Structures

Buntara S. Gan

CRC Press
Taylor & Francis Group
Boca Raton London New York

CRC Press is an imprint of the
Taylor & Francis Group, an informa business

CRC Press
Taylor & Francis Group
6000 Broken Sound Parkway NW, Suite 300
Boca Raton, FL 33487-2742

Printed on acid-free paper

International Standard Book Number-13: 978-0-367-02348-5 (Hardback)

Library of Congress Cataloging-in-Publication Data

Names: Gan, Buntara S. (Buntara Sthenly), author.
Title: Condensed isogeometric analysis for plates and shell structures / authored by Buntara S. Gan.
Description: First edition. | Boca Raton, FL : CRC Press/Taylor & Francis Group, 2019. | Includes bibliographical references and index.
Identifiers: LCCN 2019028538 (print) | LCCN 2019028539 (ebook) | ISBN 9780367023485 (hardback ; acid-free paper) | ISBN 9780429399824 (ebook)
Subjects: LCSH: Elastic plates and shells—Data processing. | Isogeometric analysis. | MATLAB.
Classification: LCC QA935 .G325 2019 (print) | LCC QA935 (ebook) | DDC 624.1/776—dc23
LC record available at https://lccn.loc.gov/2019028538
LC ebook record available at https://lccn.loc.gov/2019028539

Visit the Taylor & Francis Web site at
http://www.taylorandfrancis.com

and the CRC Press Web site at
http://www.crcpress.com

To

Rie, Kenta, Suchan-Shichan, our parents, families, and friends.

Thank you for the love and encouragement over the years.

Contents

Preface

This book is dedicated to bridging the gap between classical plate and shell theories with the advanced numerical simulation using the nonuniform rational basis spline (NURBS) concept. If we observe carefully at the state of the art of the well-recognized finite element analysis (FEA) and computer-aided design (CAD), there are two different disciplines in engineering and science that do not "go along with" each other. T.J.R. Hughes and coworkers in 2005 introduced the isogeometric analysis to bridge the gap between the CAD and FEA approaches in which it gained significant thrusts to all research studies on this topic.

This book is intended for engineering and science students who are beginning a series of courses in structures and mechanics, and practitioners in companies who are starting to learn how to analyze the NURBS curves and surfaces. This book could also be a reference for researchers who are going to pursue or extend their research in plate and shell elements using NURBS curves.

This book was written as a textbook or workbook with tutorials on how to get the problems solved. All the plate and shell theories and NURBS concepts are accompanied by MATLAB® program lists, which are intended for easy understanding and learning how the formulas work. Just like computer science, we know that the modern editor software is now adopting the "WYSIWYG" system in which the displayed text can be edited in a form exactly resembling its appearance when printed on a paper.

The organization of the book is as follows: In Chapter 1, the representation of curves on a plane and surface in space is presented. It starts from the parametric modeling of polynomial curves and surfaces through several standard curves and surfaces, which are the foundation of the NURBS. All the methods described are accompanied by the MATLAB program and function lists. Some important concepts, such as the control point, control mesh, knot vector, weight, and rationalization, are discussed. In Chapter 2, the numerical integration technique called Gauss–Legendre quadrature is introduced by using examples of area calculation of a surface for clarity purpose. The Jacobian operator, gradient, and curvature of a surface are also explained by using examples with accompanying program lists. Chapter 3 reviews the classical theory of thin plate (classical plate theory [CPT]) and shell, how to get the governing equations, and how to formulate the plate and shell in the finite element context by using a variational approach. The applications of NURBS in the theory of thick plate (first-order shear deformation theory [FSDT]) and how to formulate the governing equation are described in Chapter 4. In Chapter 5, the finite element formulations for both CPT and FSDT are presented with accompanying program lists. Chapter 6 discusses the methods of implementation of isogeometric analysis (IGA) into both CPT and FSDT plate and shell elements. In Chapter 7, the "sandwich" condensation is introduced. The sandwich condensation condenses the increasing degrees of freedom in the midspan of the plate and shell element due to the adoption of the NURBS, to a four-node 12-degree-of-freedom standard rectangular plate and shell element. The program and function lists are provided. By condensing the NURBS-based plate and shell element to the standard plate and shell element, several merits are gained. First, when we deal with nonlinear plate and shell analyses where iterative calculations are necessary, the computing cost can be reduced significantly. Second, program and function lists can be combined easily to the existing finite element codes either the in-house or commercial customizable software. In Chapters 8 and 9, flat and free surface Mindlin thick plate examples are given to demonstrate the advantage of the present condensed IGA.

The writing of this book was a pleasant experience. I received benefits from my professional works, encouragement, and support from many colleagues as well as my students who have taught me how to explain to them complicated concepts in simple terms. Although it is not possible to name all of them, without their help and support, it would not have been possible for me to finish the writing of this book.

My sincere thanks are due to my respected former Emeritus Professor Fumio Nishino in the University of Tokyo, for his philosophies in teaching and research which have been useful in my professional life. Great appreciations are to Professor Mitsuharu Kurata in Nihon University, for constantly motivating me to find the "truth" in science. I thank Dr Nguyen Dinh Kien, my research colleague in Vietnam Academy of Science and Technology, for his fruitful ideas, encouragement, and support in my research works. I am indebted to Professor Han Ay Lie in Universitas Diponegoro for her continuous encouragement and inspiring ideas in doing my research works. I am also grateful to Professors Lee Jaehong, Lee Seunghye, and Kim Nam Il at Sejong University for their hospitality, collaboration, and many stimulating discussions concerning several topics during my sabbatical leave. I appreciate much for the support, help, and kind assistance from the publishing editor Mr Joseph Clements, editorial assistant Ms Lisa Wilford and project manager Ms Sofia Buono during the preparation, editing, proofing, handling, and production of this book.

The readers are welcomed to use the MATLAB program and function lists freely. However, the author does not guarantee that the program and function lists are without any error. The author is also not responsible for any damage or loss caused while using the program or function lists. The reader is fully responsible for verifying the correctness of the formulas and equations before using the program or function lists.

Most books are not free from errors, especially those with many mathematical equations and numbers. I would like to thank in advance the readers who are willing to draw attention to any typos and errors, using the e-mail address: buntara@arch.ce.nihon-u.ac.jp.

Buntara S. Gan
Koriyama, May 2019
Fukushima, Japan

Author

Buntara S. Gan is currently a professor in the architecture department, College of Engineering, Nihon University, Koriyama, Japan. He is registered as a professional engineer in Oregon State. He received his BEng degree from the department of civil engineering, Bandung Institute of Technology in 1988, West Java, Indonesia, and his PhD in structural engineering from the University of Tokyo in 1994. Previously, he worked and practiced in a general construction company in Tokyo, Japan.

His research interests include structural engineering, structural dynamics, computational mechanics, structural optimization, soil–structure interaction, numerical analysis, solid mechanics, engineering, and applied and computational mathematics.

1

Representing a Surface Using Nonuniform Rationalized B-Spline

1.1 Parametric Modeling of a Surface

In geometric modeling, parametric presentation of polynomial functions is one of many methods to create surfaces in a three-dimensional (3D) space. The parametric presentation of the polynomial is not very common in a real application; however, the understanding of parametric presentation is necessary to deal with the isogeometric approach that we are going to learn.

In the parametric form, the coordinate of a point on a surface is represented explicitly by a common parameter that controls the curvatures on the surface. Therefore, by varying the value of the parameter, we can have a family of surfaces within the same order of the polynomial. A surface is known as a bi-parametric geometry. Function of a surface having two of second-order parametric polynomial of two parameters (ξ and η) can be written as

$$\text{function} = \left(\text{coef}_1 + \text{coef}_2\xi + \text{coef}_3\xi^2\right) \times \left(\text{coef}_4 + \text{coef}_5\eta + \text{coef}_6\eta^2\right) \tag{1.1}$$

Defining the coordinates x, y, and z in 3D space by expanding Equation (1.1), the second order of a bi-parametric polynomial can be written as

$$\begin{aligned} x_{i,j} &= a_1 + a_2\xi_i + a_3\eta_j + a_4\xi_i\eta_j + a_5\xi_i^2 + a_6\eta_j^2 + a_7\xi_i^2\eta_j + a_8\xi_i\eta_j^2 + a_9\xi_i^2\eta_j^2 \\ y_{i,j} &= b_1 + b_2\xi_i + b_3\eta_j + b_4\xi_i\eta_j + b_5\xi_i^2 + b_6\eta_j^2 + b_7\xi_i^2\eta_j + b_8\xi_i\eta_j^2 + b_9\xi_i^2\eta_j^2 \\ z_{i,j} &= c_1 + c_2\xi_i + c_3\eta_j + c_4\xi_i\eta_j + c_5\xi_i^2 + c_6\eta_j^2 + c_7\xi_i^2\eta_j + c_8\xi_i\eta_j^2 + c_9\xi_i^2\eta_j^2 \end{aligned} \tag{1.2}$$

where

$$i = 1,2,3$$

$$j = 1,2,3$$

Now we want to represent a surface as shown in Figure 1.1 as a parametric polynomial form of Equation (1.2). First, we assume that the second order of the polynomial function represents the surface. Second, we need to define the common parameters, denoted by $\xi = \{ \xi_1 \quad \xi_2 \quad \xi_3 \}$ and $\eta = \{ \eta_1 \quad \eta_2 \quad \eta_3 \}$, to represent the parametric position of those points. We can see here by defining different values of positioning those points, we can have different surfaces of polynomial functions. Now, we need to obtain the values of ξ and η parameters to construct Equation (1.2). To do this, we need to take the coordinates (**x**,**y**,**z**) of arbitrary three points ($P_1, P_2, \ldots, P_9$) on the surface. The corner points of the surface are assigned to the P_1, P_3, P_7, and P_9 points. By substituting the coordinates of the nine points given, Equation (1.2) results in

$$
\begin{aligned}
x_1 &= a_1 + a_2\xi_1 + a_3\eta_1 + a_4\xi_1\eta_1 + a_5\xi_1^2 + a_6\eta_1^2 + a_7\xi_1^2\eta_1 + a_8\xi_1\eta_1^2 + a_9\xi_1^2\eta_1^2 \\
&\cdot \\
x_9 &= a_1 + a_2\xi_3 + a_3\eta_3 + a_4\xi_3\eta_3 + a_5\xi_3^2 + a_6\eta_3^2 + a_7\xi_3^2\eta_3 + a_8\xi_3\eta_3^2 + a_9\xi_3^2\eta_3^2 \\
y_1 &= b_1 + b_2\xi_1 + b_3\eta_1 + b_4\xi_1\eta_1 + b_5\xi_1^2 + b_6\eta_1^2 + b_7\xi_1^2\eta_1 + b_8\xi_1\eta_1^2 + b_9\xi_1^2\eta_1^2 \\
&\cdot \\
y_9 &= b_1 + b_2\xi_3 + b_3\eta_3 + b_4\xi_3\eta_3 + b_5\xi_3^2 + b_6\eta_3^2 + b_7\xi_3^2\eta_3 + b_8\xi_3\eta_3^2 + b_9\xi_3^2\eta_3^2 \\
z_1 &= c_1 + c_2\xi_1 + c_3\eta_1 + c_4\xi_1\eta_1 + c_5\xi_1^2 + c_6\eta_1^2 + c_7\xi_1^2\eta_1 + c_8\xi_1\eta_1^2 + c_9\xi_1^2\eta_1^2 \\
&\cdot \\
z_9 &= c_1 + c_2\xi_3 + c_3\eta_3 + c_4\xi_3\eta_3 + c_5\xi_3^2 + c_6\eta_3^2 + c_7\xi_3^2\eta_3 + c_8\xi_3\eta_3^2 + c_9\xi_3^2\eta_3^2
\end{aligned}
\tag{1.3}
$$

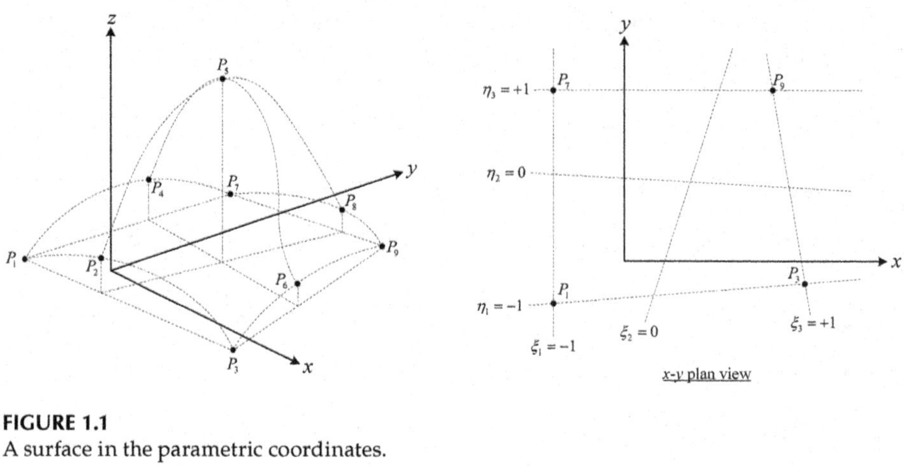

FIGURE 1.1
A surface in the parametric coordinates.

which can be written in the form of vectors and matrices as

$$\mathbf{x} = \mathbf{T}\alpha \tag{1.4}$$

where

$$\mathbf{x} = \begin{bmatrix} x_1 & . & x_9 & y_1 & . & y_9 & z_1 & . & z_9 \end{bmatrix}^T$$

$$\alpha = \begin{bmatrix} \mathbf{a} & \mathbf{b} & \mathbf{c} \end{bmatrix}^T$$

$$\mathbf{a} = \begin{bmatrix} a_1 & a_2 & a_3 & a_4 & a_5 & a_6 & a_7 & a_8 & a_9 \end{bmatrix}$$

$$\mathbf{b} = \begin{bmatrix} b_1 & b_2 & b_3 & b_4 & b_5 & b_6 & b_7 & b_8 & b_9 \end{bmatrix}$$

$$\mathbf{c} = \begin{bmatrix} c_1 & c_2 & c_3 & c_4 & c_5 & c_6 & c_7 & c_8 & c_9 \end{bmatrix}$$

$$\mathbf{T} = \begin{bmatrix} \mathbf{t} & 0 & 0 \\ 0 & \mathbf{t} & 0 \\ 0 & 0 & \mathbf{t} \end{bmatrix}$$

$$\mathbf{t} = \begin{bmatrix} 1 & \xi_1 & \eta_1 & \xi_1\eta_1 & \xi_1^2 & \eta_1^2 & \xi_1^2\eta_1 & \eta_1^2\xi_1 & \xi_1^2\eta_1^2 \\ 1 & \xi_2 & \eta_1 & \xi_2\eta_1 & \xi_2^2 & \eta_1^2 & \xi_2^2\eta_1 & \eta_1^2\xi_2 & \xi_2^2\eta_1^2 \\ 1 & \xi_3 & \eta_1 & \xi_3\eta_1 & \xi_3^2 & \eta_1^2 & \xi_3^2\eta_1 & \eta_2^2\xi_3 & \xi_3^2\eta_1^2 \\ 1 & \xi_1 & \eta_2 & \xi_1\eta_2 & \xi_1^2 & \eta_2^2 & \xi_1^2\eta_2 & \eta_2^2\xi_1 & \xi_1^2\eta_2^2 \\ 1 & \xi_2 & \eta_2 & \xi_2\eta_2 & \xi_2^2 & \eta_2^2 & \xi_2^2\eta_2 & \eta_2^2\xi_2 & \xi_2^2\eta_2^2 \\ 1 & \xi_3 & \eta_2 & \xi_2\eta_2 & \xi_3^2 & \eta_2^2 & \xi_3^2\eta_2 & \eta_2^2\xi_3 & \xi_3^2\eta_2^2 \\ 1 & \xi_1 & \eta_3 & \xi_1\eta_3 & \xi_1^2 & \eta_3^2 & \xi_1^2\eta_3 & \eta_3^2\xi_1 & \xi_1^2\eta_3^2 \\ 1 & \xi_2 & \eta_3 & \xi_2\eta_3 & \xi_2^2 & \eta_3^2 & \xi_2^2\eta_3 & \eta_3^2\xi_2 & \xi_2^2\eta_3^2 \\ 1 & \xi_3 & \eta_3 & \xi_3\eta_3 & \xi_3^2 & \eta_3^2 & \xi_3^2\eta_3 & \eta_3^2\xi_3 & \xi_3^2\eta_3^2 \end{bmatrix}$$

The vector $\boldsymbol{\alpha}$ can be obtained by multiplying the inverse of the matrix **T** with the vector **X**. Now let us consider, for example, the arbitrary coordinates of nine points to create a surface.

The common parametric values are initially assigned as $\xi_1 = \eta_1 = -1$, $\xi_2 = \eta_2 = 0$, and $\xi_3 = \eta_3 = -1$. By substituting the common parametric values into the matrix **t**, then solving for vector α as shown in Table 1.1 together with the coordinate points. By using the value of α, we can draw a surface as shown in Figure 1.1 by varying the parameters ξ and η with a small interval.

TABLE 1.1

The Coordinates of Nine Points on the Surface

Points	x	y	z	a	b	c
P_1	–5	–1	0	3	2	10
P_2	0	–1	3	7	0	0.5
P_3	9	–1	0	0	2.5	–0.5
P_4	–5	2	4	0	0	0
P_5	3	2	10	–1	0	–5.5
P_6	9	2	–5	–3	–0.5	–7.5
P_7	–5	4	0	0	0	0.5
P_8	0	4	2	0	0	–0.5
P_9	9	4	0	3	0	3

Note: In this book, the intervals of parameters ξ and η are decided within the range of $-1 \le \xi \le +1$ and $-1 \le \eta \le +1$, respectively, so that it is consistent with the range of Gauss quadrature integration scheme that will be discussed in the following subsequent chapters. The Gauss quadrature integration scheme will be used for integrating all the matrix and vector quantities when we deal with finite element formulations for a beam developed by the isogeometric approach.

MATLAB® code PolynomialSurface.m solves the values of vector $\boldsymbol{\alpha}$, constructs the parametric polynomial equation in Equation (1.3), and draws the polynomial curve as shown in Figure 1.2. Now, if we continue drawing all the curves by varying the value of parameter ξ, in the range of –1 to +1, with 0.1 intervals, then we can get a second-order polynomial family that passes through those nine points given.

1.1.1 PolynomialSurface Program List

```
% PolynomialSurface.m
% Parametric creation of a second-order polynomial curve
% Nodal coordinates: xi, yi and zi
xi=[-5  0  9 -5  3  9 -5  0  9];
yi=[-1 -1 -1  2  2  2  4  4  4];
zi=[ 0  3  0  4 10  5  0  2  0];
ik=size(xi,2);
xyzi=transpose([xi yi zi]);
% Parameter s(xi) and t(eta)
s=[-1 0 1];
t=[-1 0 1];
% Transformation matrix [tran]
ti = [ ...
1 s(1) t(1) s(1)*t(1) s(1)^2 t(1)^2 s(1)^2*t(1) s(1)*t(1)^2
s(1)^2*t(1)^2;
1 s(2) t(1) s(2)*t(1) s(2)^2 t(1)^2 s(2)^2*t(1) s(2)*t(1)^2
s(2)^2*t(1)^2;
1 s(3) t(1) s(3)*t(1) s(3)^2 t(1)^2 s(3)^2*t(1) s(3)*t(1)^2
s(3)^2*t(1)^2;
```

```
1 s(1) t(2) s(1)*t(2) s(1)^2 t(2)^2 s(1)^2*t(2) s(1)*t(2)^2
s(1)^2*t(2)^2;
1 s(2) t(2) s(2)*t(2) s(2)^2 t(2)^2 s(2)^2*t(2) s(2)*t(2)^2
s(2)^2*t(2)^2;
1 s(3) t(2) s(3)*t(2) s(3)^2 t(2)^2 s(3)^2*t(2) s(3)*t(2)^2
s(3)^2*t(2)^2;
1 s(1) t(3) s(1)*t(3) s(1)^2 t(3)^2 s(1)^2*t(3) s(1)*t(3)^2
s(1)^2*t(3)^2;
1 s(2) t(3) s(2)*t(3) s(2)^2 t(3)^2 s(2)^2*t(3) s(2)*t(3)^2
s(2)^2*t(3)^2;
1 s(3) t(3) s(3)*t(3) s(3)^2 t(3)^2 s(3)^2*t(3) s(3)*t(3)^2
s(3)^2*t(3)^2];
tran=[ti         zeros(9,9) zeros(9,9);
      zeros(9,9) ti         zeros(9,9);
      zeros(9,9) zeros(9,9) ti];
% Coefficients a's, b's and c's calculation
abc=tran\xyzi;
a=abc(1:ik,1);
b=abc(ik+1:2*ik,1);
c=abc(2*ik+1:3*ik,1);
% Drawing graph
```

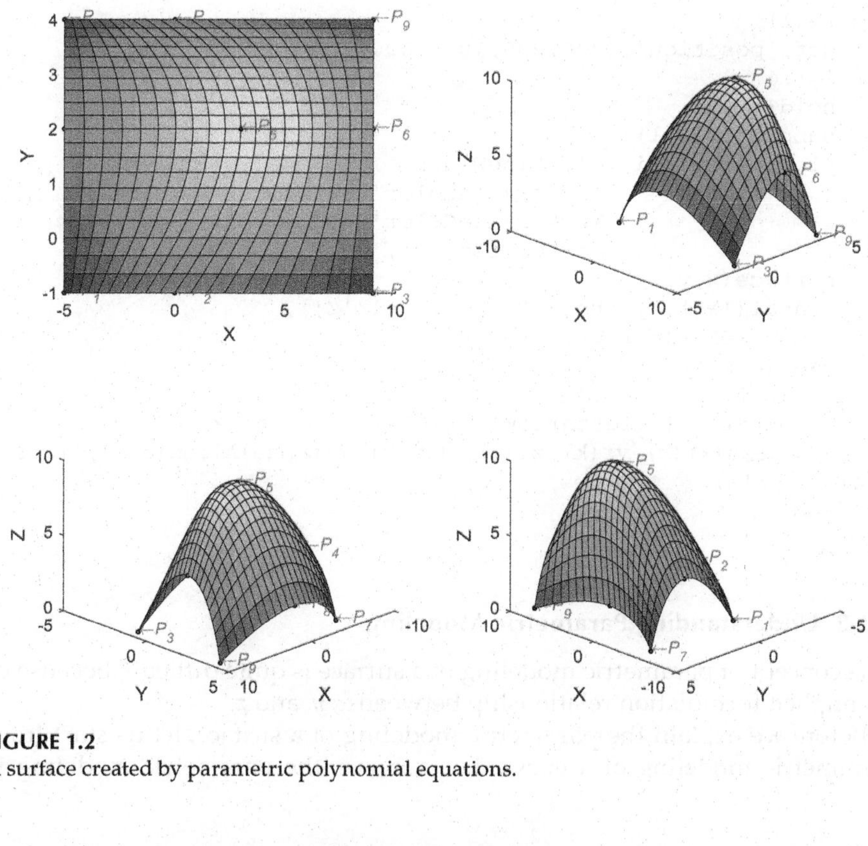

FIGURE 1.2
A surface created by parametric polynomial equations.

```
s=-1:0.1:1;
t=-1:0.1:1;
m=size(s,2);
n=size(t,2);
x=zeros(m,n);
y=zeros(m,n);
z=zeros(m,n);
for i=1:m
  for j=1:n
    x(i,j)=a(1)+a(2)*s(i)+a(3)*t(j)+a(4)*s(i)*t(j)+a(5)*
s(i)^2+ ...
        a(6)*t(j)^2+a(7)*s(i)^2*t(j)+a(8)*s(i)*t(j)^2+a(9)
*s(i)^2*t(j)^2;
    y(i,j)=b(1)+b(2)*s(i)+b(3)*t(j)+b(4)*s(i)*t(j)+b(5)*
s(i)^2+ ...
        b(6)*t(j)^2+b(7)*s(i)^2*t(j)+b(8)*s(i)*t(j)^2+b(9)
*s(i)^2*t(j)^2;
    z(i,j)=c(1)+c(2)*s(i)+c(3)*t(j)+c(4)*s(i)*t(j)+c(5)*
s(i)^2+ ...
        c(6)*t(j)^2+c(7)*s(i)^2*t(j)+c(8)*s(i)*t(j)^2+c(9)
*s(i)^2*t(j)^2;
  end
end
Figure(1)
set(gcf,'position',[200,200,1000,800])
for l=1:4
    hold;
    subplot(2,2,l);
    plot3(xi,yi,zi,'o','LineWidth',1,...
                       'MarkerEdgeColor','k',...
                       'MarkerFaceColor','w',...
                       'MarkerSize',2);
    surface(x,y,z)
    xlabel('X'); ylabel('Y'); zlabel('Z');
    azm = -45+90*(l-1);
    view(azm,30)
    for k=1:ik
        itxt = ['\leftarrow' '\itP_{' num2str(k) '}'];
        text(xi(k),yi(k),zi(k),itxt,'HorizontalAlignment','left',
'Color','r');
    end
end
```

1.1.2 Understanding Parametric Modeling

The concept of parametric modeling of a surface is quite difficult, because of its implicit formulation relationship between x, y, and z.

Before we explain the parametric modeling of a surface, let us start from parametric modeling of a curve. For a curve, the parametric modeling is quite easy to understand if we think a common parameter formulates both the abscissa x and the ordinate y in terms of ξ only. Suppose on the x–y plane, the third order of parametric polynomial takes the following form:

$$\begin{aligned} x_i &= a_1 + a_2\xi_i + a_3\xi_i^2 + a_4\xi_i^3 \\ y_i &= b_1 + b_2\xi_i + b_3\xi_i^2 + b_4\xi_i^3 \end{aligned} \tag{1.5}$$

where

$$i=1,2,3,4$$

The values of x and y are connected with one common parameter ξ_i at a time. Hence, we can have a relationship between x and y, after we assign a value to the parameter ξ_i. It is difficult to find a value of the parameter ξ_i that can give precisely a pair of x and y that we want if we do not have all the coefficients of functions in Equation (1.5). We will face difficulties if we have to transform Equation (1.5) into the "y as a function of x" form of equation. Especially for higher order polynomial functions, where the elimination of parameter ξ_i from both equations becomes a tedious work by hand, indeed. However, in the geometric computation using a computer, the parametric modeling suits well the representation of curves numerically.

Let us consider a third-order nonparametric polynomial as given below:

$$y = 0.18x^3 + 0.03x^2 - 1.18x + 0.97 \tag{1.6}$$

Now let us approximate the above function by using the parametric formulation described in the Section 1.1.1. First, by using Equation (1.6), let us take four arbitrary coordinates x and compute their corresponding coordinates y to get four arbitrary points as shown in Table 1.2. These four points are lying on the curve with two points at both ends and two points in between both ends of the curve. From the four values of coordinates x and y, we can obtain four simultaneous equations by using Equation (1.5). By solving the coefficient vector α in Equation (1.4), we obtain all the coefficients a's and b's which are tabulated in Table 1.2.

TABLE 1.2

The Coordinates of Four Arbitrary Points on a Plane

Points	x	y	A	b
P_1	–3	–0.08	–0.0625	0.6144
P_2	–1	2	2.9375	–3.8956
P_3	1	0	0.5625	3.4706
P_4	4	8.25	0.5625	8.0606

To show how the parameter ξ_i works, we can draw both graphs of $\xi - y$ and $\xi - x$ from the constructed x–y curve as shown in Figure 1.3.

Substituting the coefficient vector $\boldsymbol{\alpha}$ into Equation (1.5) and assuming the parameter $\xi = \begin{bmatrix} -1 & -\frac{1}{3} & +\frac{1}{3} & +1 \end{bmatrix}$, we can draw the function y with respect to x as shown in Figure 1.3. In Figure 1.4, Equation (1.6) together with its approximated parametric graph is shown.

As we can see in Figure 1.4, the graphs from Equations (1.5) and (1.6) do not coincide, only the locations of the four assumed points are coinciding. The question now is, why don't we get the same graphs? What happened if we guessed the parametric vector not as $\xi = \begin{bmatrix} -1 & -\frac{1}{3} & +\frac{1}{3} & +1 \end{bmatrix}$? Will we get the same answer? The answer is "Yes." In fact, we made a wrong guess in determining the parameter ξ! The two parametric values of ξ which were guessed at both extremities of the curve are correct. However, the second and third values which are the equal-distance points from both ends of the curve in the parametric modeling were not the correct values. Actually, each parameter ξ_i must be interpolated along the length of the path of the curve; unless the curve is a linear one, the relationship between parameter ξ and coordinate x or y is not linear. To find the correct values of ξ, an additional algorithm that can calculate the length of the curve to select a correct sampling point is the only choice to have the real curve that we wanted to construct. Integrating the length of the curve can give the exact position of the points.

Although the curve construction by using parametric polynomial modeling has some disadvantages to represent a curve in geometric modeling, it is still very useful in generating various forms of polynomial curves because the polynomial forms are simple and mathematically well understood. The parametric form of a curve is suitable for designing and representing

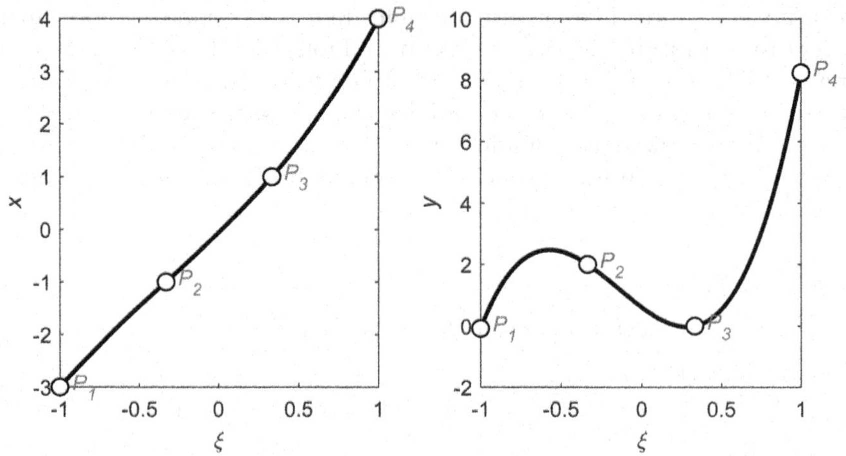

FIGURE 1.3
Parametric representation of a curve.

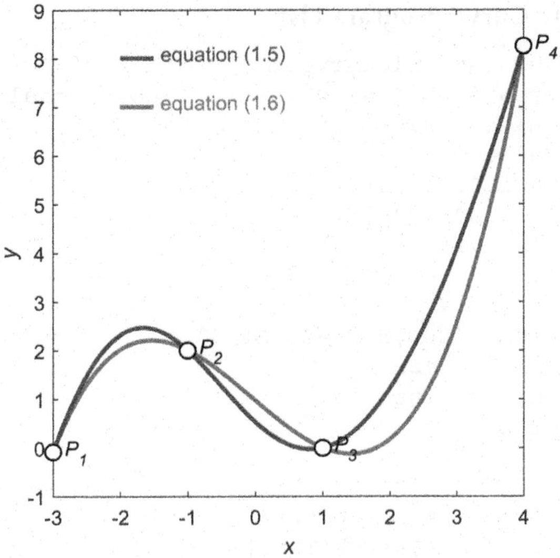

FIGURE 1.4
Difference between parametric and polynomial equations.

free geometric form by using a computer. In the remaining parts of this chapter, we will deal mostly with parametric modeling.

To be more familiar with the parametric concept, let us vary the value of the parameter ξ_2 within the limit of $\xi_1 \leq \xi_2 \leq \xi_3$ while keeping the value of parameter ξ_3 constant and vice versa. Figure 1.5 shows the results of varying the values of parameters ξ_2 and ξ_3 using ten intervals. Hence, an infinite number of graphs can be drawn using the concept of parametric representation.

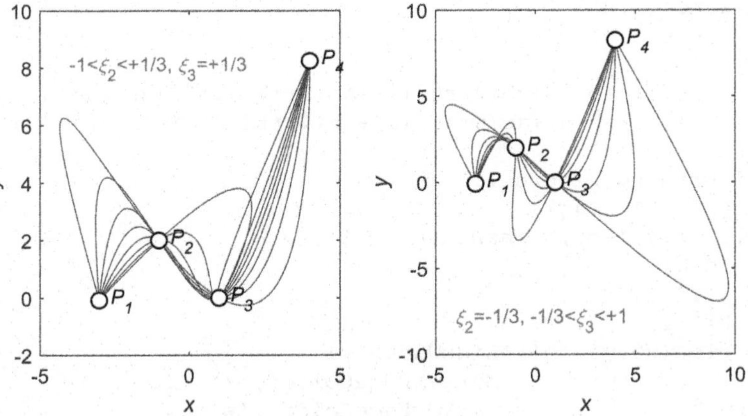

FIGURE 1.5
Variation of curves by changing the values of ξ_2 and ξ_3.

1.1.3 Parametric Curve Program List

```
% UnderstandingParametricCurve.m
% Parametric creation of a third-order polynomial curve
% Point coordinates: xi and yi
clear variables; clc;
xi=[-3.00 -1.00 1.00 4.00];
yi=[-0.08  2.00 0.00 8.25];
xyi=transpose([xi yi]);
ijn = 10;
Figure(1);
set(gcf,'position',[200,200,600,300]);
for jj = 1:2    % P2 and P3
  for ii =1:ijn-2  % interval of xi
    % Parameter t
    l2 = -1/3;
    l3 = +1/3;
    if jj==1, l2 = -1+(1/3+1)*ii/ijn; end
    if jj==2, l3 = -1/3+(1+1/3)*ii/ijn; end
    t=[-1 l2 l3 1];
    % Transformation matrix [tran]
    tran=[1 t(1) t(1)^2 t(1)^3 0 0 0 0;      % x1
          1 t(2) t(2)^2 t(2)^3 0 0 0 0;      % x2
          1 t(3) t(3)^2 t(3)^3 0 0 0 0;      % x3
          1 t(4) t(4)^2 t(4)^3 0 0 0 0;      % x4
          0 0 0 0 1 t(1) t(1)^2 t(1)^3;      % y1
          0 0 0 0 1 t(2) t(2)^2 t(2)^3;      % y2
          0 0 0 0 1 t(3) t(3)^2 t(3)^3;      % y3
          0 0 0 0 1 t(4) t(4)^2 t(4)^3];     % y4
    % Coefficients a's and b's calculation
    ab=tran\xyi;
    % Drawing curve
    ti=-1:0.001:1; n=size(ti,2);
    x=zeros(1,n);
    y=zeros(1,n);
    for i=1:n
       x(1,i)=ab(1)+ab(2)*ti(i)+ab(3)*ti(i)^2+ab(4)*ti(i)^3;
       y(1,i)=ab(5)+ab(6)*ti(i)+ab(7)*ti(i)^2+ab(8)*ti(i)^3;
    end % of i
    %% Drawing Figure 1.5
    subplot(1,2,jj)
    plot(x,y,'-b','LineWidth',0.5);
    hold all;
  end % of ii
  for k=1:4
     plot(xi,yi,'o','LineWidth',1,...
                    'MarkerEdgeColor','k',...
                    'MarkerFaceColor','w',...
                    'MarkerSize',8);
     hold all;
```

```
        itxt = ['  \itP_{' num2str(k) '}'];
        text(xi(k),yi(k),itxt,'HorizontalAlignment','left','Color'
,'k');
        hold all;
    end % of k
    xlabel('{\itx}'); ylabel('{\ity}');
    hold off;
    if jj==1
        itxt = ['-1<{\it\xi}_2<+1/3, {\it\xi}_3=+1/3'];
        text(-4,8,itxt,'HorizontalAlignment','left','Color','r');
    end
    if jj==2
        itxt = ['{\it\xi}_2=-1/3, -1/3<{\it\xi}_3<+1'];
        text(-4,-8,itxt,'HorizontalAlignment','left','Color','r');
    end
    hold all;
end % of jj
```

Figure 1.6 shows a few variations of surfaces by changing the values of the parameters ξ_2 and η_2 on the x–y plane. The height of the surface in the z-axis can be seen from the contouring level in colors.

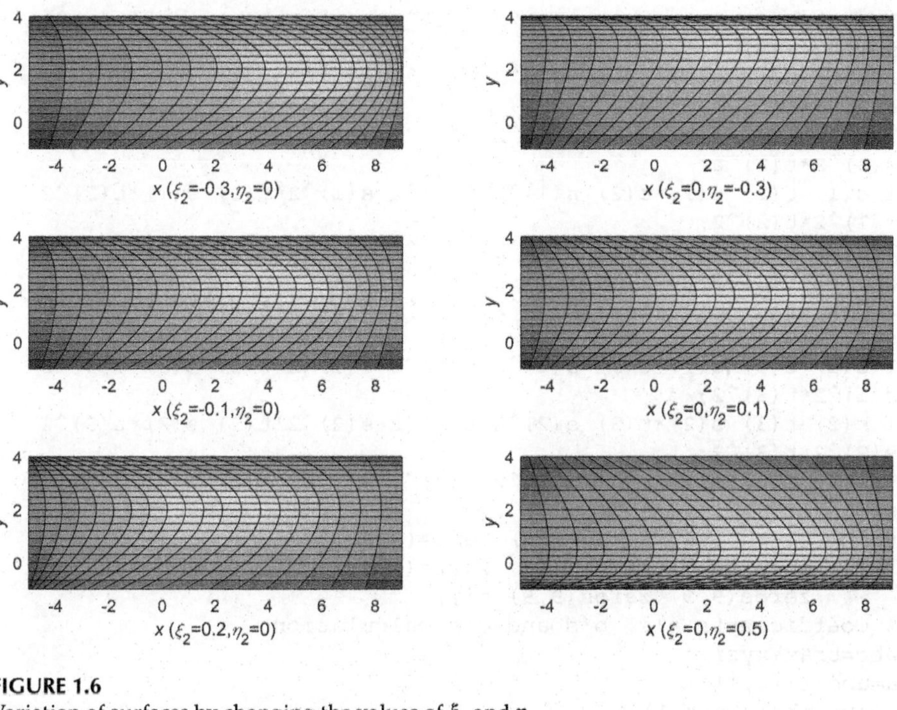

FIGURE 1.6
Variation of surfaces by changing the values of ξ_2 and η_2.

1.1.4 Parametric Surface Program List

```
% UnderstandingParametricSurface.m
% Parametric creation of a second-order polynomial surface
% Point coordinates: xi, yi and zi
clear variables; clc;
xi=[-5  0  9 -5  3  9 -5  0  9];
yi=[-1 -1 -1  2  2  2  4  4  4];
zi=[ 0  3  0  4 10  5  0  2  0];
ik=size(xi,2);
xyzi=transpose([xi yi zi]);
st2 = [-0.3  -0.1 0.2;
       -0.3   0.1 0.5];
Figure(1)
set(gcf,'position',[200,200,900,600]);
hold;
for is = 1:2
for it = 1:3
% Parameter s(xi) and t(eta)
    s=[-1 0 1];
    t=[-1 0 1];
    if is==1, s(2)=st2(is,it); t(2)=0; end
    if is==2, t(2)=st2(is,it); s(2)=0; end
% Transformation matrix [tran]
ti = [ ...
1 s(1) t(1) s(1)*t(1) s(1)^2 t(1)^2 s(1)^2*t(1) s(1)*t(1)^2
s(1)^2*t(1)^2;
1 s(2) t(1) s(2)*t(1) s(2)^2 t(1)^2 s(2)^2*t(1) s(2)*t(1)^2
s(2)^2*t(1)^2;
1 s(3) t(1) s(3)*t(1) s(3)^2 t(1)^2 s(3)^2*t(1) s(3)*t(1)^2
s(3)^2*t(1)^2;
1 s(1) t(2) s(1)*t(2) s(1)^2 t(2)^2 s(1)^2*t(2) s(1)*t(2)^2
s(1)^2*t(2)^2;
1 s(2) t(2) s(2)*t(2) s(2)^2 t(2)^2 s(2)^2*t(2) s(2)*t(2)^2
s(2)^2*t(2)^2;
1 s(3) t(2) s(3)*t(2) s(3)^2 t(2)^2 s(3)^2*t(2) s(3)*t(2)^2
s(3)^2*t(2)^2;
1 s(1) t(3) s(1)*t(3) s(1)^2 t(3)^2 s(1)^2*t(3) s(1)*t(3)^2
s(1)^2*t(3)^2;
1 s(2) t(3) s(2)*t(3) s(2)^2 t(3)^2 s(2)^2*t(3) s(2)*t(3)^2
s(2)^2*t(3)^2;
1 s(3) t(3) s(3)*t(3) s(3)^2 t(3)^2 s(3)^2*t(3) s(3)*t(3)^2
s(3)^2*t(3)^2];
tran=[ti          zeros(9,9) zeros(9,9);
      zeros(9,9) ti          zeros(9,9);
      zeros(9,9) zeros(9,9) ti];
% Coefficients a's, b's and c's calculation
abc=tran\xyzi;
a=abc(1:ik,1);
b=abc(ik+1:2*ik,1);
```

```
c=abc(2*ik+1:3*ik,1);
% Drawing graph on plane with z as the height contour
s=-1:0.1:1;
t=-1:0.1:1;
m=size(s,2);
n=size(t,2);
x=zeros(m,n);
y=zeros(m,n);
z=zeros(m,n);
for i=1:m
  for j=1:n
    x(i,j)=a(1)+a(2)*s(i)+a(3)*t(j)+a(4)*s(i)*t(j)+a(5)*
s(i)^2+ ...
        a(6)*t(j)^2+a(7)*s(i)^2*t(j)+a(8)*s(i)*t(j)^2+a(9)
*s(i)^2*t(j)^2;
    y(i,j)=b(1)+b(2)*s(i)+b(3)*t(j)+b(4)*s(i)*t(j)+b(5)*
s(i)^2+ ...
        b(6)*t(j)^2+b(7)*s(i)^2*t(j)+b(8)*s(i)*t(j)^2+b(9)
*s(i)^2*t(j)^2;
    z(i,j)=c(1)+c(2)*s(i)+c(3)*t(j)+c(4)*s(i)*t(j)+c(5)*
s(i)^2+ ...
        c(6)*t(j)^2+c(7)*s(i)^2*t(j)+c(8)*s(i)*t(j)^2+c(9)
*s(i)^2*t(j)^2;
  end
end
ip=(it-1)*2+is;
subplot(3,2,ip);
surface(x,y,z);
if is==1
    txt = ['{\itx} (' '{\it\xi}_2=' num2str(st2(is,it))
',{\it\eta}_2=0)'];
end
if is==2
    txt = ['{\itx} ({\it\xi}_2=0,' '{\it\eta}_2=' num2str
(st2(is,it)) ')'];
end
xlabel(txt); ylabel('{\ity}');
axis equal
axis tight
end % for it
end % for is
```

In this book, we are not going into much detail on the geometry modeling of curves. For more informative journals and standard textbooks of the topic, the reader can find more detail and other aspects in geometric modeling in Bajaj et al. (1995), Bernstein (1912), Faux and Pratt (1981), Mortenson (1985), Hoffmann (1989) and Beach (1991), Lei and Wang (2009), Piegl and Tiller (1997, 2003), Lorentz (1986).

1.2 Bézier Curve

In Section 1.1, we found that the parametric polynomial is widely used as a class of functions in geometric modeling. Although they are easy, efficient, and accurately processed in a computer, there are some important curve types that cannot be precisely represented by using the parametric polynomials, like a circle. Those unordinary types of curves must be represented by a system that consists of polynomials. In this section, we learn a method so-called the Bézier curve. Our discussion of the Bézier curve is only the essential part that is related to the parametric modeling. For more details, the reader should consult other references (Bajaj et al. 1995; Bézier 1972, 1986; Chang and Wu 1981; Farin 1993; Gordon and Riesenfeld 1974; Piegl and Tiller 2003; Hoschek and Lasser 1993; Rogers and Adams 1990; Yamaguchi 1988).

For a curve, the nth-order Bézier curve is given by

$$\begin{aligned} x(\xi) &= \sum_{i=0}^{n} B_{i,n}(\xi) P_{x(\xi)} \\ y(\xi) &= \sum_{i=0}^{n} B_{i,n}(\xi) P_{y(\xi)} \end{aligned} \qquad -1 \le \xi \le 1 \tag{1.7}$$

In the preceding equation, $P_{x(\xi)}$ and $P_{y(\xi)}$ are the coordinates which are called *control points* in the x–y plane. $B_{i,n}(\xi)$ is a collection of ξ-parametric-based basis functions which are also known as the nth-order Bernstein polynomials (Bernstein 1912; Lorentz 1986). The Bernstein polynomials are defined by

$$B_{i,p}(\xi) = \frac{n!}{i!(n-i)!} \times \left(\frac{1+\xi}{2} \right)^{i} \times \left(\frac{1-\xi}{2} \right)^{p-i} \tag{1.8}$$

MATLAB code BernsteinBasis.m uses function Bernstein.m (Section 1.2.2) and draws the Bernstein curves for $n = 2$ and $n = 4$ as shown in Figure 1.7 which are based on Equations (1.7) and (1.8). It is worth noting that we can have more polynomial forms to compose a curve by following the Bernstein polynomials rules because we have $(n+1)$ curves as the basis functions corresponding to each $(n+1)$ control points to draw a polynomial curve with an order of n.

1.2.1 BernsteinBasis Program List

```
% BernsteinBasisDraw.m
% Drawing the Bernstein basis functions
```

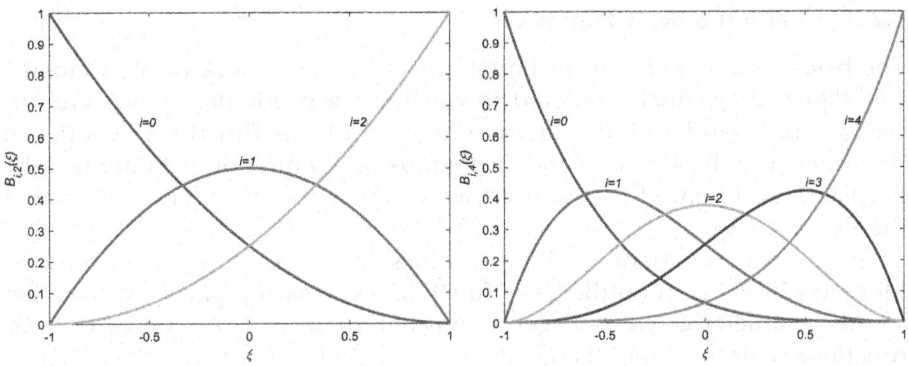

FIGURE 1.7
Bézier curves of second (a) and fourth (b) orders of Bernstein basis functions.

```
% n = degree of the polynomial
% t = -1 to +1, incremented by dt
n=2;  % second/fourth-order polynomial
dt=1/1000;
t=-1:dt:1;
y=zeros(size(t));
for i=0:n
  for dt=1:size(t,2)
    y(dt) = Bernstein(n,i,t(dt));
  end
  plot(t,y,'LineWidth',2);
  hold all;
end
xlabel('\it{\xi}'); ylabel('\itB_{i,2}({\xi})');
text(-0.55,0.65,'\iti=0');
text(-0.05,0.525,'\iti=1');
text(0.5,0.65,'\iti=2');
```

MATLAB code Berstein.m is the function which is present inside the BernsteinBasis.m code.

1.2.2 Bernstein Function List

```
function Brnbasis = Bernstein(n,i,t)
% Bernstein.m
% Bernstein basis functions
% n = degree of Bezier curve
% i = 0 to n
% t = -1 to +1, parameter
Brnbasis=factorial(n)/factorial(i)/...
         factorial(n-i)*(1/2+t/2)^i*(1/2-t/2)^(n-i);
return
```

1.2.3 All about a Basis Function

The basis function shown in Equation (1.8) is a series of curve elements (see Figure 1.8) which are used to construct a particular curve. We can see that in Figure 1.10 (n = 2), the number of basis functions is equal to the order of polynomial of the curve plus one. Equation (1.8) defines the Bernstein polynomial that is used as the basis function for constructing the Bézier curve.

These basis functions will give values less than 1 along the ξ-axis. Moreover, if we sum all the basis functions' values at a particular location of the parameter ξ, we will get a constant value of 1. All types of basis functions with this kind of characteristic are said to have a *partition of unity.* The partition of unity is an important property that is required in selecting a shape function to be used in the finite element formulations.

Each curve element of the basis functions is associated with a corresponding control point in calculating Equations (1.7) and (1.9) for a curve and surface, respectively.

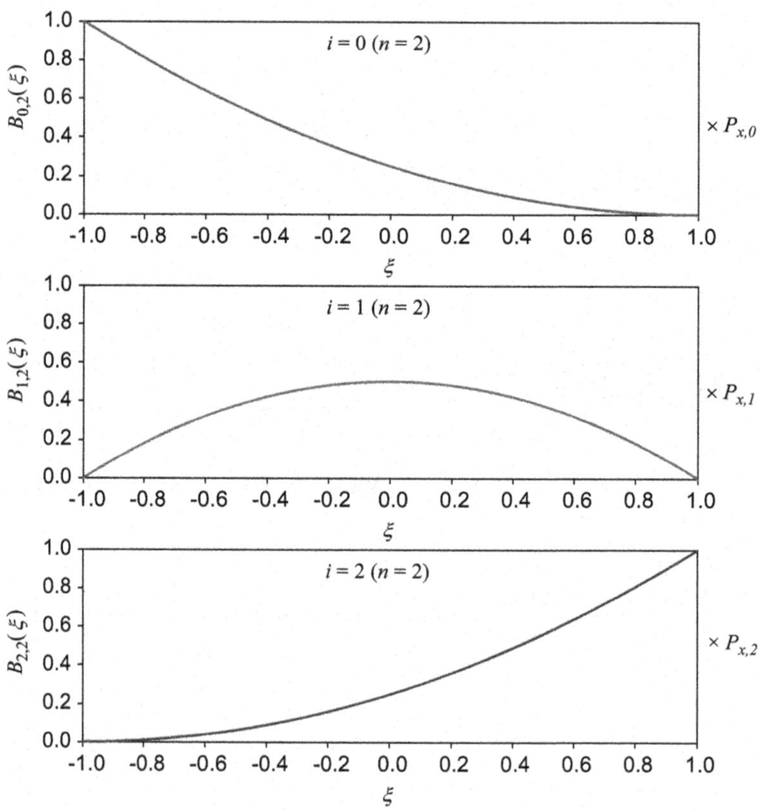

FIGURE 1.8
Second-order Bernstein basis functions with the associated control points.

1.2.4 The Roles of a Control Point

A control point plays a major role in constructing a Bézier curve or surface. If we observe the relationship between a basis function and a control point in Equation (1.7) or (1.9), the ξ_i or (ξ_i, η_j) control point is a scalar multiplier for the corresponding basis function. If one control point acts as a "controller" to its corresponding basis function, then we need (n+1) and (m+1) × (n+1) number of control points to construct a Bézier curve or surface.

We can see in Figure 1.9 that some of the control points of a Bézier surface are located altered from their sampling points taken from the surface. All of these control points and their associated basis functions should be used to determine a position on the Bézier surface given in Equation (1.9).

To illustrate the roles of control points, Figure 1.9 shows that by arranging the control points, we can obtain various shapes of interesting curves easily.

a.	b.
c.	d.

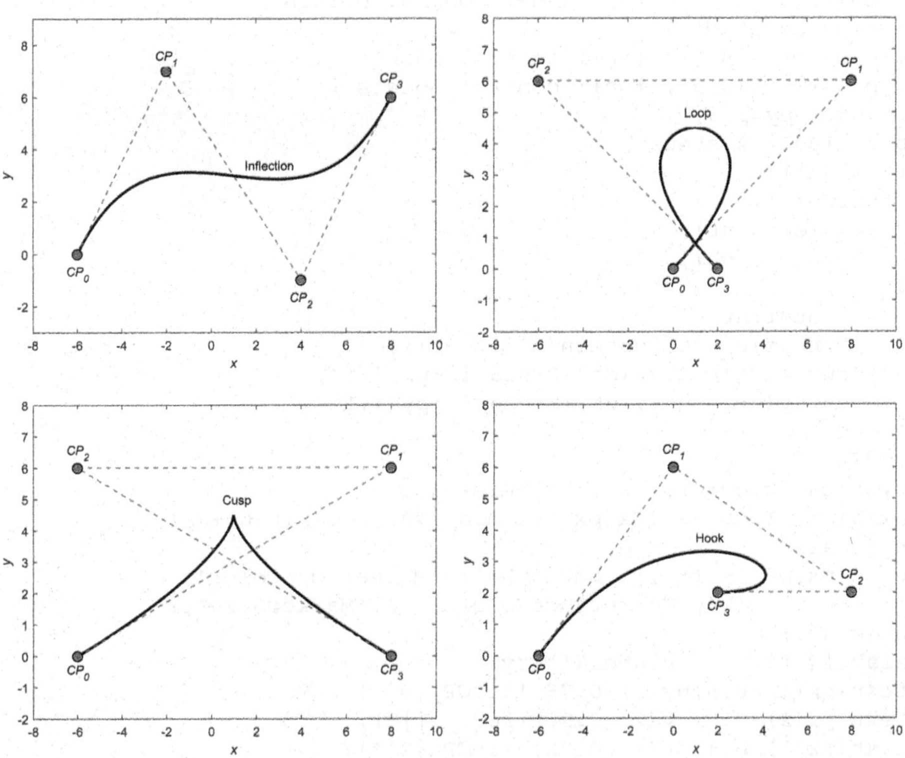

FIGURE 1.9
Third-order Bézier curve examples.

MATLAB code BezierCurveCP.m uses the function Bernstein.m (Section 1.2.2) and draws the Bézier curves as shown in Figure 1.9.

1.2.5 BezierCurveCP Program List

```
% BezierCurveCP.m (a)
% Drawing the Bezier curve from given control points
% n = degree of Bezier curve
% t = -1 to +1, incremented by dt
clear all;
n=3;
ndiv=10;
% Figure 1.9(a)
px0=[ -6 -2  4 8]; % n+1 control points
py0=[  0  7 -1 6]; % n+1 control points
% Figure 1.9(b)
% px0=[ 0 8 -6 2]; % n+1 control points
% py0=[ 0 6  6 0]; % n+1 control points
% Figure 1.9(c)
% px0=[ -6  8  -6  8]; % n+1 control points
% py0=[  0  6   6  0]; % n+1 control points
% Figure 1.9(d)
% px0=[-6 0 8 2]; % n+1 control points
% py0=[ 0 6 2 2]; % n+1 control points
px=px0; py=py0;
dt=1/1000; % steps
t=-1:dt:1;
nt=size(t');
curvx=zeros(nt);
curvy=zeros(nt);
for i=0:n
  for dt=1:nt
    Brnbasis = Bernstein(n,i,t(dt));
    curvx(dt)=curvx(dt)+Brnbasis*px(i+1);
    curvy(dt)=curvy(dt)+Brnbasis*py(i+1);
  end
end
plot(curvx,curvy,'-k','LineWidth',2);
axis([min(px0)-2 max(px0)+2 min(py0)-2 max(py0)+2]);
hold all;
plot(px,py,'--ro','LineWidth',1,'MarkerEdgeColor',...
           'k','MarkerFaceColor','r','MarkerSize',8);
hold all;
xlabel('\itx'); ylabel('\ity');
text(px(1)-0.5,py(1)-0.75,'\itCP_{0}');
text(px(2)-0.5,py(2)+0.5,'\itCP_{1}');
text(px(3)-0.5,py(3)-0.75,'\itCP_{2}');
text(px(4)-0.5,py(4)+0.65,'\itCP_{3}');
text(1.5,3.4,'Inflection');
```

1.2.6 Rational Bézier Curve

Many applications require the family of functions of curves which have been normalized. Bézier curves can be represented by rational functions which are defined as the ratio between the family of the polynomials. We can construct such curves by dividing each series function by the summation of all the series of functions as a denominator.

The *n*th-order rational Bézier curve is defined by

$$x(\xi) = \sum_{i=0}^{n} R_{i,n}(\xi) P_{x(\xi)}$$
$$y(\xi) = \sum_{i=0}^{n} R_{i,n}(\xi) P_{y(\xi)} \qquad -1 \le \xi \le 1 \tag{1.9}$$
$$z(\xi) = \sum_{i=0}^{n} R_{i,n}(\xi) P_{z(\xi)}$$

where

$$R_{i,n}(\xi) = \frac{B_{i,n}(\xi) w_i}{\sum_{j=0}^{n} B_{j,n}(\xi) w_j} \qquad w_i > 0$$

w_i, w_j are the weighting parameters of the control points, and $R_{i,n}(\xi)$ is the rational basis function of the Bézier curve.

Let us see if we use different weights of $w_0 = 2$, $w_1 = 4$, and $w_2 = 1$ for the Bernstein basis functions shown in Figure 1.7a.

MATLAB code RationalBezierBasis.m uses the function Bernstein.m (Section 1.2.2) and draws the rational Bézier basis functions for $n = 2$ Bézier curve as shown in Figure 1.12 which is based on Equation (1.12). We can observe from the figure that by changing the weighting parameters of a particular basis function, we can adjust either "stronger" ($w_1 = 4$) or "weaker" ($w_2 = 1$) influence of that basis function to the curve (Figure 1.10).

1.2.7 RationalBezierBasis Program List

```
text(px(4)-0.5,py(4)-0.5,'CP3');
text(0.5,5,'Loop');
% RationalBezierBasis.m
% Drawing the weighted Rational Bezier basis functions
% n = degree of polynomial
% t = -1 to +1, incremented by dt
```

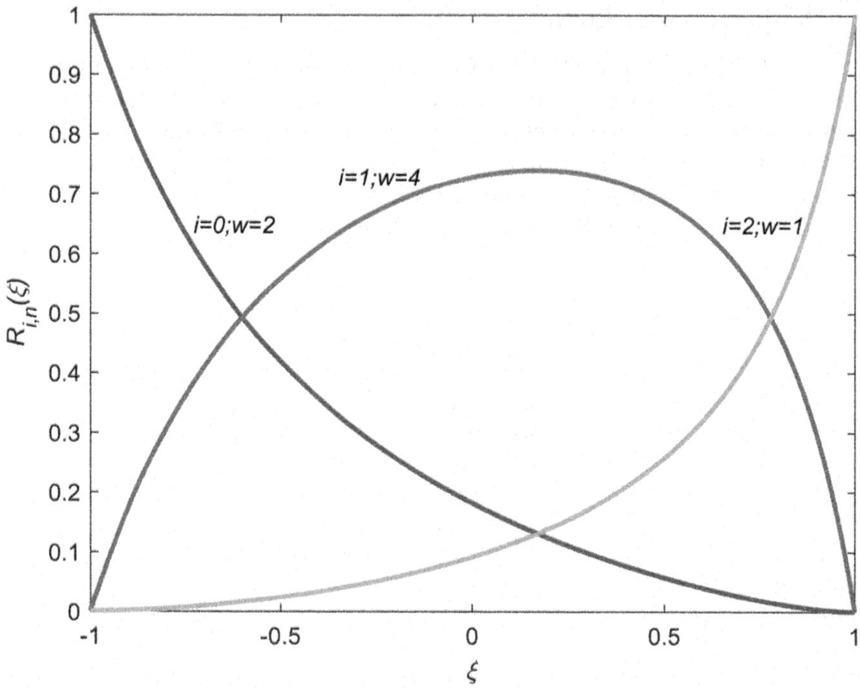

FIGURE 1.10
Weighted second-order Bernstein basis functions.

```
n=2;  % second degree polynomial
dt=1/1000;
t=-1:dt:1;
nt=size(t,2);
wt=[2 4 1];
rb=zeros(nt,1);
for i=0:n
  for dt=1:nt
    Brnbasis = Bernstein(n,i,t(dt));
    rb(dt)=Brnbasis*wt(i+1);
    ydenom=0;
    for j=0:n
      Brnbasis = Bernstein(n,j,t(dt));
      ydenom=ydenom+Brnbasis*wt(j+1);
    end
    rb(dt)=rb(dt)/ydenom;
  end
  plot(t',rb,'LineWidth',2);
  hold all;
end
xlabel('\it{\xi}'); ylabel('\itR_{i,n}({\xi})');
```

1.2.8 Weighting Parameter

As we can see in Figure 1.11, by using higher weight values more than one, we can increase the "influence" of the corresponding control points. The influence can be noticed from the changing curvature of the curve approaching the coordinates of the control points that "pull" the curve toward a particular control point. The higher the weight given to the corresponding control point, the closer the curve to the control point.

In practice, some curves such as circle, ellipse, and hyperbola cannot be constructed perfectly by using a simple Bézier curve; however, with the help of weighting values, these curves can be created.

1.2.9 Rationalization of Basis Functions

In the upcoming chapter 5 regarding the finite element analysis, we will deal with the so-called shape functions. The shape function is an important factor to be considered in the finite element method that leads to the accuracy and efficiency of the formulations and analyses. The shape function for the finite element formulation has to satisfy the partition of unity. These shape functions will be used for constructing the element stiffness matrix, mass matrix, loading vector, and other properties of a plate. Therefore, the rationalization or normalization of the curve will be necessary to satisfy the requirements

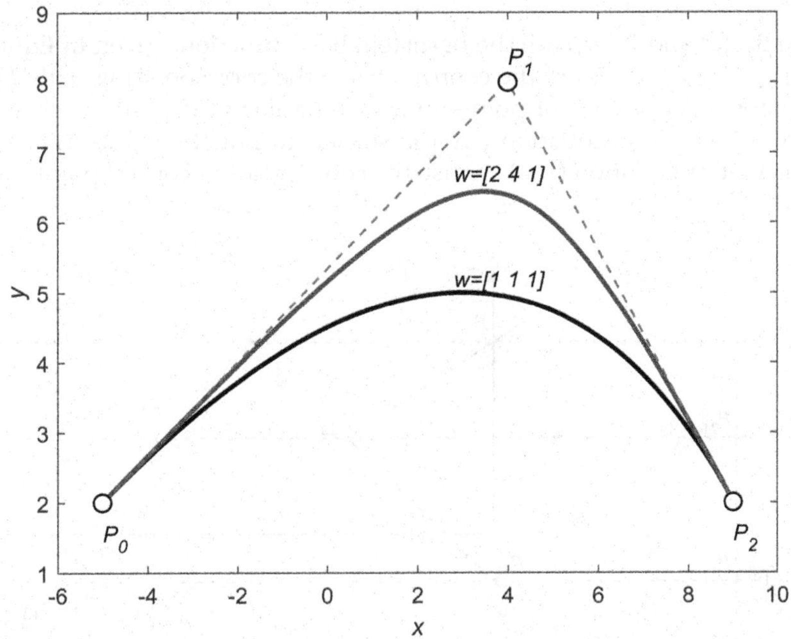

FIGURE 1.11
Example of giving weight to the rational Bézier curve.

of shape function. The rationalization of Bernstein basis functions using a weighting parameter w_i in Equation (1.9) results in the basis functions that satisfy the condition of the partition of unity.

1.3 Bézier Surface

The Bézier polynomials of a curve given in Equation (1.4) can be expanded to create a surface by using the following equation:

$$\begin{aligned} x(\xi,\eta) &= \sum_{i=0}^{m}\sum_{j=0}^{n} B_{i,m}(\xi)B_{j,n}(\eta)P_{x(\xi,\eta)} \\ y(\xi,\eta) &= \sum_{i=0}^{m}\sum_{j=0}^{n} B_{i,m}(\xi)B_{j,n}(\eta)P_{y(\xi,\eta)}, \quad \begin{cases} -1 \le \xi \le 1 \\ -1 \le \eta \le 1 \end{cases} \\ z(\xi,\eta) &= \sum_{i=0}^{m}\sum_{j=0}^{n} B_{i,m}(\xi)B_{j,n}(\eta)P_{z(\xi,\eta)} \end{aligned} \tag{1.10}$$

where $B_{i,m}(\xi)$ and $B_{j,n}(\eta)$ are the Bernstein basis functions given in Equation (1.8). $P_{x(\xi,\eta)}, P_{y(\xi,\eta)}, P_{z(\xi,\eta)}$ are the coordinates of the corresponding *control mesh* that maps a set of control points in a rectangular grid (ξ,η) as shown in Figure 1.12. In fact, Equation (1.10) is similar to Equation (1.2). The difference is that in Equation (1.10) we use the coordinates of control points as the

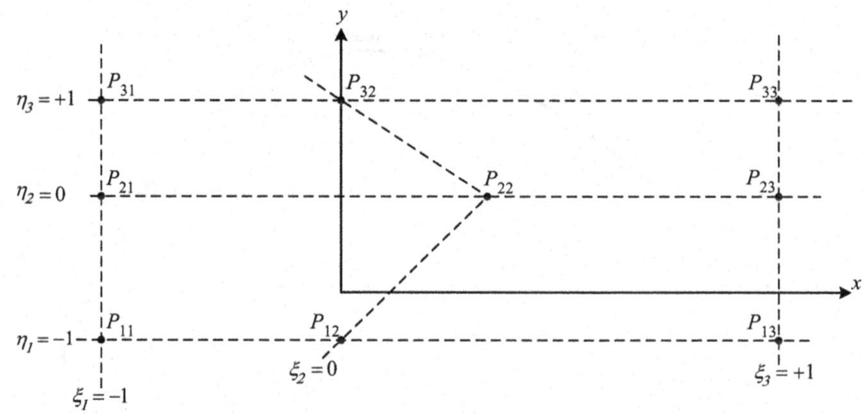

FIGURE 1.12
Illustration of the control mesh on the x–y plane.

coefficients in Equation (1.2). Therefore, the control points have more important physical meaning than the coefficients in a polynomial as discussed in Section 1.2.4.

Suppose we want to represent a second-order polynomial surface as shown in Figure 1.2 using the Bézier surface. First, we need to obtain the control points to construct Equation (1.10). The Bernstein basis functions should be generated in $m = 2$ and $n = 2$ mesh of points. The coordinates (x, y, z) of nine control points $(P_{11}, P_{12}, \ldots, P_{33})$ as shown in Figure 1.12 are substituted into Equation (1.9), resulting in

$$
\begin{aligned}
x_{11} &= B_{0,2}(\xi_1)B_{0,2}(\eta_1)P_{x,11} + \cdots + B_{2,2}(\xi_1)B_{2,2}(\eta_1)P_{x,33} \\
&\;\;\cdot \\
x_{33} &= B_{0,2}(\xi_3)B_{0,2}(\eta_3)P_{x,11} + \cdots + B_{2,2}(\xi_3)B_{2,2}(\eta_3)P_{x,33} \\
y_{11} &= B_{0,2}(\xi_1)B_{0,2}(\eta_1)P_{y,11} + \cdots + B_{2,2}(\xi_1)B_{2,2}(\eta_1)P_{y,33} \\
&\;\;\cdot \\
y_{33} &= B_{0,2}(\xi_3)B_{0,2}(\eta_3)P_{y,11} + \cdots + B_{2,2}(\xi_3)B_{2,2}(\eta_3)P_{y,33} \\
z_{11} &= B_{0,2}(\xi_1)B_{0,2}(\eta_1)P_{z,11} + \cdots + B_{2,2}(\xi_1)B_{2,2}(\eta_1)P_{z,33} \\
&\;\;\cdot \\
z_{33} &= B_{0,2}(\xi_3)B_{0,2}(\eta_3)P_{z,11} + \cdots + B_{2,2}(\xi_3)B_{2,2}(\eta_3)P_{z,33}
\end{aligned}
\tag{1.11}
$$

The nine control points $(P_{11}, P_{12}, \ldots, P_{33})$ can be computed from the sampling data of $(m+1)\times(n+1)$ points taken from the surface. Equation (1.11) can be written in a matrix form similar to Equation (1.4) as follows:

$$
\mathbf{x} = \begin{bmatrix} x_{11} & y_{11} & z_{11} \\ \cdot & \cdot & \cdot \\ x_{33} & y_{33} & z_{33} \end{bmatrix}_{(m+1)\times(n+1),3}
$$

$$
\alpha = \begin{bmatrix} P_{x,11} & P_{y,11} & P_{z,11} \\ \cdot & \cdot & \cdot \\ P_{x,33} & P_{y,33} & P_{z,33} \end{bmatrix}_{(m+1)\times(n+1),3}
$$

$$
\mathbf{T} = \begin{bmatrix} B_{0,2}(\xi_1)B_{0,2}(\eta_1) & \cdot & \cdot & \cdot & B_{2,2}(\xi_1)B_{2,2}(\eta_1) \\ \cdot & \cdot & \cdot & \cdot & \cdot \\ \cdot & \cdot & B_{1,1}(\xi_2)B_{1,1}(\eta_2) & \cdot & \cdot \\ \cdot & \cdot & \cdot & \cdot & \cdot \\ B_{0,2}(\xi_3)B_{0,2}(\eta_3) & \cdot & \cdot & \cdot & B_{2,2}(\xi_3)B_{2,2}(\eta_3) \end{bmatrix}_{(m+1)\times(n+1),(m+1)\times(n+1)}
\tag{1.12}
$$

The computed control points in the control mesh are tabulated in Table 1.3.

It can be seen in Table 1.3 that the resulting nine control points of the control mesh can be computed from the nine sampling points on the surface. Now we can use these control points to generate the surface. The matrices **T** and $\boldsymbol{\alpha}$ in Equation (1.12) can be constructed by dividing the range of ξ and η into 20 intervals. The result of mesh points **x** which are generated on the surface is depicted in Figure 1.13.

The Bézier curves and surfaces are suitable for intuitive design to create free-form curves and surfaces. This is because the control points and mesh give the designer more controls handles to harness the curve and surface creations systematically to be solved numerically by using a computer.

MATLAB code BezierSurface.m uses the function Bernstein.m (Section 1.2.2) and solves the control points coordinates vector $\boldsymbol{\alpha}$, and uses the control points to construct the Bézier surface in Equation (1.11) and draws the surface as shown in Figure 1.13.

1.3.1 BezierSurface Program List

```
% BezierSurface.m
% Drawing the Bezier surface by calculating the control points
% m,n = order of Bezier surface in two directions
clear variables; clc;
% Nodal coordinates of nine points: xi, yi and zi
xi=[-5   0  9 -5  3  9 -5  0  9];
yi=[-1 -1 -1  2  2  2  4  4  4];
zi=[ 0  3  0  4 10  5  0  2  0];
m=2; n=2;
xyzi=transpose([xi' yi' zi'])';
% Parameter ksi=s and eta=t
s=[-1 0 1];
t=[-1 0 1];
```

TABLE 1.3

The Control Mesh of Nine Points on the Surface

Sampling Points	x	y	z	Control Points	x	y	z
P_1	–5	–1	0	P_{11}	–5	–1	0
P_2	0	–1	3	P_{12}	–2	–1	6
P_3	9	–1	0	P_{13}	9	–1	0
P_4	–5	2	4	P_{21}	–5	2.5	8
P_5	3	2	10	P_{22}	10	2.5	26
P_6	9	2	–5	P_{23}	9	2.5	10
P_7	–5	4	0	P_{31}	–5	4	0
P_8	0	4	2	P_{32}	–2	4	4
P_9	9	4	0	P_{33}	9	4	0

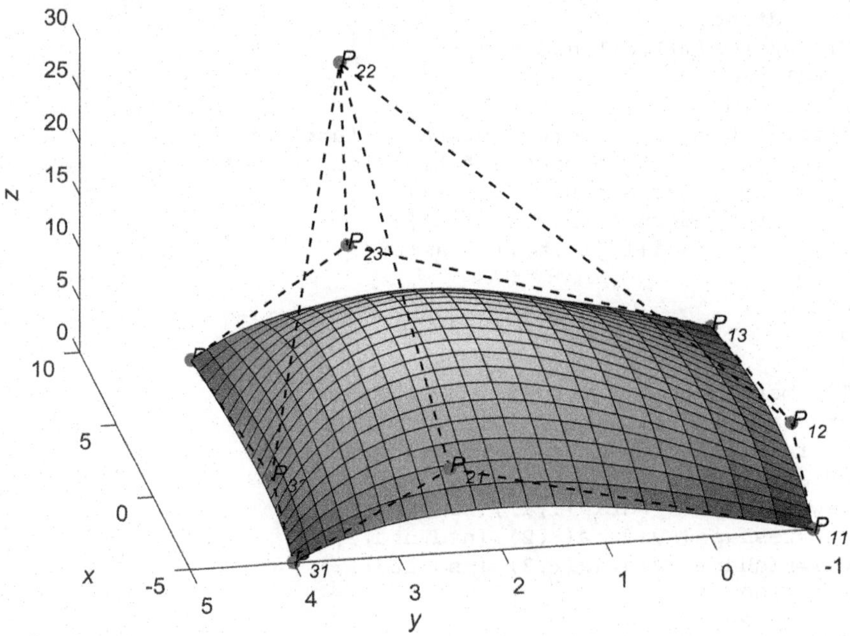

FIGURE 1.13
Control mesh with its control points.

```
ns=size(s,2);
nt=size(t,2);
nst=ns*nt;
stt=zeros(m+1,n+1,ns,nt);
%
for k=1:ns
for l=1:nt
    for i=0:m       % order of basis function
    for j=0:n       % order of basis function
        ss = Bernstein(m,i,s(k));
        tt = Bernstein(n,j,t(l));
        stt(i+1,j+1,k,l) = ss*tt ;
    end
    end
end
end
% [T] matrix
st=reshape(stt,[(m+1)*(n+1),nst])';
% Control Points Px's, Py's and Pz's calculation
pxyz=st\xyzi;
% Drawing the Bezier Surface
is=2/20;          it=is;         % equal interval
si=-1:is:+1;    ti=-1:it:+1;
nsd=size(si,2);  ntd=size(ti,2);
```

```
nstd=nsd*ntd;
stdraw=zeros(m+1,n+1,nsd,ntd);
for k=1:nsd
for l=1:ntd
    for i=0:m       % order of basis function
    for j=0:n       % order of basis function
        ss = Bernstein(m,i,si(k));
        tt = Bernstein(n,j,ti(l));
        stdraw(i+1,j+1,k,l) = ss*tt ;
    end
    end
end
end
% [T] matrix
st=reshape(stdraw,[(m+1)*(n+1),nstd])';
% Coordinates of points in the control mesh
xyzdraw=st*pxyz;
xdraw=reshape(xyzdraw(:,1),[nsd,ntd]);
ydraw=reshape(xyzdraw(:,2),[nsd,ntd]);
zdraw=reshape(xyzdraw(:,3),[nsd,ntd]);
azm = -100;
view(azm,35)
surface(xdraw,ydraw,zdraw)
xlabel('\itx'); ylabel('\ity'); zlabel('\itz');
hold all;
% Plotting the control points
scatter3(pxyz(:,1)',pxyz(:,2)',pxyz(:,3)','filled','MarkerFace
Color','r');
% Plotting x-direction control mesh
for j = 1:nt
    lxyz=[];
    for i = 1:ns
        lxyz=[lxyz pxyz(i+(j-1)*nt,:)'];
    end
    hold all;
    plot3(lxyz(1,:),lxyz(2,:),lxyz(3,:),'--k','LineWidth',1);
end
% Plotting y-direction control mesh
for j = 1:ns
    lxyz=[];
    for i = 1:nt
        lxyz=[lxyz pxyz(j+(i-1)*ns,:)'];
    end
    hold all;
    plot3(lxyz(1,:),lxyz(2,:),lxyz(3,:),'--k','LineWidth',1);
end
% Plotting labels
for i = 1:nt
    for j = 1:ns
        itxt=['\itP_{' num2str(i) num2str(j) '}'];
```

```
        ij=j+(i-1)*ns;
        text(pxyz(ij,1),pxyz(ij,2),pxyz(ij,3),itxt);
    end
end
```

1.4 B-Spline Curve

There are occasions that we want to increase the number of control points for the accuracy of a curve. To represent a curve that has control points more than the $(n+1)$ order of the polynomial, we need the corresponding basis functions to accommodate these control points. To do this, we need a B-spline curve. The B-spline curve can do the task by using a series of basis functions which can have a different type of polynomial elements.

The pth order of B-spline curve is defined by

$$\begin{aligned} x(\xi) &= \sum_{i=0}^{n} N_{i,n}(\xi) P_{x(\xi)} \\ y(\xi) &= \sum_{i=0}^{n} N_{i,n}(\xi) P_{y(\xi)} \end{aligned} \qquad -1 \le \xi \le 1 \tag{1.13}$$

where $N_{i,n}(\xi)$ are the ξ parameter-based basis functions (De Boor 1978, 1987). The basis functions of the B-spline are defined by

$$\begin{aligned} N_{i,0}(\xi) &= \begin{cases} 1 & \text{if } \xi_i \le \xi < \xi_{i+1} \\ 0 & \text{otherwise} \end{cases} \\ N_{i,p}(\xi) &= \frac{\xi - \xi_i}{\xi_{i+p} - \xi_i} N_{i,p-1}(\xi) + \frac{\xi_{i+p+1} - \xi}{\xi_{i+p+1} - \xi_{i+1}} N_{i+1,p-1}(\xi) \end{aligned} \tag{1.14}$$

$N_{i,n}(\xi)$ are the pth-degree B-spline basis functions defined on the nonperiodic *knot* vector Ξ. Next, will discuss the concept of the knot vector. The knot vector Ξ consists of the parameter ξ in the range of $-1 \le \xi \le +1$. The ξ's in Equation (1.14) with the subscript notation i are fixed values which belong to the knot vector, whereas the ξ's without any subscript notation are the variables. When we discussed the Bernstein basis functions to construct the Bézier curve, the number of basis functions is equal to $(n+1)$, where n is the order of polynomial of the curve (see Figure 1.8). However, in the B-spline basis functions, the number of basis functions is not the same as the order of the polynomial of the curve. The standard ξ parametric vector for $p = 2$ order B-spline curve is $\Xi = \begin{bmatrix} -1 & -1 & -1 & +1 & +1 & +1 \end{bmatrix}$, but this will give the same basis functions like the Bézier curve. Here, let us add a zero value inside the ξ parametric vector so that it becomes $\Xi = \begin{bmatrix} -1 & -1 & -1 & 0 & +1 & +1 & +1 \end{bmatrix}$. The B-spline curve can be defined by four basis functions as shown in Figure 1.14.

We can learn that by increasing the amount of parametric vector ξ, the basis function will increase without increasing the order of the polynomial, thus increasing the accuracy of the curve.

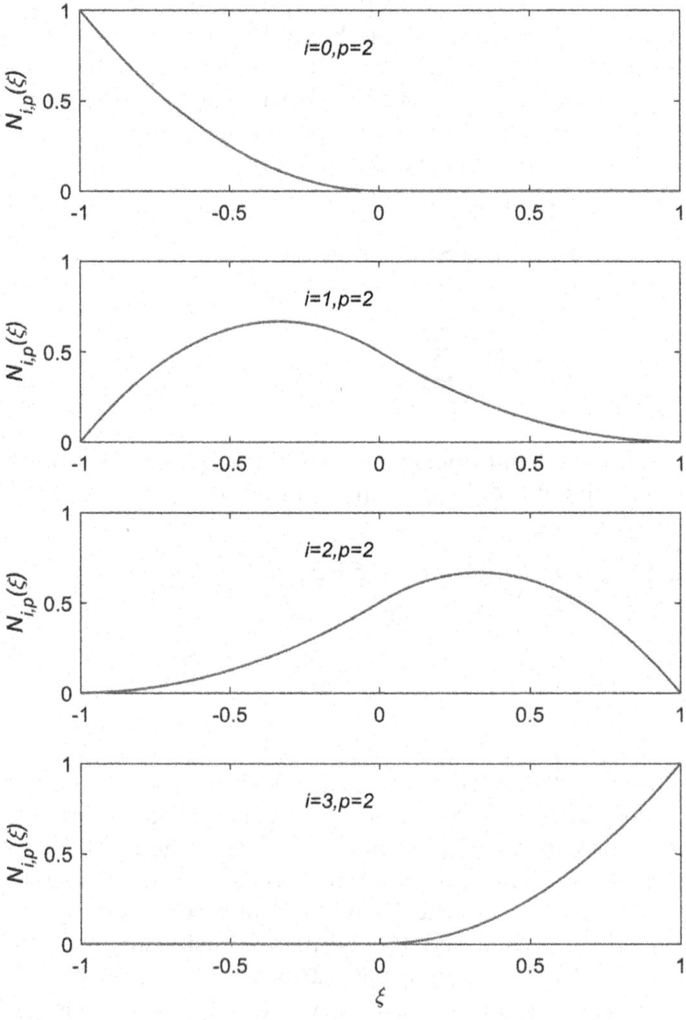

FIGURE 1.14
Four B-spline basis functions of the $p = 2$ order curve.

MATLAB code BsplineBasis.m (Section. 1.4.1) and function Bspline.m (Section 1.4.2) draw the B-spline basis curves for $p = 2$ as shown in Figure 1.14 which are based on the basis functions in Equation (1.14).

1.4.1 BsplineBasis Program List

```
% BsplineBasis.m
% Drawing the Bspline basis functions
% p = degree of Bspline curves
clear;clc;
p=2;
knot=[-1 -1 -1 0 +1 +1 +1];
knot=[knot ones(1,p)];
nb=size(knot,2)-p-1;                   % max number of basis
functions
ndt=1000;                              % t steps
t1=min(knot); t2=max(knot);
t=t1:(t2-t1)/ndt:t2;
nt=size(t,2);
y=zeros(nb,p+1,nt);                    % BsplineBasisPlot
(order,basis)
for it=1:nt
   Bspln = Bspline(p,knot,t(it));
   y(1:nb,1:(p+1),it) = Bspln(1:nb,1:(p+1));
end
set(0,'defaultFigurePosition',[0 0 10*50 15*50]);
for ip=p:p
 Figure(ip+1);
 for ib=1:(nb-ip)
   subplot(nb-ip,1,ib);
   yC(1:nt)=y(ib,ip+1,1:nt);
   plot(t,yC,'LineWidth',1);
   ylabel('\itN_{i,p}({\xi})');
   atext=['\iti=',int2str(ib-1),',\itp=2'];
   text(-0.25,0.8,atext);
 end
end
xlabel('\it{\xi}');
```

1.4.2 Bspline Basis Function List

```
function Bspln = Bspline(p,knot,t)
% Bspline.m
% p = degree of Bspline curve
% knot = knot vector
% t = parameter
%--------------------------------------------------------------
nb=size(knot,2)-p-1;             % number of basis functions
Bspln=zeros(nb,p+1);             % B spline (order,basis)
```

```
for ip=0:p                        % degree of polynomial
 for i=1:(nb-ip)                  % i-th basis function
  iknot=knot(i);
  jknot=knot(i+1);
  kknot=knot(i+ip);
  lknot=knot(i+ip+1);
  if ip==0                        % ip = 0 degree
    Bspln(i,ip+1)=1;
    if t>=iknot && t<jknot
      Bspln(i,ip+1)=1;            % unity
    else
      Bspln(i,ip+1)=0;            % zero
    end
    if i==(nb-p) && t==jknot
      Bspln(i,ip+1)=1;            % unity
    end
  end
  if ip>0                         % ip > 0 degree
    if kknot==iknot
       fract1=0;
    else
       fract1=(t-iknot)/(kknot-iknot);
    end
    if lknot==jknot
       fract2=0;
    else
       fract2=(lknot-t)/(lknot-jknot);
    end
       Bspln(i,ip+1)=fract1*Bspln(i,ip)+fract2*Bspln(i+1,ip);
  end
 end
end
return
```

1.4.3 Rational B-Spline Curve

In the subsequent sections, the *nonuniform rationalized B-spline* (*NURBS*) requires the formulation of rationalized B-spline curve. The procedure for rationalizing the B-spline curve is the same with the Bézier curve as previously discussed.

The nth order of the rational B-spline curve is defined by

$$\begin{aligned} x(\xi) &= \sum_{i=0}^{n} S_{i,n}(\xi) P_{x(\xi)} \\ y(\xi) &= \sum_{i=0}^{n} S_{i,n}(\xi) P_{y(\xi)} \end{aligned} \qquad -1 \le \xi \le 1 \tag{1.15}$$

where

$$S_{i,p}(\xi) = \frac{N_{i,p}(\xi)w_i}{\sum_{k=0}^{n} N_{k,p}(\xi)w_k} \quad w_i > 0$$

Here, w_i are the weighting parameters of the control points, and $S_{i,p}(\xi)$ are the rationalized $N_{i,p}(\xi)$ basis functions of the B-spline curve.

1.5 NURBS Curve

The term NURBS is an abbreviation of nonuniform rational B-spline. NURBS is a special case of the B-spline curve where the knot vector is not uniform. The nonuniformity of the knot vector is necessary to draw a system of polynomial curves with the desired level of continuity (see Section 1.4.3) at several knots. To demonstrate the capability of NURBS in combining two polynomial forms into one formulation, the following examples are given.

The second order of NURBS basis functions $p = 2$ with a given knot vector is as follows:

$\Xi = \begin{bmatrix} -1 & -1 & -1 & -0.4 & 0.2 & 0.2 & 0.4 & 0.6 & 0.7 & +1 & +1 & +1 \end{bmatrix}$ with uniform unit weight parameters $\mathbf{w} = \begin{bmatrix} 1 & 1 & 1 & 1 & 1 & 1 & 1 & 1 & 1 \end{bmatrix}$.

1.5.1 The Interpretation of a Knot

A typical knot vector consists of a sequence of ξ parameters which are arranged as follows:

$$\Xi = \Bigg[\underbrace{-1 \quad . \quad . \quad -1}_{pknots} \quad \underbrace{-1 \quad (\xi ...)_{m_i} \quad . \quad . \quad (\xi ...)_{m_{m+1}} \quad 1}_{(p+1+m)\,basis\ functions} \qquad (1.16)$$

$$\underbrace{1 \quad . \quad . \quad 1}_{pknots} \Bigg]^T$$

where m_i is the inclusion of knots in the range of [–1 +1]. The continuity condition of the curve at the repeated knots can be defined as

$$C^{p-m} \qquad (1.17)$$

Various orders of parametric level of continuity between curves can be described as follows:

- C^{-1}: the curves have discontinuities.
- C^0: the curves are connected.
- C^1: the first derivative of the curves is continuous.
- C^2: the first and second derivatives of the curves are continuous.
- C^n: the first through nth derivatives of the curves are continuous.

As we can see the knot vector in Equation (1.15), each knot can be considered as a breakpoint of several polynomial curves. Therefore, one polynomial will be represented between two nonrepeating knots. For a $p = 2$ B-spline curve, if the two of three control points are to be fixed at the extremities of the curve, the first and last knots have to be repeated $p+1$ times as the knots. This will enforce a level of discontinuity at both extremity points. Hence, both points will be "clamped" at both extremities of the curve.

The simplest knot vector of the $p = 2$ B-spline curve representation can be given as

$$\Xi = \Bigg[\underbrace{-1 \quad -1}_{p=2\ knots} \quad \underbrace{-1 \qquad\qquad 1}_{3\ basis\ functions} \quad \underbrace{1 \quad 1}_{p=2\ knots} \Bigg]^T \tag{1.18}$$

From Equation (1.17), we can observe there are six knots; however, because three knots are the repetition of two numbers, that is, −1 and +1, essentially there are only two knots that are unique. The first two and the last two numbers of p basis functions will not generate any basis function. The control points required within the range of $\begin{bmatrix} -1 & +1 \end{bmatrix}$ are three. Because of this, we have three corresponding basis functions which are exactly similar to the Bernstein polynomial basis functions as shown in Figure 1.8.

Implicitly, by inserting a knot inside the interval, we can increase the number of basis functions, and it also means that we can increase the number of control points required to draw a B-spline curve without increasing the order of polynomial of the curve.

For example, the $\Xi = \begin{bmatrix} -1 & -1 & -1 & 1 & 1 & 1 \end{bmatrix}$ knot vector is a standard $p = 2$ NURBS curve which has three $(p+1=3)$ basis functions, and the three control points correspond to each basis function. The $\Xi = \begin{bmatrix} -1 & -1 & -1 & -0.2 & 0.5 & 1 & 1 & 1 \end{bmatrix}$ knot vector is a nonstandard $p = 2$ NURBS curve which has five $(m+p+1=5)$ basis functions, and the five control points correspond to each basis function.

MATLAB code NurbsBasis.m uses function Nurbs.m (Section 1.5.3) to draw NURBS basis functions which are shown in Figure 1.15.

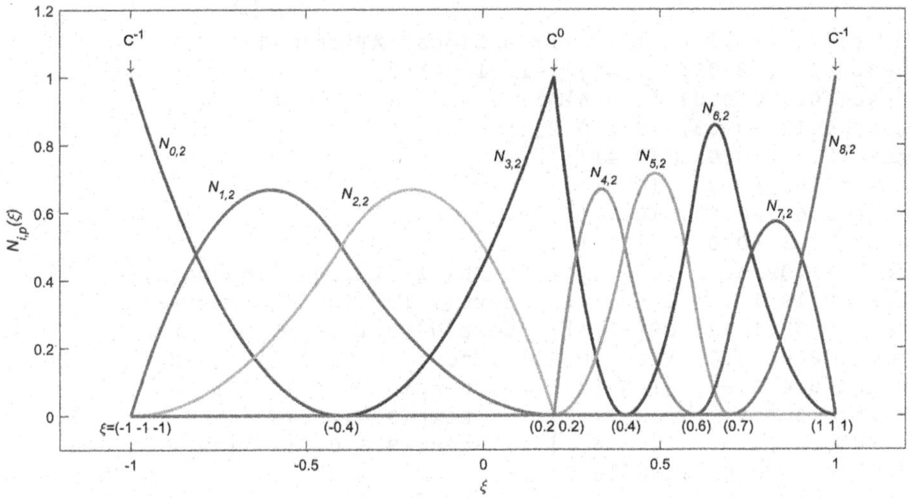

FIGURE 1.15
The second-order NURBS basis functions example.

1.5.2 NurbsBasis Program List

```
% NurbsBasis.m
% Drawing the NURBS basis functions
% p = degree of Bspline curves
clear;clc;
p=2;
knot=[-1 -1 -1 -0.4 0.2 0.2 0.4 0.6 0.7 1 1 1]; %Nonuniform
Bspline
knot=[knot ones(1,p)];
nb=size(knot,2)-p-1;                % max number of basis
functions
ndt=1000;
t1=min(knot); t2=max(knot);
t=t1:(t2-t1)/ndt:t2;
nt=size(t,2);
wt=ones(nb-p);                      % Uniform weights
Figure(1);
Sb=zeros(nt,1);                     % BsplineBasisPlot
(order,basis)
set(0,'defaultFigurePosition',[0 0 20*50 10*50]);
for j=1:(nb-p)
  for it=1:nt
    Nurbsbasis = Nurbs(p,knot,wt,t(it));
    Sb(it,1)=Nurbsbasis(j,p+2);
  end
  plot(t',Sb,'LineWidth',2);
  hold all;
```

```
end
axis([-1.2 1.2 -0.1 1.2]); set(gca,'XTick',-1:0.5:1);
text(-1.1,-0.03,' {\xi}=(-1 -1 -1)');
text(-0.45,-0.03,'(-0.4)');
text(0.13,-0.03,'(0.2 0.2)');
text(0.36,-0.03,'(0.4)');
text(0.56,-0.03,'(0.6)');
text(0.68,-0.03,'(0.7)');
text(0.93,-0.03,'(1 1 1)');
text(-1.02,1.12,'C^{-1}'); text(-1.01,1.04,'\downarrow');
text(0.18,1.12,'C^{0}');   text(0.19,1.04,'\downarrow');
text(0.98,1.12,'C^{-1}');  text(0.99,1.04,'\downarrow');
text(-0.92,0.8,'\itN_{0,2}'); text(-0.78,0.67,'\itN_{1,2}');
text(-0.4,0.65,'\itN_{2,2}'); text(0.03,0.76,'\itN_{3,2}');
text(0.3,0.71,'\itN_{4,2}');  text(0.44,0.76,'\itN_{5,2}');
text(0.62,0.9,'\itN_{6,2}');  text(0.8,0.61,'\itN_{7,2}');
text(0.98,0.8,'\itN_{8,2}');
xlabel('\it{\xi}'); ylabel('\itN_{i,p}(\it{\xi})');
```

1.5.3 Nurbs Function List

```
function Nurbsbasis = Nurbs(p,knot,wt,t)
% Nurbs.m
% p = degree of Bspline curve
% knot = knot vector
% w = weight vector
% t = parameter
%-----------------------------------------------------------
nb=size(knot,2)-p-1;             % number of basis functions
% Constructing NURBS basis functions
Nurbsbasis=zeros(nb,p+1+1);               % Rational NURBS
(order,basis)
for ip=0:p                       % degree of polynomial
 for i=1:(nb-ip)                 % i-th basis function
  iknot=knot(i);
  jknot=knot(i+1);
  kknot=knot(i+ip);
  lknot=knot(i+ip+1);
  if ip==0                       % ip = 0 degree
    Nurbsbasis(i,ip+1)=1;
    if t>=iknot && t<jknot
      Nurbsbasis(i,ip+1)=1;           % unity
    else
      Nurbsbasis(i,ip+1)=0;           % zero
    end
    if i==(nb-p) && t==jknot
      Nurbsbasis(i,ip+1)=1;           % unity
    end
  end
  if ip>0                        % ip > 0 degree
```

```
    if kknot==iknot
       fract1=0;
    else
       fract1=(t-iknot)/(kknot-iknot);
    end
    if lknot==jknot
       fract2=0;
    else
       fract2=(lknot-t)/(lknot-jknot);
    end
    Nurbsbasis(i,ip+1)=fract1*Nurbsbasis(i,ip)+fract2*Nurbsbas
is(i+1,ip);
  end
 end
end
% Constructing Rational NURBS basis functions
for j=1:(nb-p)
    Nurbsbasis(j,p+2)=Nurbsbasis(j,p+1)*wt(j);
    ydenom=0;
    for i=1:(nb-p)
      ydenom=ydenom+Nurbsbasis(i,p+1)*wt(i);
    end
    Nurbsbasis(j,p+2)=Nurbsbasis(j,p+2)/ydenom;
end
return
```

MATLAB code NurbsCurveDrawCP.m uses the function Nurbs.m (Section 1.5.3) to construct the NURBS curve by using the following control points: (0,0), (1,5), (3,0), (5,5), (7,0), (10,0), (13,5), (15,–1), and (17,7). By using the given control points, we can draw the curve created by the NURBS as shown in Figure 1.16.

From the figure, we can see the capability of the NURBS to connect several polynomial curves into a single formulation. By inserting the $\begin{bmatrix} -0.4 & 0.2 & 0.2 & 0.4 & 0.6 & 0.7 \end{bmatrix}$ knots into the standard $\Xi = \begin{bmatrix} -1 & -1 & -1 & 1 & 1 & 1 \end{bmatrix}$ knot vector of $p = 2$ NURBS curve, we can get a continuity condition that connects all the polynomial curves.

1.5.4 NurbsCurveDrawCP Program List

```
% NurbsCurveDrawCP.m
% Drawing the NURBS curve using Control Points
% p = degree of Bspline curves
clear var;clc;
p=2;
knot=[-1 -1 -1 -0.4 0.2 0.2 0.4 0.6 0.7 1 1 1]; % Nonuniform
knots
px0=[ 0 1 3 5 7 10 13  15 17];% Control points
py0=[ 0 5 0 5 0  0  5  -1  7];
```

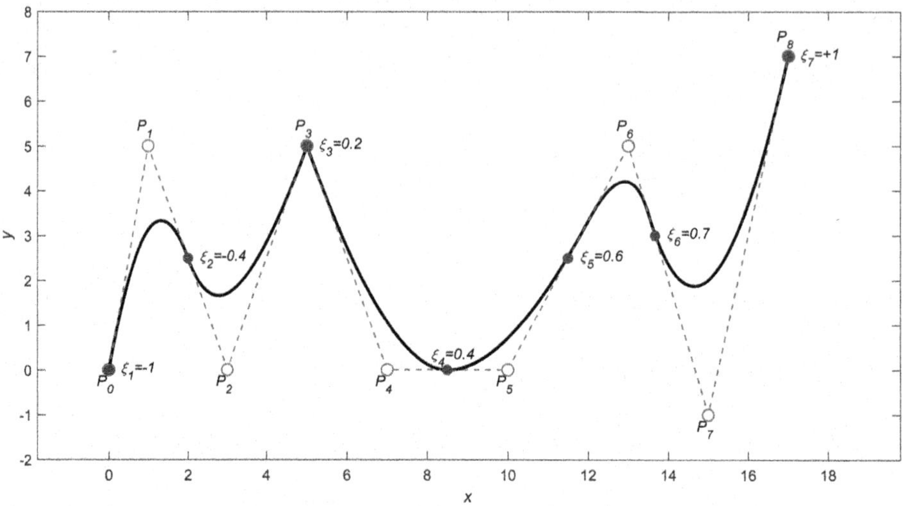

FIGURE 1.16
The NURBS curve example.

```
knot=[knot ones(1,p)];
px=px0; py=py0;
%-----------------------------------------------------------------
nb=size(knot,2)-p-1;                    % max number of basis
functions
ndt=1000;                                % steps of parameter t
t1=min(knot); t2=max(knot);
t=t1:(t2-t1)/ndt:t2;
nt=size(t,2);
wt=ones(nb-p);                          % Uniform weights
curvx=zeros(nt,1);                      % Nurbs variables
curvy=zeros(nt,1);
for ib=1:nb-p                           % Number of basis function
  for it=1:nt
    Nurbsbasis = Nurbs(p,knot,wt,t(it));
    curvx(it)=curvx(it)+Nurbsbasis(ib,p+2)*px(ib);
    curvy(it)=curvy(it)+Nurbsbasis(ib,p+2)*py(ib);
  end
end
set(0,'defaultFigurePosition',[0 0 20*50 10*50]);
Figure(2);
plot(curvx,curvy,'-k','LineWidth',2);
hold all;
plot(px,py,'--ro','LineWidth',1,'MarkerEdgeColor',...
          'r','MarkerFaceColor','w','MarkerSize',8);
hold all;
xknot=[0 2.0 5 8.5 11.5 13.6667 17];
```

```
yknot=[0 2.5 5 0.0  2.5  3          7];
plot(xknot,yknot,'bo','LineWidth',1,'MarkerEdgeColor',...
            'b','MarkerFaceColor','b','MarkerSize',6);
hold all; axis([-1.8 19.8 -2 8]);
text(px(1)-0.3,py(1)-0.3,'\itP_{0}'); text(px(2)-0.3,
py(2)+0.4,'\itP_{1}');
text(px(3)-0.3,py(3)-0.3,'\itP_{2}'); text(px(4)-0.3,
py(4)+0.4,'\itP_{3}');
text(px(5)-0.3,py(5)-0.3,'\itP_{4}'); text(px(6)-0.3,
py(6)-0.3,'\itP_{5}');
text(px(7)-0.3,py(7)+0.4,'\itP_{6}'); text(px(8)-0.3,
py(8)-0.3,'\itP_{7}');
text(px(9)-0.3,py(9)+0.4,'\itP_{8}');
text(xknot(1)+0.3,yknot(1),'\it{\xi}_1=-1');
text(xknot(2)+0.3,yknot(2),'\it{\xi}_2=-0.4');
text(xknot(3)+0.3,yknot(3),'\it{\xi}_3=0.2');
text(xknot(4)-0.4,yknot(4)+0.3,'\it{\xi}_4=0.4');
text(xknot(5)+0.3,yknot(5),'\it{\xi}_5=0.6');
text(xknot(6)+0.3,yknot(6),'\it{\xi}_6=0.7');
text(xknot(7)+0.3,yknot(7),'\it{\xi}_7=+1');
xlabel('\itx'); ylabel('\ity');
```

1.6 NURBS Surface

The rational B-spline polynomials of a curve which are given in Equation (1.15) can be extended to create a surface by using the following equation:

$$\begin{aligned} x(\xi,\eta) &= \sum_{i=0}^{m}\sum_{j=0}^{n} R_{i,j}^{p,q}(\xi,\eta) P_{x(\xi,\eta)} \\ y(\xi,\eta) &= \sum_{i=0}^{m}\sum_{j=0}^{n} R_{i,j}^{p,q}(\xi,\eta) P_{y(\xi,\eta)}, \quad \left\{ \begin{array}{l} -1 \le \xi \le 1 \\ -1 \le \eta \le 1 \end{array} \right. \\ z(\xi,\eta) &= \sum_{i=0}^{m}\sum_{j=0}^{n} R_{i,j}^{p,q}(\xi,\eta) P_{z(\xi,\eta)} \end{aligned} \tag{1.19}$$

where

$$R_{i,j}^{p,q}(\xi,\eta) = \frac{S_{i,p}(\xi) S_{j,q}(\eta) w_{i,j}}{\sum_{k=0}^{m}\sum_{l=0}^{n} S_{k,p}(\xi) S_{l,q}(\eta) w_{k,l}} \qquad w_{i,j} > 0$$

Here, $w_{i,j}$ is the weighting parameter of points in the *control mesh*. $R_{i,j}^{p,q}(\xi,\eta)$ are the rationalized $S_{i,p}(\xi)S_{j,q}(\eta)$ basis functions of the rational B-spline curves. Similar to the Bézier surface formulation, the points $P_{x(\xi,\eta)}, P_{y(\xi,\eta)}, P_{z(\xi,\eta)}$ are the coordinates of corresponding control mesh that maps a set of control points in a rectangular grid (ξ,η) as illustrated in Figure 1.12.

In the matrix form, Equation (1.19) can be written as

$$\mathbf{x} = \begin{bmatrix} x_{11} & y_{11} & z_{11} \\ \cdot & \cdot & \cdot \\ x_{33} & y_{33} & z_{33} \end{bmatrix}_{(p+1+m)\times(q+1+n),3}$$

$$\alpha = \begin{bmatrix} P_{x,11} & P_{y,11} & P_{z,11} \\ \cdot & \cdot & \cdot \\ P_{x,33} & P_{y,33} & P_{z,33} \end{bmatrix}_{(p+1+m)\times(q+1+n),3}$$

$$\mathbf{T} = \begin{bmatrix} R_{0,0}^{2,2}(\xi_1,\eta_1) & \cdot & \cdot & \cdot & R_{2,2}^{2,2}(\xi_1,\eta_1) \\ \cdot & \cdot & \cdot & \cdot & \cdot \\ \cdot & \cdot & R_{1,1}^{2,2}(\xi_2,\eta_2) & \cdot & \cdot \\ \cdot & \cdot & \cdot & \cdot & \cdot \\ R_{0,0}^{2,2}(\xi_3,\eta_3) & \cdot & \cdot & \cdot & R_{2,2}^{2,2}(\xi_3,\eta_3) \end{bmatrix}_{\substack{(p+1+m)\times(q+1+n),\\ (p+1+m)\times(q+1+n)}} \tag{1.20}$$

Consider a second degree of NURBS basis functions of a surface with $p=2$ having knot vector as follow:

$\Xi = \begin{bmatrix} -1 & -1 & -1 & -0.2 & 0.5 & 1 & 1 & 1 \end{bmatrix}$ moreover, $\mathbf{H} = \begin{bmatrix} -1 & -1 & -0.2 & 0.5 & 1 & 1 & 1 \end{bmatrix}$ with uniform unit weight parameters $\mathbf{w}_{i,j} = \mathbf{1}$ is considered.

MATLAB code NurbsBasisSurface.m uses function NurbsSurface.m (Section 1.8.1). The sampling points are created from a 5×5 equal interval control. The basis functions that are used to create the NURBS surface are shown in Figure 1.17.

1.6.1 NurbsBasisSurface Program List

```
% NurbsBasisSurface.m
% Drawing the NURBS basis functions of a surface
% p,q = order of NURBS surface in two directions
clear variables; clc;
%
```

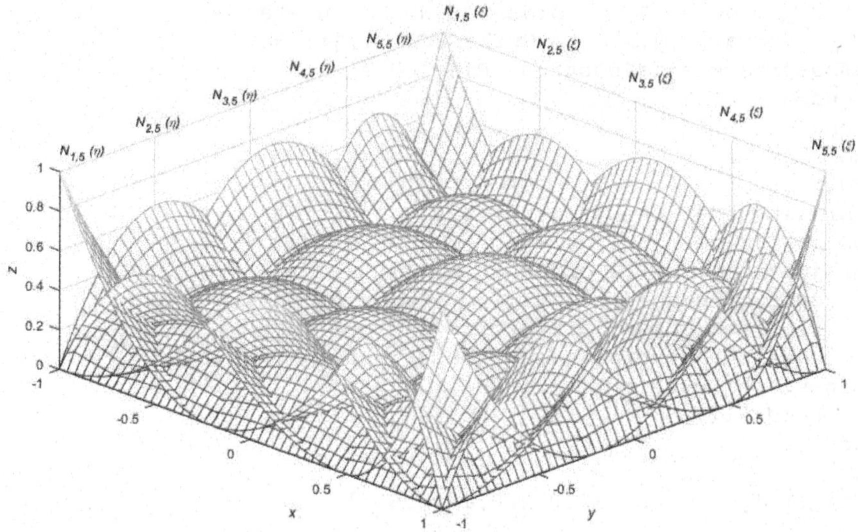

FIGURE 1.17
The basis functions of a NURBS surface.

```
p=2; m=0;
q=2; n=0;
% Parameter ksi=s and eta=t
knotksi=[-1 -1 -1 -0.2 0.5 1 1 1 ones(1,p)]; % ones p-dummy
knoteta=[-1 -1 -1 -0.2 0.5 1 1 1 ones(1,q)]; % ones q-dummy
ns=size(knotksi,2)-p-1;
nt=size(knoteta,2)-q-1;
ws=ones(ns-p);
wt=ones(nt-q);
%
% Drawing the NURBS Surface
is=2/50;          it=is;         % 50 equal intervals
si=-1:is:+1;      ti=-1:it:+1;
nsd=size(si,2);   ntd=size(ti,2);
xdraw=zeros(nsd,ntd);
ydraw=zeros(nsd,ntd);
for k=1:nsd   % ksi refinement data
for l=1:ntd   % eta refinement data
    xdraw(k,l)=si(k);
    ydraw(k,l)=ti(l);
end
end
stdraw=zeros(ns-p,nt-q,nsd,ntd);
for k=1:nsd   % ksi refined
for l=1:ntd   % eta refined
  NurbsSFC = NurbsSurface(p,knotksi,q,knoteta,si(k),ti(l));
  for i=1:ns-p          % order of NURBS in ksi
```

```
  for j=1:nt-q      % order of NURBS in eta
      stdraw(i,j,k,l) = NurbsSFC(1,i,j); %
NurbsSurfaceBasisFunctions Rij
  end
  end
end
end
% Drawing the (ns-p)X(nt-q) basis functions
for i=1:ns-p       % order of NURBS in ksi
for j=1:nt-q       % order of NURBS in eta
    zdraw=reshape(stdraw(i,j,:,:),[nsd,ntd]);
    %surface(xdraw,ydraw,zdraw)
    mesh(xdraw,ydraw,zdraw,'FaceLighting','gouraud',
'LineWidth',0.5)
    hold all;
end
end
set(gcf,'position',[200,200,1000,600]);
azm = 45;
view(azm,45);
hold all;
% Plotting labels
for i=1:ns-p       % order of NURBS in ksi
        itxt=['\itN_{' num2str(i) ',' num2str(ns-p) '} (\xi)' ];

text(xdraw(1+(i-1)*12,51),ydraw(1+(i-1)*12,51),1.1,itxt);
        hold all;
end
for i=1:nt-q       % order of NURBS in eta
        itxt=['\itN_{' num2str(i) ',' num2str(nt-q) '} (\eta)' ];

text(xdraw(1,1+(i-1)*10),ydraw(1,1+(i-1)*10),1.1,itxt);
        hold all;
end
xlabel('\itx'); ylabel('\ity'); zlabel('\itz');
```

MATLAB code NurbsSurfaceCP.m uses function NurbsSurface.m (Section 1.8.1). The sampling points are created from a control mesh of 5×5 with the heights of the points randomly determined. The program then solves the control points matrix α in Equation (1.20) to compute the other coordinates of points on the surface in 50×50 drawing mesh. The surface created by using the NURBS surface is shown in Figure 1.18.

1.6.2 NurbsSurfaceCP Program List

```
% NURBSSurfaceCP.m
% Drawing the NURBS surface by calculating the control points
first
% p,q = order of NURBS surface in two directions
```

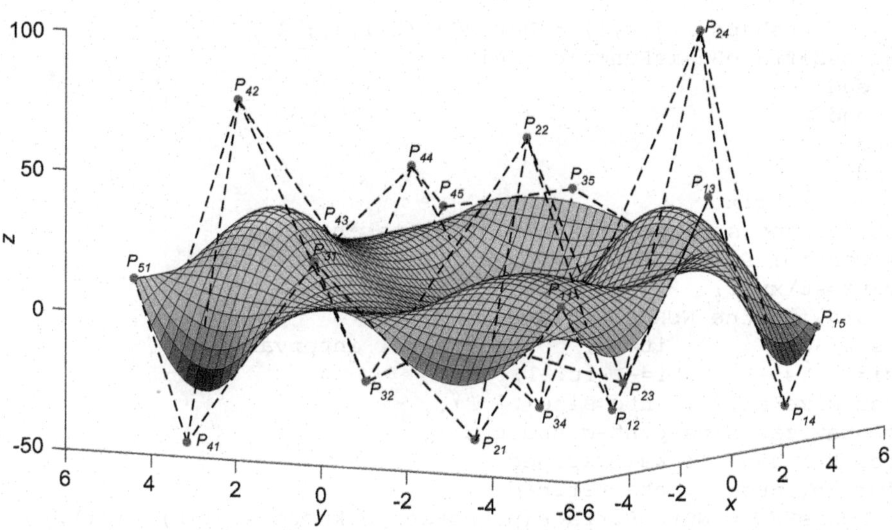

FIGURE 1.18
A surface created by using NURBS.

```
clear variables; clc;
% Coordinates of sampling points: xi, yi and zi
%%
xi = [linspace(-5,5,5) linspace(-5,5,5) linspace(-5,5,5) ...
      linspace(-5,5,5) linspace(-5,5,5)];
yi = [linspace(-5,-5,5) linspace(-2.5,-2.5,5) linspace(0,0,5)
...
      linspace(2.5,2.5,5) linspace(5,5,5)];
rng(400); zi=randi([-20 25],1,25);
p=2; m=0;
q=2; n=0;
xyzi=transpose([xi' yi' zi'])';
% Parameter ksi=s and eta=t
knotksi=[-1 -1 -1 -0.2 0.5 1 1 1 ones(1,p)]; % ones p-dummy
knoteta=[-1 -1 -1 -0.2 0.5 1 1 1 ones(1,q)]; % ones q-dummy
ns=size(knotksi,2)-p-1;
nt=size(knoteta,2)-q-1;
ws=ones(ns-p); ss=[-1 -0.5 0 0.5 1]; nss=size(ss,2);
wt=ones(nt-q); tt=[-1 -0.5 0 0.5 1]; ntt=size(tt,2);
wst=ones(ns-p,nt-q);
stshape=zeros(ns-p,nt-q,nss,ntt);
for k=1:nss   % ksi
for l=1:ntt   % eta
  NurbsSFC = NurbsSurface(p,knotksi,q,knoteta,ss(k),tt(l));
  for i=1:ns-p       % order of NURBS in ksi
  for j=1:nt-q       % order of NURBS in eta
```

```
      stshape(i,j,k,l) = NurbsSFC(1,i,j); %
NurbsSurfaceBasisFunctions Rij
  end
  end
end
end
% [T] matrix
st=reshape(stshape,[(ns-p)*(nt-q),nss*ntt])';
% Control Points Px's, Py's and Pz's calculation
pxyz=st\xyzi;
% Drawing the NURBS Surface
is=2/50;        it=is;        % equal interval
si=-1:is:+1;    ti=-1:it:+1;
nsd=size(si,2);  ntd=size(ti,2);
stdraw=zeros(ns-p,nt-q,nsd,ntd);
for k=1:nsd    % ksi refined
for l=1:ntd    % eta refined
  NurbsSFC = NurbsSurface(p,knotksi,q,knoteta,si(k),ti(l));
  for i=1:ns-p      % order of NURBS in ksi
  for j=1:nt-q      % order of NURBS in eta
      stdraw(i,j,k,l) = NurbsSFC(1,i,j); %
NurbsSurfaceBasisFunctions Rij
  end
  end
end
end
% [T] matrix
st=reshape(stdraw,[(ns-p)*(nt-q),nsd*ntd])';
% Coordinates of points in the control mesh
xyzdraw=st*pxyz;
xdraw=reshape(xyzdraw(:,1),[nsd,ntd]);
ydraw=reshape(xyzdraw(:,2),[nsd,ntd]);
zdraw=reshape(xyzdraw(:,3),[nsd,ntd]);
set(gcf,'position',[200,200,1000,600]);
azm = -59;
view(azm,9)
surface(xdraw,ydraw,zdraw)
xlabel('\itx'); ylabel('\ity'); zlabel('\itz');
hold all;
% Plotting the control points
scatter3(pxyz(:,1)',pxyz(:,2)',pxyz(:,3)','filled',
'MarkerFaceColor','r');
% Plotting x-direction control mesh
for j = 1:nt-q
    lxyz=[];
    for i = 1:ns-p
        lxyz=[lxyz pxyz(i+(j-1)*(nt-q),:)'];
    end
    hold all;
    plot3(lxyz(1,:),lxyz(2,:),lxyz(3,:),'--k','LineWidth',1);
```

```
end
% Plotting y-direction control mesh
for j = 1:ns-p
    lxyz=[];
    for i = 1:nt-q
        lxyz=[lxyz pxyz(j+(i-1)*(ns-p),:)'];
    end
    hold all;
    plot3(lxyz(1,:),lxyz(2,:),lxyz(3,:),'--k','LineWidth',1);
end
% Plotting labels
for i = 1:nt-q
    for j = 1:ns-p
        itxt=['\itP_{' num2str(i) num2str(j) '}'];
        ij=j+(i-1)*(ns-p);
        text(pxyz(ij,1),pxyz(ij,2),pxyz(ij,3),itxt);
    end
end
```

1.7 Derivatives of NURBS Curve

Because a NURBS curve is originated from the family of rational B-spline curves, the derivatives of the NURBS can also be obtained from the derivatives of the rational B-spline curve formula. The derivatives of the basis functions of the rational B-spline curves can be obtained by applying the quotient rule to Equation (1.15) as

$$\frac{dS_{i,p}(\xi)}{d\xi} = w_i \frac{\Psi(\xi)N'_{i,p}(\xi) - \Psi'(\xi)N_{i,p}(\xi)}{\left(\Psi(\xi)\right)^2} \tag{1.21}$$

where

$$(\cdot)' = \frac{d(\cdot)}{d\xi}$$

and

$$\Psi(\xi) = \sum_{j=0}^{n} N_{j,p}(\xi)w_j \tag{1.22}$$

The derivatives of the basis functions of the B-spline curves can be obtained from

$$\frac{dN_{i,p}(\xi)}{d\xi} = \frac{p}{\xi_{i+p} - \xi_i} N_{i,p-1}(\xi) - \frac{p}{\xi_{i+p+1} - \xi_{i+1}} N_{i+1,p-1}(\xi) \tag{1.23}$$

We can generalize for the higher derivatives as

$$\frac{d^k N_{i,p}(\xi)}{d\xi^k} = \frac{p!}{(p-k)!} \sum_{j=0}^{k} \alpha_{k,j} N_{i+j,p-k}(\xi) \tag{1.24}$$

where

$$\alpha_{0,0} = 1,\ \alpha_{k,0} = \frac{\alpha_{k-1,0}}{\xi_{i+p-k+1} - \xi_i},\ \alpha_{k,j} = \frac{\alpha_{k-1,j} - \alpha_{k-1,j-1}}{\xi_{i+p+j-k+1} - \xi_{i+j}},\ \alpha_{k,k} = \frac{-\alpha_{k-1,k-1}}{\xi_{i+p+1} - \xi_{i+k}}$$

where $j = 1, \ldots, k-1$.

The coefficient is defined to be zero when the denominator of these coefficients becomes zero in the repeated knots (Piegl and Tiller 1997).

Generalization of the derivatives of the rational B-spline curve in Equation (1.15) is given as

$$\begin{aligned} \frac{d^k x(\xi)}{d\xi^k} &= \sum_{i=0}^{n} \frac{d^k S_{i,p}(\xi)}{d\xi^k} P_{x(\xi)} \\ \frac{d^k y(\xi)}{d\xi^k} &= \sum_{i=0}^{n} \frac{d^k S_{i,p}(\xi)}{d\xi^k} P_{y(\xi)} \end{aligned} \qquad -1 \le \xi \le 1 \tag{1.25}$$

where

$$\frac{d^k S_{i,p}(\xi)}{d\xi^k} = \frac{\Lambda_i^k(\xi) - \sum_{j=1}^{k} \binom{k}{j} \Psi^j(\xi) \frac{d^{k-j} S_{i,p}(\xi)}{d\xi^{k-j}}}{\Psi(\xi)^{2k}}$$

$$\binom{k}{j} = \frac{k!}{j!(k-j)!} \text{ and } \Lambda_i^k(\xi) = w_i \frac{d^k S_{i,p}(\xi)}{d\xi^k}$$

MATLAB code DNurbsBasis.m uses function DNurbsLeibnitz.m (Section 1.7.2) to draw the first derivative of NURBS curve in Section 1.5; the results are shown in Figure 1.19.

1.7.1 DNurbsBasis Program List

```
% DNurbsBasis.m
% Drawing the first derivative NURBS basis functions
```

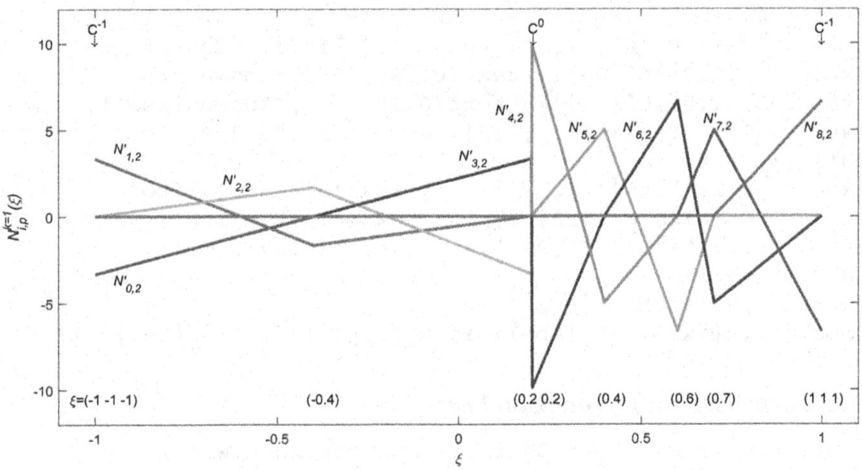

FIGURE 1.19
The derivatives of NURBS basis functions.

```
% p = order of Bspline curves
clear var;clc;
p=2;
knot=[-1 -1 -1 -0.4 0.2 0.2 0.4 0.6 0.7 1 1 1]; % Nonuniform
Knots
knot=[knot ones(1,p)];
nb=size(knot,2)-p-1;                    % max number of basis
functions
ndt=1000;                                % t steps
t1=min(knot); t2=max(knot);
t=t1:(t2-t1)/ndt:t2;
nt=size(t,2);
wt=ones(nb-p);                          % uniform weights
Sb=zeros(nt,1);
set(0,'defaultFigurePosition',[0 0 20*50 10*50]);
Figure(1);
for j=1:(nb-p)
  for it=1:nt
    DNurbsbasis = DNurbsLeibnitz(p,knot,wt,t(it));
    Sb(it,1)=DNurbsbasis(j,p+7); % first derivative of the
function
  end
  plot(t',Sb,'LineWidth',2);
  hold all;
end
axis([-1.1 1.1 -12 12]); % first derivative
set(gca,'XTick',-1:0.5:1);
text(-1.08,-10.5,' {\xi}=(-1 -1 -1)');
text(-0.42,-10.5,'(-0.4)'); text(0.15,-10.5,'(0.2 0.2)');
text(0.38,-10.5,'(0.4)');   text(0.58,-10.5,'(0.6)');
```

```
text(0.68,-10.5,'(0.7)');    text(0.95,-10.5,'(1 1 1)');
text(-1.02,11.0,'C^{-1}'); text(-1.01,10.3,'\downarrow');
text(0.19,11.0,'C^{0}'); text(0.195,10.3,'\downarrow');
text(0.98,11.0,'C^{-1}'); text(0.99,10.3,'\downarrow');
text(-0.95,-3.8,'\itN''_{0,2}'); text(-0.95,3.8,'\
itN''_{1,2}');
text(-0.65,2,'\itN''_{2,2}');      text(0.0,3.4,'\itN''_{3,2}');
text(0.1,6,'\itN''_{4,2}');       text(0.3,5,'\itN''_{5,2}');
text(0.45,5,'\itN''_{6,2}');        text(0.67,5.5,'\
itN''_{7,2}');
text(0.95,5,'\itN''_{8,2}');
xlabel('\it{\xi}'); ylabel('\itN_{i,p}^{k=1}(\it{\xi})');
```

1.7.2 DNurbsLeibnitz Function List

```
function DNurbsbasis = DNurbsLeibnitz(p,knot,wt,t)
% DNurbsLeibnitz.m for Nonuniform knots
% p = degree of Bspline curves, knot = knot vector, t =
parameter
% storage info: N_i^0 (1 to p+1); N_i' (p+2 to p+4)
% storage info: R_i^0 (p+5);    R_i' and R_i'' (p+7 to p+8)
%---------------------------------------------------------------
nb=size(knot,2)-p-1;                    % number of basis
functions
% Constructing Bspline basis functions (p+1)
DNurbsbasis=zeros(nb,nb-p+11);              % B spline
(order,basis)
for ip=0:p                              % degree of polynomial
 for i=1:(nb-ip)                        % number of basis function
  iknot=knot(i);
  jknot=knot(i+1);
  kknot=knot(i+ip);
  lknot=knot(i+ip+1);
  if ip==0
    if t>=iknot && t<jknot              % ip = 0 degree
      DNurbsbasis(i,ip+1)=1;            % unity
    else
      DNurbsbasis(i,ip+1)=0;            % zero
    end
    if i==(nb-p) && t==jknot
      DNurbsbasis(i,ip+1)=1;            % unity at the right end
to close
    end
  end
  if ip>0                               % ip = 0 degree
    if kknot==iknot
       fract1=0;
    else
       fract1=(t-iknot)/(kknot-iknot);
    end
```

```
    if lknot==jknot
       fract2=0;
    else
       fract2=(lknot-t)/(lknot-jknot);
    end
       DNurbsbasis(i,ip+1)=fract1*DNurbsbasis(i,ip)+fract2*
DNurbsbasis(i+1,ip);
  end
 end
end
% Constructing Bspline basis functions (p+2~2p+1)
% Derivative Basis Functions (only for the k's first and
second order the p basis)
kmin=min(knot);
kmax=max(knot);
%dknot=p/(kmax-kmin);  % for drawing curves using NURBS
dknot=1;               % as the basis function
for k=1:3                  % first order for p-1,p and second
order for p only
  if k<3
     ip=p+k-2;             % p=p-1
     npi=p+1+k-3;          % storage array
     if ip == 0
        ip=1;
        npi=1;
     end
  else
     ip=p;                 % p=p
     npi=p+1+1;            % storage array
  end
  for i=1:nb-ip            % corresponding basis function
i=1:p+1
     k_i_knot=(knot(i+ip)-knot(i))*dknot;
     m_j_knot=(knot(i+ip+1)-knot(i+1))*dknot;
     if k_i_knot < 1.E-10 && m_j_knot < 1.E-10
        DNurbsbasis(i,p+1+k)=0;
     elseif k_i_knot < 1.E-10

DNurbsbasis(i,p+1+k)=-ip*DNurbsbasis(i+1,npi)/m_j_knot;
     elseif m_j_knot < 1.E-10
        DNurbsbasis(i,p+1+k)=+ip*DNurbsbasis(i,npi)/k_i_knot;
     else
        DNurbsbasis(i,p+1+k)=+ip*DNurbsbasis(i,npi)/k_i_knot...
                             -ip*DNurbsbasis(i+1,npi)/m_j_knot;
     end
  end
end
%
% Constructing NURBS p-th basis functions(stored in 2p+2
array)
```

```
for j=1:(nb-p)
    DNurbsbasis(j,p+5)=DNurbsbasis(j,p+1)*wt(j);
    ydenom=0;
    for i=1:(nb-p)
      ydenom=ydenom+DNurbsbasis(i,p+1)*wt(i);
    end
    DNurbsbasis(j,p+5)=DNurbsbasis(j,p+5)/ydenom;
end
% Constructing k-th Derivative NURBS basis functions (stored
in 2p+2+k array)
  SumNw=zeros(p+1);
  for i=1:p+1    % i=1 non-derivation
   for j=1:nb-p
    SumNw(i)=SumNw(i)+DNurbsbasis(j,p+i)*wt(j);   % Sigma(N(t)
xw(t)) term
   end
  end
  for i=1:nb-p
   for k=1:3
    SumRhs=0;
    for m=1:k
        Binomial=factorial(k)/factorial(m)/factorial(k-m);
        SumRhs=Binomial*SumNw(1+k)*DNurbsbasis(i,p+4+m);
    end
    DNurbsbasis(i,p+5+k)=(DNurbsbasis(i,p+1+k)*wt(i)-SumRhs)/
SumNw(1);
   end
  end
return
```

1.8 Derivatives of NURBS Surface

Suppose the NURBS surface is defined as

$$Q(\xi,\eta)=\sum_{i=0}^{m}\sum_{j=0}^{n}R_{i,j}^{p,q}(\xi,\eta)P_{x(\xi,\eta)}, \quad \begin{cases} -1\le\xi\le1 \\ -1\le\eta\le1 \end{cases} \tag{1.26}$$

where

$$R_{i,j}^{p,q}(\xi,\eta)=\frac{S_{i,p}(\xi)S_{j,q}(\eta)w_{i,j}}{\sum_{k=0}^{m}\sum_{l=0}^{n}S_{k,p}(\xi)S_{l,q}(\eta)w_{k,l}} \qquad w_{i,j}>0$$

$w_{i,j}$ are the weighting parameters of points in the *control mesh*. $R_{i,j}^{p,q}(\xi,\eta)$ are the rationalized $S_{i,p}(\xi)S_{j,q}(\eta)$ basis functions of the rationalized B-spline curves.

Similar to the Bézier surface formulation, the points $P_{x(\xi,\eta)}, P_{y(\xi,\eta)}, P_{z(\xi,\eta)}$ are the coordinates of corresponding control mesh that maps a set of control points in a rectangular grid (ξ,η) as illustrated in Figure 1.12.

Symbolizing the rationalized basis functions of the NURBS surface as

$$R_{i,j}^{p,q}(\xi,\eta) = \frac{S_{i,p}(\xi)S_{j,q}(\eta)w_{i,j}}{\sum_{k=0}^{m}\sum_{l=0}^{n} S_{k,p}(\xi)S_{l,q}(\eta)w_{k,l}} = \frac{T}{Z} \tag{1.27}$$

Several derivatives of the NURBS surface in Equation (1.26) can be obtained using the quotient rule for derivatives as follows:

1. The first derivatives of the NURBS surface are given by

$$\begin{aligned} Q_{,x} &= \frac{dQ(\xi,\eta)}{dx} = \frac{T_{,x}}{Z} - \frac{TZ_{,x}}{Z^2} \\ Q_{,y} &= \frac{dQ(\xi,\eta)}{dy} = \frac{T_{,y}}{Z} - \frac{TZ_{,y}}{Z^2} \end{aligned} \tag{1.28}$$

2. The second derivatives of the NURBS surface are given by

$$\begin{aligned} Q_{,xx} &= \frac{d^2Q(\xi,\eta)}{dx^2} = \frac{T_{,xx}}{Z} - 2\frac{T_{,x}Z_{,x}}{Z^2} + 2\frac{TZ_{,x}^2}{Z^3} - \frac{TZ_{,xx}}{Z^2} \\ Q_{,xy} &= \frac{d^2Q(\xi,\eta)}{dx\,dy} = \frac{T_{,xy}}{Z} - \frac{T_{,x}Z_{,y}}{Z^2} - \frac{T_{,y}Z_{,x}}{Z^2} + 2\frac{TZ_{,x}Z_{,y}}{Z^3} - \frac{TZ_{,xy}}{Z^2} \\ Q_{,yy} &= \frac{d^2Q(\xi,\eta)}{dy^2} = \frac{T_{,yy}}{Z} - 2\frac{T_{,y}Z_{,y}}{Z^2} + 2\frac{TZ_{,y}^2}{Z^3} - \frac{TZ_{,yy}}{Z^2} \end{aligned} \tag{1.29}$$

where

$$T_{,x} = \frac{dT}{dx},\quad T_{,y} = \frac{dT}{dy},\quad T_{,xx} = \frac{d^2T}{dx^2},\quad T_{,xy} = \frac{d^2T}{dxdy},\quad T_{,yy} = \frac{d^2T}{dy^2}$$

$$Z_{,x} = \frac{dZ}{dx},\quad Z_{,y} = \frac{dZ}{dy},\quad Z_{,xx} = \frac{d^2Z}{dx^2},\quad Z_{,xy} = \frac{d^2Z}{dxdy},\quad Z_{,yy} = \frac{d^2Z}{dy^2}$$

The derivatives of T are given by

$$\begin{aligned}
T &= S_{i,p}(\xi)S_{j,q}(\eta)w_{i,j} \\
T_{,x} &= \frac{dS_{i,p}(\xi)}{dx}S_{j,q}(\eta)w_{i,j} \\
T_{,y} &= S_{i,p}(\xi)\frac{dS_{j,q}(\eta)}{dx}w_{i,j} \\
T_{,xx} &= \frac{d^2S_{i,p}(\xi)}{dx^2}S_{j,q}(\eta)w_{i,j} \\
T_{,xy} &= \frac{dS_{i,p}(\xi)}{dx}\frac{dS_{j,q}(\eta)}{dx}w_{i,j} \\
T_{,yy} &= S_{i,p}(\xi)\frac{d^2S_{j,q}(\eta)}{dx^2}w_{i,j}
\end{aligned} \tag{1.30}$$

The derivatives of Z are given by

$$\begin{aligned}
Z &= \sum_{k=0}^{m}\sum_{l=0}^{n}\left(S_{k,p}(\xi)S_{l,q}(\eta)w_{k,l}\right) \\
Z_{,x} &= \sum_{k=0}^{m}\sum_{l=0}^{n}\left(\frac{dS_{i,p}(\xi)}{dx}S_{j,q}(\eta)w_{i,j}\right) \\
Z_{,y} &= \sum_{k=0}^{m}\sum_{l=0}^{n}\left(S_{i,p}(\xi)\frac{dS_{j,q}(\eta)}{dx}w_{i,j}\right) \\
Z_{,xx} &= \sum_{k=0}^{m}\sum_{l=0}^{n}\left(\frac{d^2S_{i,p}(\xi)}{dx^2}S_{j,q}(\eta)w_{i,j}\right) \\
Z_{,xy} &= \sum_{k=0}^{m}\sum_{l=0}^{n}\left(\frac{dS_{i,p}(\xi)}{dx}\frac{dS_{j,q}(\eta)}{dx}w_{i,j}\right) \\
Z_{,yy} &= \sum_{k=0}^{m}\sum_{l=0}^{n}\left(S_{i,p}(\xi)\frac{d^2S_{j,q}(\eta)}{dx^2}w_{i,j}\right)
\end{aligned} \tag{1.31}$$

MATLAB code NurbsSurface.m uses function DNurbsLeibnitz.m (Section 1.7.2) to compute the first and second derivatives of NURBS surface.

1.8.1 NurbsSurface Function List

```
function NurbsSFC = NurbsSurface(p,knotp,q,knotq,ss,tt)
% NurbsSurface.m using DNurbsLeibnitzR7
% p,q = degree of Bspline curves, knot = knot vector, ss,tt =
parameter
%-----------------------------------------------------------------
% Parameter ksi=s and eta=t
ns=size(knotp,2)-p-1;
nt=size(knotq,2)-q-1;
ws=ones(ns-p);
wt=ones(nt-q);
wst=ones(ns-p,nt-q);
NurbsSFC=zeros(6,ns-p,nt-q);
% 1 for NurbsSurface
% 2 for first derivative R,x of NurbsSurface
% 3 for first derivative R,y of NurbsSurface
% 4 for second derivative R,xx of NurbsSurface
% 5 for second derivative R,yx of NurbsSurface
% 6 for second derivative R,yy of NurbsSurface
Nurbsksi = DNurbsLeibnitz(p,knotp,ws,ss);
SX0(:) = Nurbsksi(:,p+5);
SX1(:) = Nurbsksi(:,p+7);
SX2(:) = Nurbsksi(:,p+8);
Nurbseta = DNurbsLeibnitz(q,knotq,wt,tt);
SY0(:) = Nurbseta(:,q+5);
SY1(:) = Nurbseta(:,q+7);
SY2(:) = Nurbseta(:,q+8);
T = zeros(6,ns-p,nt-q); % Numerator   :T,Tx,Ty,Txx,Txy,Tyy
Z = zeros(6,1);         % Denominator :Z,Zx,Zy,Zxx,Zxy,Zyy
for i=1:ns-p            % order of NURBS in ksi
  for j=1:nt-q          % order of NURBS in eta
    T(1,i,j) = SX0(i)*SY0(j)*wst(i,j); % NurbsSurface
    T(2,i,j) = SX1(i)*SY0(j)*wst(i,j);
    T(3,i,j) = SX0(i)*SY1(j)*wst(i,j);
    T(4,i,j) = SX2(i)*SY0(j)*wst(i,j);
    T(5,i,j) = SX1(i)*SY1(j)*wst(i,j);
    T(6,i,j) = SX0(i)*SY2(j)*wst(i,j);
    Z(1,1) = Z(1,1) + SX0(i)*SY0(j)*wst(i,j); % 0th-der
    Z(2,1) = Z(2,1) + SX1(i)*SY0(j)*wst(i,j);
    Z(3,1) = Z(3,1) + SX0(i)*SY1(j)*wst(i,j);
    Z(4,1) = Z(4,1) + SX2(i)*SY0(j)*wst(i,j);
    Z(5,1) = Z(5,1) + SX1(i)*SY1(j)*wst(i,j);
    Z(6,1) = Z(6,1) + SX0(i)*SY2(j)*wst(i,j);
  end
end
% Rationalized NurbsSurface
for i=1:ns-p           % order of NURBS in ksi
    for j=1:nt-q       % order of NURBS in eta
```

```
        NurbsSFC(1,i,j) = Nurbsksi(i,p+5)*Nurbseta(j,q+5)*wst
(i,j)/Z(1);
        % Derivative of NurbsSurface 2:R,x 3:R,y 4:R,xx 5:R,xy
6:R,yy
        NurbsSFC(2,i,j) = T(2,i,j)/Z(1)
- T(1,i,j)*Z(2)/Z(1)/Z(1);
        NurbsSFC(3,i,j) = T(3,i,j)/Z(1)
- T(1,i,j)*Z(3)/Z(1)/Z(1);
        NurbsSFC(4,i,j) = T(4,i,j)/Z(1)
- 2*T(2,i,j)*Z(2)/Z(1)^2 ...
            +2*T(1,i,j)*Z(2)^2/Z(1)^3 - T(1,i,j)*Z(4)/Z(1)^2;
        NurbsSFC(5,i,j) = T(5,i,j)/Z(1) - T(2,i,j)*Z(3)/Z(1)^2
...
           -T(3,i,j)*Z(2)/Z(1)^2 + 2*T(1,i,j)*Z(2)*Z(3)/Z(1)^3 ...
            -T(1,i,j)*Z(5)/Z(1)^2; NurbsSFC(6,i,j) =
T(6,i,j)/Z(1) - 2*T(3,i,j)*Z(3)/Z(1)^2 ...
            +2*T(1,i,j)*Z(3)^2/Z(1)^3 - T(1,i,j)*Z(6)/Z(1)^2;
    end
end
return
```

References

Bajaj C, Chen J, Xu G (1995) Modeling with cubic A-patches. *ACM Trans on Graphics* 14:103–133.

Beach RC (1991) *An Introduction to the Curves and Surfaces of Computer-Aided Design*. Van Nostrand Reinhold, New York.

Bernstein SN (1912) Démonstration du théorèm de Weirstrass fondée sûr le calcul dès proba-bilités. *Commun Soc Math Khrakow* 12(2):1–2.

Bézier PE (1972) *Numerical Control: Mathematics and Applications*. John Wiley, New York.

Bézier PE (1986) *The Mathematical Basis of the UNISURF CAD System*. Butterworth, London.

Chang G, Wu J (1981) Mathematical foundations of Bézier's technique. *CAD* 13(3):133–136.

De Boor C (1978) *A Practical Guide to Splines*. Springer-Verlag, New York.

De Boor C (1987) Cutting corners always works. *Comput Aid Geom Des* 4(1–2):125–131.

Farin GE (1993) *Curves and Surfaces for Computer Aided Geometric Design—A Practical Guide*, 3rd edition. Academic Press, Boston.

Faux ID, Pratt MJ (1981) *Computational Geometry for Design and Manufacture*. Ellis Horwood Ltd., Chichester.

Hoffmann CM (1989) *Geometric & Solid Modeling*. Morgan Kaufmann, San Mateo, CA.

Gordon WJ, Riesenfeld RF (1974) Berstein-Bézier methods for the computer-aided design of free-form curves and surfaces. *J Assoc Comp Mach* 21(2):293–310.

Hoschek J, Lasser D (1993) *Fundamentals of Computer Aided Geometric Design*. AK Peters Ltd., Wellesley.

Lei WT, Wang SB (2009) Robust real-time NURBS path interpolators. *Int J Mach Tools & Manuf* 49:625–633.

Lorentz GG (1986) *Bernstein Polynomials*. Chelsea Publishing Co., New York.

Mortenson ME (1985) *Geometric Modeling*. John Wiley, New York.

Piegl L, Tiller W (1997) *The NURBS Book*. Springer-Verlag, Berlin.

Piegl LA, Tiller W (2003) Circle approximation using integral B-splines. *Computer-Aided Des* 35:601–607.

Rogers DF, Adams JA (1990) *Mathematical Elements for Computer Graphics*, 2nd edition. McGraw-Hill, New York.

Yamaguchi F (1988) *Curves and Surfaces in Computer Aided Geometric Design*. Springer-Verlag, New York.

2

Numerical Integrations on a Surface

2.1 Numerical Integration

In the isogeometric approach, the vector and matrix formulations are necessitating numerical integration scheme because the high order polynomial presentations of the nonuniform rational basis spline (NURBS) functions are too complex. The NURBS functions that will be adopted as the shape functions in the formulations of a plate/shell element are difficult to be expressed by using formulas explicitly. Our discussion on the numerical integration and its related methods is only related to the plate/shell element formulation; for other purposes of interests, the reader should consult to other books (Gradshteyn and Ryzhik 2000; Hildebrand 1956; Kaplan 1984; Ralston and Rabinowitz 1978).

2.2 Gauss–Legendre Quadrature

The Gauss–Legendre numerical integration can achieve the exact solution of an n order of polynomials or less by using the weighted sum of function series at specified $2n-1$ points within the integration domain. The method estimates the double integration of the following two-variable function without the explicit expression of the solution:

$$\int_{-1}^{+1}\int_{-1}^{+1} f(\xi,\eta)d\xi d\eta \approx \sum_{i=1}^{m}\sum_{j=1}^{n} f(\xi_i,\eta_j)w_i w_j \tag{2.1}$$

where n is the order of polynomial function inside the integral form, and ξ_i,η_i and w_i,w_j are the local coordinates and weighting parameters of the Gauss points i and j, respectively.

The Legendre polynomials for the ξ coordinate can be defined by a recursive relation as follows:

$$P_n(\xi) = \frac{1}{n}\{(2n-1)\xi P_{n-1}(\xi) - (n-1)P_{n-2}(\xi)\} \quad n \geq 2 \tag{2.2}$$

where

$$P_0(\xi) = 1, \quad P_1(\xi) = \xi$$

The points ξ_i $(i = 1,\ldots,n)$ are the roots of Equation (2.2) within the range of [−1 1] by equating the polynomial equation to zero as

$$P_n(\xi) = 0 \quad \text{for} \quad n \geq 1 \tag{2.3}$$

The weights w_i are calculated from the corresponding ξ_i points as

$$w_i = \frac{2\left(1-\xi_i^2\right)}{n^2\ P_{n-1}^2(\xi_i)} \quad \text{for} \quad n \geq 1 \tag{2.4}$$

The points and weights of the nth-order Legendre polynomial up to the fifth order are listed in Table 2.1. The Gauss–Legendre formulations for the η coordinate are similar to those for the ξ coordinate given above.

As an illustrative example, let us calculate the volume under a given function (see Figure 2.1) of the following:

$$f(x,y) = 1 - x^2 - y^2,$$

within the range of $-1 \leq x \leq +1$, $-1 \leq y \leq +1$ and $0 \leq f(x,y) \leq +1$, by using the Gauss–Legendre numerical scheme.

The volume inside the spherical function can be calculated by using the following formula:

$$Volume = V = \int_{-1}^{+1}\int_{-1}^{+1} f(x,z)dx\,dy$$

To integrate the above equation by using the Gauss–Legendre numerical scheme, the variables x and y must be transformed to the local coordinates ξ and η. The function $f(x,y)$ must also be transformed into the local coordinates ξ and η. In this example, since the lower and upper limits of coordinates x and y are within the range of Gaussian coordinates (ξ,η) of $[\ -1 \quad +1\ ]$, then the relationships between $x = \xi$ and $y = \eta$ are held. Hence, the function and its volume can be written as

$$f(\xi,\eta) = 1 - \xi^2 - \eta^2$$

TABLE 2.1
Legendre Polynomial Points and Weights up to $n = 5$

n	Point, ξ_i / Weight, w_i
1	0
	2
2	$-\sqrt{\frac{1}{3}}\ldots,\ldots+\sqrt{\frac{1}{3}}$
	$1\ldots,\ldots1$
3	$-\sqrt{\frac{3}{5}}\cdots0\cdots+\sqrt{\frac{3}{5}}$
	$\frac{5}{9}\cdots,\frac{8}{9}\cdots,\frac{5}{9}$
4	$-\sqrt{\frac{3}{7}+\frac{2}{7}\sqrt{\frac{6}{5}}}\ldots,\ldots-\sqrt{\frac{3}{7}-\frac{2}{7}\sqrt{\frac{6}{5}}}\ldots,\ldots+\sqrt{\frac{3}{7}-\frac{2}{7}\sqrt{\frac{6}{5}}}\ldots,\ldots+\sqrt{\frac{3}{7}+\frac{2}{7}\sqrt{\frac{6}{5}}}$
	$\frac{1}{2}+\frac{1}{6}\sqrt{\frac{5}{6}}\ldots,\ldots\frac{1}{2}-\frac{1}{6}\sqrt{\frac{5}{6}}\ldots,\ldots\frac{1}{2}-\frac{1}{6}\sqrt{\frac{5}{6}}\ldots,\ldots\frac{1}{2}+\frac{1}{6}\sqrt{\frac{5}{6}}$
5	$-\frac{1}{3}\sqrt{5+2\sqrt{\frac{10}{7}}}\ldots,\ldots-\frac{1}{3}\sqrt{5-2\sqrt{\frac{10}{7}}}\ldots,\ldots+\frac{1}{3}\sqrt{5-2\sqrt{\frac{10}{7}}}\ldots,\ldots+\frac{1}{3}\sqrt{5+2\sqrt{\frac{10}{7}}}$
	$\frac{322+13\sqrt{70}}{900}\ldots,\ldots\frac{322-13\sqrt{70}}{900}\ldots,\ldots\frac{322-13\sqrt{70}}{900}\ldots,\ldots\frac{322+13\sqrt{70}}{900}$

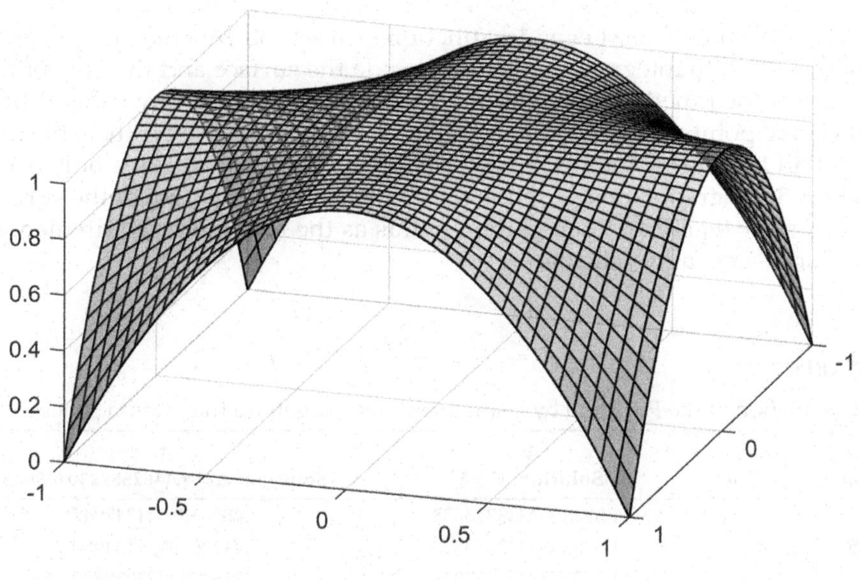

FIGURE 2.1
Surface of a sphere.

$$V = \int_{-1}^{+1}\int_{-1}^{+1}\left(1-\xi^2-\eta^2\right)d\xi\, d\eta$$

By substituting the lower and upper limits of both ξ and η variables, we obtain $V = 4/3$.

If we use the Gauss–Legendre integration scheme to do the integration, it is not necessary to find the solution by integrating the function, but we can use the original function, instead. This is the advantage of using a numerical integration scheme with the help of a computer, where finding the solution of integration might be clumsy and sometimes cannot be solved by hand.

The Gauss–Legendre integration scheme is used to calculate the area of the surface by using the following:

$$A = \int_{-1}^{+1}\int_{-1}^{+1}\sqrt{1+\left(\frac{\partial f(\xi,\eta)}{\partial \xi}\right)^2+\left(\frac{\partial f(\xi,\eta)}{\partial \eta}\right)^2}\, d\xi\, d\eta$$

$$= \sum_{i=1}^{m}\sum_{j=1}^{n}\sqrt{1+\left(\frac{\partial f(\xi,\eta)}{\partial \xi}\right)^2+\left(\frac{\partial f(\xi,\eta)}{\partial \eta}\right)^2}\, w_j w_i$$

$$= \sum_{i=1}^{m}\sum_{j=1}^{n}\sqrt{1+(-2\xi)^2+(-2\eta)^2}\, w_j w_i$$

MATLAB® code GaussLegendreFunction.m uses the function Legendre.m (Section 2.2.2) to integrate the volume inside the surface and the area of the surface of the function numerically. The computed V and A by using different Gauss' points are tabulated in Table 2.2. It can be seen from the table that to get an accurate solution, the computation of volume requires only $m = 3$ and $n = 3$ integration points, while the computation of the area of the surface requires $m = 9$ and $n = 9$ integration points as the minimum requirement in the Gauss–Legendre scheme.

TABLE 2.2

Calculation of the Function by Using the Gauss–Legendre Integration Scheme

m	n	V **Solution:** $V = 4/3$	A **Solution:** $A = 7.4462567230123635$
3	3	1.3333333333333333	7.4056458544347947
5	5	1.3333333333333333	7.4443650539134980
7	7	1.3333333333333333	7.4461243073059835
9	9	1.3333333333332493	7.4462454144412638

2.2.1 GaussLegendreFunction Program List

```
% GaussLegendreFunction.m
% Volume and Area calculation of a surface example
% Gauss point is stored in Gpw(n,1)
% Gauss weight is stored in Gpw(n,2)
clear variables;clc;
mg = 9;
ng = 9;
Gpwksi = Legendre(mg);
Gpweta = Legendre(ng);
% Plot the surface
[x,y] = meshgrid(-1:0.05:1,-1:0.05:1);
z = 1 - x .* x .* y .* y;
h=surf(x,y,z);
h.FaceAlpha = 0.5;
h.LineWidth = 1.0;
h.EdgeColor = [0.0 0.0 0.0];
ax = gca;
axis equal;
view(108,18);
ax.ZLim = [0 1];
V = 0;  % volume
A = 0;  % area
for i = 1:mg
    eta = Gpweta(i,1);
    wi  = Gpweta(i,2);
    for j = 1:ng
     ksi  = Gpwksi(j,1);
      wj  = Gpwksi(j,2);
      fz = 1 - ksi*ksi - eta*eta;
      V = V + fz * wj * wi;
      fa = sqrt(1+(-2*ksi)^2+(-2*eta)^2);
      A = A + fa * wj * wi;
    end
end
disp('Volume inside the surface = ');
disp(vpa(sym(real(V)),17));
disp('Area of the surface = ');
disp(vpa(sym(real(A)),17));
```

2.2.2 Legendre Function List

```
function Gpw = Legendre(ng)
% ng = number of Gauss points
% Gauss point is stored in Gpw(n,1)
% Gauss weight is stored in Gpw(n,2)
%---------------------------------
Gpw=zeros(ng,2,1);
```

```
switch ng
  case 1
    Gpw(1,1) =  0.000000000000000000000;
    Gpw(1,2) =  2.000000000000000000000;
  case 2
    Gpw(1,1) = -0.577350269189625764509;
    Gpw(2,1) =  0.577350269189625764509;
    Gpw(1,2) =  1.000000000000000000000;
    Gpw(2,2) =  1.000000000000000000000;
  case 3
    Gpw(1,1) = -0.774596669241483377036;
    Gpw(2,1) =  0.000000000000000000000;
    Gpw(3,1) =  0.774596669241483377036;
    Gpw(1,2) =  0.555555555555555555556;
    Gpw(2,2) =  0.888888888888888888889;
    Gpw(3,2) =  0.555555555555555555556;
  case 4
    Gpw(1,1) = -0.861136311594052575224;
    Gpw(2,1) = -0.339981043584856264803;
    Gpw(3,1) =  0.339981043584856264803;
    Gpw(4,1) =  0.861136311594052575224;
    Gpw(1,2) =  0.347854845137453857373;
    Gpw(2,2) =  0.652145154862546142627;
    Gpw(3,2) =  0.652145154862546142627;
    Gpw(4,2) =  0.347854845137453857373;
  case 5
    Gpw(1,1) = -0.906179845938663992798;
    Gpw(2,1) = -0.538469310105683091036;
    Gpw(3,1) =  0.000000000000000000000;
    Gpw(4,1) =  0.538469310105683091036;
    Gpw(5,1) =  0.906179845938663992798;
    Gpw(1,2) =  0.236926885056189087514;
    Gpw(2,2) =  0.478628670499366468041;
    Gpw(3,2) =  0.568888888888888888889;
    Gpw(4,2) =  0.478628670499366468041;
    Gpw(5,2) =  0.236926885056189087514;
  case 6
    Gpw(1,1) = -0.932469514203151938982;
    Gpw(2,1) = -0.661209386466264592509;
    Gpw(3,1) = -0.238619186083196932469;
    Gpw(4,1) =  0.238619186083196932469;
    Gpw(5,1) =  0.661209386466264592509;
    Gpw(6,1) =  0.932469514203151938982;
    Gpw(1,2) =  0.171324492379171867455;
    Gpw(2,2) =  0.360761573048138139704;
    Gpw(3,2) =  0.467913934572691007877;
    Gpw(4,2) =  0.467913934572691007877;
    Gpw(5,2) =  0.360761573048138139704;
    Gpw(6,2) =  0.171324492379171867455;
```

```
  case 7
    Gpw(1,1) = -0.94910791234275848621 4;
    Gpw(2,1) = -0.74153118559939468207 4;
    Gpw(3,1) = -0.40584515137739718415 6;
    Gpw(4,1) =  0.00000000000000000000;
    Gpw(5,1) =  0.40584515137739718415 6;
    Gpw(6,1) =  0.74153118559939468207 4;
    Gpw(7,1) =  0.94910791234275848621 4;
    Gpw(1,2) =  0.12948496616887045273 2;
    Gpw(2,2) =  0.27970539148927488222 1;
    Gpw(3,2) =  0.38183005050511889374 9;
    Gpw(4,2) =  0.41795918367346938775 5;
    Gpw(5,2) =  0.38183005050511889374 9;
    Gpw(6,2) =  0.27970539148927488222 1;
    Gpw(7,2) =  0.12948496616887045273 2;
  case 8
    Gpw(1,1) = -0.96028985649753650916 2;
    Gpw(2,1) = -0.79666647741362694995 6;
    Gpw(3,1) = -0.52553240991632899076 3;
    Gpw(4,1) = -0.18343464249564980783 6;
    Gpw(5,1) =  0.18343464249564980783 6;
    Gpw(6,1) =  0.52553240991632899076 3;
    Gpw(7,1) =  0.79666647741362694995 6;
    Gpw(8,1) =  0.96028985649753650916 2;
    Gpw(1,2) =  0.10122853629037002200 5;
    Gpw(2,2) =  0.22238103445337263424 6;
    Gpw(3,2) =  0.31370664587788726706 1;
    Gpw(4,2) =  0.36268378337836197937 5;
    Gpw(5,2) =  0.36268378337836197937 5;
    Gpw(6,2) =  0.31370664587788726706 1;
    Gpw(7,2) =  0.22238103445337263424 6;
    Gpw(8,2) =  0.10122853629037002200 5;
case 9
    Gpw(1,1) = -0.96816023950762508659 8;
    Gpw(2,1) = -0.83603110732663610255 2;
    Gpw(3,1) = -0.61337143270059257810 4;
    Gpw(4,1) = -0.32425342340380897132 6;
    Gpw(5,1) =  0.00000000000000000000;
    Gpw(6,1) =     0.32425342340380897132 6;
    Gpw(7,1) =     0.61337143270059257810 4;
    Gpw(8,1) =     0.83603110732663610255 2;
    Gpw(9,1) =     0.96816023950762508659 8;
    Gpw(1,2) =     0.08127438836159960639 6;
    Gpw(2,2) =     0.18064816069485431126 8;
    Gpw(3,2) =     0.26061069640292428513 8;
    Gpw(4,2) =     0.31234707704000274434 8;
    Gpw(5,2) =     0.33023935500125976316 5;
    Gpw(6,2) =     0.31234707704000274434 8;
    Gpw(7,2) =     0.26061069640292428513 8;
```

```
        Gpw(8,2) =     0.18064816069485431126 8;
        Gpw(9,2) =     0.08127438836159960639 6;
    case 10
        Gpw(1,1) = -0.9739065285171691898 64;
        Gpw(2,1) = -0.8650633666889867567 38;
        Gpw(3,1) = -0.6794095682990239914 46;
        Gpw(4,1) = -0.4333953941292473799 33;
        Gpw(5,1) = -0.1488743389816312157 06;
        Gpw(6,1) =  0.1488743389816312157 06;
        Gpw(7,1) =  0.4333953941292473799 33;
        Gpw(8,1) =  0.6794095682990239914 46;
        Gpw(9,1) =  0.8650633666889867567 38;
        Gpw(10,1)=  0.9739065285171691898 64;
        Gpw(1,2) =  0.0666713443087583118 81;
        Gpw(2,2) =  0.1494513491505552092 79;
        Gpw(3,2) =  0.2190863625159845668 33;
        Gpw(4,2) =  0.2692667193099957572 14;
        Gpw(5,2) =  0.2955242247147528654 02;
        Gpw(6,2) =  0.2955242247147528654 02;
        Gpw(7,2) =  0.2692667193099957572 14;
        Gpw(8,2) =  0.2190863625159845668 33;
        Gpw(9,2) =  0.1494513491505552092 79;
        Gpw(10,2) = 0.0666713443087583118 81;
end
return
```

2.3 Jacobian Operator of a Surface

In subsequent chapters, when we deal with the plate/shell problems in the finite element formulations, the numerical integration on the surface of plate/shell is necessary. The plate/shell elements such as stiffness matrix, mass matrix, and loading vector are constructed by integrating the material properties, cross section, or distributed mass over the area of a surface.

In the previous example, we integrated the functions of x and y variables both within the range of $\begin{bmatrix} -1 & +1 \end{bmatrix}$, respectively. What happens if we need to integrate an equation by using the real Cartesian coordinates of different ranges? Apparently, we will find in most of the cases when we deal with the real problems. Therefore, we need an operator that can map the relationships (*geometry mapping*) between both variables ξ,η with their associated real values of x,y.

In calculus, a scaling factor that relates the incremental variable of both variables ξ and η is called the "Jacobian" operator. We denote it by J.

In case of geometry mapping of one variable $x \rightarrow \xi$, the derivative of the variable can be obtained by invoking the chain rule as follows:

$$\frac{d}{d\xi} = \frac{dx}{d\xi}\frac{d}{dx} = J\frac{d}{dx} \tag{2.5}$$

where

$$\frac{dx}{d\xi} = J \quad \text{or} \quad \frac{d\xi}{dx} = \frac{1}{J} = J^{-1}$$

which requires that J is a nonzero value. The numerical integration becomes

$$\int_{x_1}^{x_2} f(x)dx = \int_{-1}^{+1} f(\xi)J\,d\xi \approx \sum_{i=1}^{n} f(\xi_i)Jw_i \tag{2.6}$$

In case of geometry mapping of two variables $x \to \xi$ and $y \to \eta$, the derivatives of both variables can be obtained from the chain rule as

$$\mathbf{J} = \begin{bmatrix} \dfrac{\partial x}{\partial \xi} & \dfrac{\partial y}{\partial \xi} \\ \dfrac{\partial x}{\partial \eta} & \dfrac{\partial y}{\partial \eta} \end{bmatrix} \tag{2.7}$$

In lieu of Equation (2.5),

$$\begin{Bmatrix} \dfrac{\partial(\cdot)}{\partial \xi} \\ \dfrac{\partial(\cdot)}{\partial \eta} \end{Bmatrix} = \mathbf{J} \begin{Bmatrix} \dfrac{\partial(\cdot)}{\partial x} \\ \dfrac{\partial(\cdot)}{\partial y} \end{Bmatrix} \tag{2.8}$$

In this case of geometry mapping, the two variables are reversed as $x \to \xi$ and $y \to \eta$, and the derivatives can be obtained by inversing the Jacobian operator to $\mathbf{J}^{-1}$, which is expressed as follows:

$$\begin{Bmatrix} \dfrac{\partial(\cdot)}{\partial x} \\ \dfrac{\partial(\cdot)}{\partial y} \end{Bmatrix} = \mathbf{J}^{-1} \begin{Bmatrix} \dfrac{\partial(\cdot)}{\partial \xi} \\ \dfrac{\partial(\cdot)}{\partial \eta} \end{Bmatrix} \tag{2.9}$$

The numerical integration becomes

$$\int_{y_1}^{y_2}\int_{x_1}^{x_2} f(x,y)dx\,dy = \int_{-1}^{+1}\int_{-1}^{+1} f(\xi,\eta)J\,d\xi\,d\eta \approx \sum_{i=1}^{m}\sum_{j=1}^{n} f(\xi_i,\eta_j)Jw_jw_i \tag{2.10}$$

where

$J = \det[\mathbf{J}]$

2.4 Area Calculation of the Parametric Surface

The surface (see Figure 2.2) constructed by using parametric modeling in Section 1.1 is considered for area calculation example using the Gauss–Legendre scheme.

In this example, the limits of integration of either x or y coordinates are not $\begin{bmatrix} -1 & +1 \end{bmatrix}$ as the previous functional surface. To use the Gauss–Legendre integration scheme, we need to map the x and y coordinates into the ξ and η coordinates. This can be done by generating the Jacobian operator J used in Equation (2.10).

From the program list in Section 1.1.1, we can compute the coefficients **a**, **b**, and **c** from which we can construct the parametric surface by substituting the x, y, and z coordinates in Table 1.1 into Equation (1.2):

$$x_{i,j} = 3 + 7\xi_i - \xi_i^2 - 3\eta_j^2 + 3\xi_i^2\eta_j^2$$

$$y_{i,j} = 2 + 2.5\eta_j - 0.5\eta_j^2$$

$$z_{i,j} = 10 + 0.5\xi_i - 0.5\eta_j - 5.5\xi_i^2 - 7.5\eta_j^2 + 0.5\xi_i^2\eta_j - 0.5\xi_i\eta_j^2 + 3\xi_i^2\eta_j^2 \tag{2.11}$$

The partial differentiations of the variables are

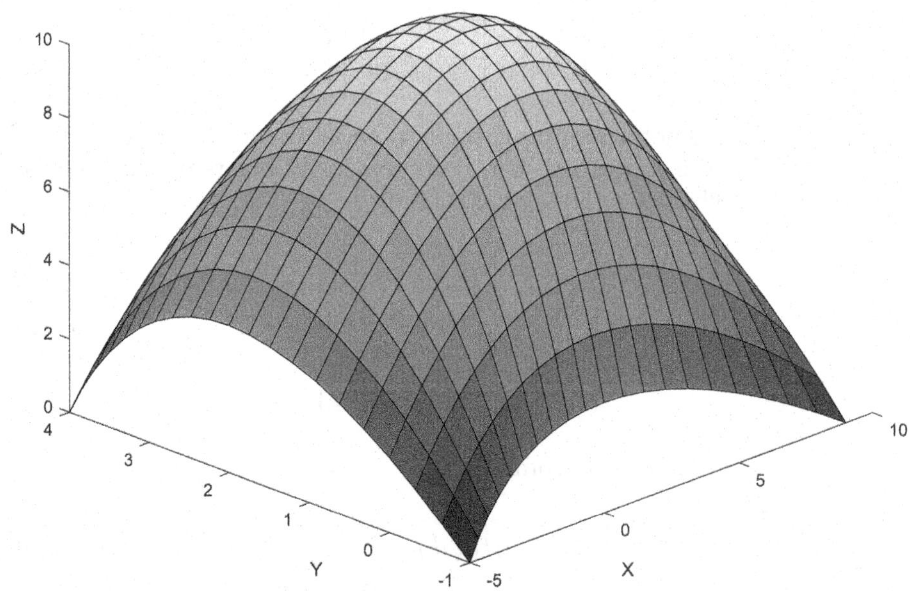

FIGURE 2.2
Area calculation of the parametric modeling of a surface.

$$\frac{\partial x}{\partial \xi} = 7 - 2\xi + 6\xi\eta^2; \quad \frac{\partial y}{\partial \xi} = 0$$

$$\frac{\partial x}{\partial \eta} = -6\eta + 6\xi^2\eta; \quad \frac{\partial y}{\partial \eta} = 2.5 - \eta$$

$$\frac{\partial z}{\partial \xi} = 0.5 - 11\xi + \xi\eta - \eta^2 + 6\xi\eta^2$$

$$\frac{\partial z}{\partial \eta} = -0.5 - 15\eta + 0.5\xi^2 - \xi\eta + 6\xi^2\eta$$

$$\frac{\partial z}{\partial x} = \frac{\partial z}{\partial \xi} \cdot \frac{\partial \xi}{\partial x} + \frac{\partial z}{\partial \eta} \cdot \frac{\partial \eta}{\partial x}$$

$$\frac{\partial z}{\partial y} = \frac{\partial z}{\partial \xi} \cdot \frac{\partial \xi}{\partial y} + \frac{\partial z}{\partial \eta} \cdot \frac{\partial \eta}{\partial y} \tag{2.12}$$

It is worth noting that the term $\partial y/\partial \xi = 0$ in Equation (2.12) means the y variable does not depend on the ξ variable. Thus, the term $\frac{\partial z}{\partial \xi} \cdot \frac{\partial \xi}{\partial y}$ in the last part of Equation (2.12) becomes $\frac{\partial z}{\partial y} = \frac{\partial z}{\partial \eta} \cdot \frac{\partial \eta}{\partial y} + \frac{\partial z}{\partial \xi} \cdot \frac{\partial \xi}{\partial y} = \frac{\partial z}{\partial \eta} \cdot \frac{1}{J}$.

The Jacobian operator at every control point (i, j) is given by

$$J = \begin{vmatrix} (7 - 2\xi + 6\xi\eta^2) & 0 \\ -6\eta + 6\xi^2\eta & 2.5 - \eta \end{vmatrix} \tag{2.13}$$

The area of the surface can be computed by

$$A = \int_{y1}^{y2} \int_{x1}^{x2} \left(\sqrt{E \times G - F^2}\right) dx\, dy \tag{2.14}$$

where

$$E = \left(\frac{dx}{d\xi}\right)^2 + \left(\frac{dy}{d\xi}\right)^2 + \left(\frac{dz}{d\xi}\right)^2$$

$$F = \frac{dx}{d\xi}\frac{dx}{d\eta} + \frac{dy}{d\xi}\frac{dy}{d\eta} + \frac{dz}{d\xi}\frac{dz}{d\eta}$$

$$G = \left(\frac{dx}{d\eta}\right)^2 + \left(\frac{dy}{d\eta}\right)^2 + \left(\frac{dz}{d\eta}\right)^2$$

TABLE 2.3

Calculation of the Parametric Surface Area by Using the Gauss–Legendre Integration Scheme

m	n	A
4	4	214.61004568130394
5	5	211.14877715433676
6	6	212.90575649555447
7	7	212.02420807746859

After mapping to the ξ and η coordinates, we can obtain the following:

$$A = \int_{-1}^{+1}\int_{-1}^{+1}\left(\sqrt{E \times G - F^2}\right) J\,d\xi\,d\eta \approx \sum_{i=1}^{m}\sum_{j=1}^{n}\left(\sqrt{E \times G - F^2}\right) J w_j w_i \tag{2.15}$$

MATLAB code ParametricSurfaceArea.m (Section 2.4.1) uses the function Legendre.m (Section 2.2.2) to calculate the area of the surface shown in Figure 2.2 which was created by parametric modeling. The area A computed by using different Gauss' points is shown in Table 2.3. It can be seen from the table that to get an accurate solution, the computation of the area of the surface requires $m = 7$ and $n = 7$ integration points as the minimum requirement in the Gauss–Legendre integration scheme. The area of the flat plate when the elevations are set to zero is 70.

We can see here that the solutions shown in Table 2.3 converged at high number of Gaussian points; hence, more integration points or smaller division of the surface area will improve the results. The more complicated the surface, the less accurate the area computed.

2.4.1 ParametricSurfaceArea Program List

```
% ParametricSurfaceArea.m
% Area calculation of the surface in Figure 1.2
% Gauss point is stored in Gpw(n,1)
% Gauss weight is stored in Gpw(n,2)
clear variables;clc;
mg = 7;
ng = 7;
Gpwksi = Legendre(mg);
Gpweta = Legendre(ng);
% Parametric Coefficients of Figure 1.2
abc(1,:) = [ 3;  7;    0;0;   -1;   -3;  0;    0;3]; %
coefficients a
abc(2,:) = [ 2;  0; 2.5;0;    0;-0.5;  0;    0;0]; %
coefficients b
```

```
abc(3,:) = [10;0.5;-0.5;0;-5.5;-7.5;0.5;-0.5;3]; %
coefficients c
%abc(3,:) = [0;0;0;0;0;0;0;0;0]; % coefficients c for flat
plate
Area = 0;
tol=1.0e-8;
for i = 1:mg
    eta = Gpweta(i,1);
    wi  = Gpweta(i,2);
    for j = 1:ng
      ksi = Gpwksi(j,1);
      wj  = Gpwksi(j,2);
      for k = 1:3     % x,y,z functions
       dksi(k,1) = abc(k,2)+abc(k,4)*eta+2*abc(k,5)*ksi+ ...
                  2*abc(k,7)*ksi*eta+abc(k,8)*eta^2+2*abc(k,9)
*ksi*eta^2;
        deta(k,1) = abc(k,3)+abc(k,4)*ksi+2*abc
(k,6)*eta+ ...
                  abc(k,7)*ksi^2+2*abc(k,8)*ksi*eta+2*abc(k,9)
*ksi^2*eta;
      end
      E=dksi(1,1)^2+dksi(2,1)^2+dksi(3,1)^2;
      F=dksi(1,1)*deta(1,1)+dksi(2,1)*deta(2,1)+dksi(3,1)*
deta(3,1);
      G=deta(1,1)^2+deta(2,1)^2+deta(3,1)^2;
      farea=sqrt(E*G-F*F);
      Area = Area + farea * wi * wj;
    end
end
disp(vpa(sym(Area),17));
```

2.5 Area Calculation of the NURBS Surface

The surface (see Figure 2.3) constructed by using NURBS in Section 1.6 is considered for the area calculation example using the Gauss–Legendre integration scheme.

Similar to the previous example, the limits of integration of x or y coordinates are not in the range of $\begin{bmatrix} -1 & +1 \end{bmatrix}$. To use the Gauss–Legendre integration scheme, we need to map the x and y coordinates into the ξ and η coordinates. This can be done by generating the Jacobian operator J used in Equation (2.10).

The NURBS equations of the surface given in Equation (1.19) are repeated here for convenience.

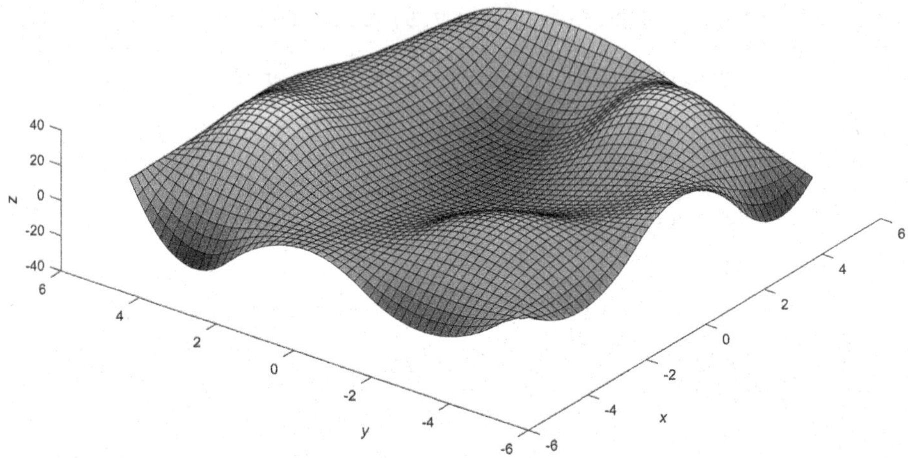

FIGURE 2.3
Area calculation of the NURBS surface.

$$
\begin{aligned}
x(\xi,\eta) &= \sum_{i=0}^{m}\sum_{j=0}^{n} S_{i,p}(\xi)S_{j,q}(\eta)P_{x(\xi,\eta)} \\
y(\xi,\eta) &= \sum_{i=0}^{m}\sum_{j=0}^{n} S_{i,p}(\xi)S_{j,q}(\eta)P_{y(\xi,\eta)}, \qquad \begin{cases} -1 \le \xi \le 1 \\ -1 \le \eta \le 1 \end{cases} \\
z(\xi,\eta) &= \sum_{i=0}^{m}\sum_{j=0}^{n} S_{i,p}(\xi_i)S_{j,q}(\eta)P_{z(\xi,\eta)}
\end{aligned}
\tag{2.16}
$$

where

$$
S_{i,p}(\xi) = \frac{N_{i,p}(\xi)w_i}{\sum_{k=0}^{n} N_{k,p}(\xi)w_k} \qquad w_i > 0
$$

$$
S_{j,q}(\eta) = \frac{N_{j,q}(\eta)w_j}{\sum_{k=0}^{m} N_{k,q}(\eta)w_k} \qquad w_j > 0
$$

The partial differentiations of the x and y coordinates with respect to the ξ and η coordinates are given in Equations (1.27) and (1.28). The partial differentiations of the z coordinate with respect to the x and y coordinates can be obtained from the chain rules as follows:

$$\frac{\partial z}{\partial x}=\frac{\partial z}{\partial \xi}\cdot\frac{\partial \xi}{\partial x}+\frac{\partial z}{\partial \eta}\cdot\frac{\partial \eta}{\partial x}$$
$$\frac{\partial z}{\partial y}=\frac{\partial z}{\partial \xi}\cdot\frac{\partial \xi}{\partial y}+\frac{\partial z}{\partial \eta}\cdot\frac{\partial \eta}{\partial y} \tag{2.17}$$

It is worth noting that if one of the terms in Equations (1.27) and (1.28) is equal to zero, it means that the origin of the mapping coordinate has no dependency. Hence, the related term in Equation (2.17) should be omitted to avoid a division by zero.

The Jacobian operator is given by

$$J=\begin{vmatrix} \dfrac{dx(\xi,\eta)}{d\xi} & \dfrac{dy(\xi,\eta)}{d\xi} \\ \dfrac{dx(\xi,\eta)}{d\eta} & \dfrac{dy(\xi,\eta)}{d\eta} \end{vmatrix} \tag{2.18}$$

The area of the surface can be computed by

$$A=\int_{y_1}^{y_2}\int_{x_1}^{x_2}\left(\sqrt{1+\left(\frac{dz}{dx}\right)^2+\left(\frac{dz}{dy}\right)^2}\right)dx\,dy \tag{2.19}$$

After mapping to the ξ and η coordinates, we can obtain the following:

$$A=\int_{-1}^{+1}\int_{-1}^{+1}\left(\sqrt{1+\left(\frac{dz}{dx}\right)^2+\left(\frac{dz}{dy}\right)^2}\right)d\xi\,d\eta$$
$$\approx\sum_{i=1}^{m}\sum_{j=1}^{n}\left(\sqrt{1+\left(\frac{dz}{d\xi}\right)^2+\left(\frac{dz}{d\xi}\right)^2}\right)Jw_jw_i \tag{2.20}$$

MATLAB code NurbsSurfaceArea.m (Section 2.5.1) uses the functions DNurbsLeibnitz.m (Section 1.7.2) and Legendre.m (Section 2.2.2) to calculate the area of the surface shown in Figure 2.3 which was created by using NURBS. The area A computed by using different Gauss' points is given in Table 2.4. It can be seen from the table that to get an accurate solution, the computation of the area of the surface requires $m=10$ and $n=10$ integration points as the minimum requirement in the Gauss–Legendre integration scheme. The area of the flat plate when the elevations are set to zero is 100.

We can see here that the solutions shown in Table 2.3 still not converged at higher NURBS polynomial order; hence, more integration points or smaller division of the surface area will improve the results.

TABLE 2.4

Calculation of the NURBS Surface Area by Using the Gauss–Legendre Integration Scheme

m	n	A
7	7	113.81798742161318
8	8	115.88946345292808
9	9	113.71726497895747
10	10	114.45956097411079

2.5.1 NurbsSurfaceArea Program List

```
% NurbsSurfaceArea.m
% Area calculation of the surface in Figure 1.17
% Gauss point is stored in Gpw(n,1)
% Gauss weight is stored in Gpw(n,2)
clear variables;clc;
% NURBS surface parameters of Figure 1.17
p=2; m=0;
q=2; n=0;
xi = [linspace(-5,5,5) linspace(-5,5,5) linspace(-5,5,5) ...
      linspace(-5,5,5) linspace(-5,5,5)];
yi = [linspace(-5,-5,5) linspace(-2.5,-2.5,5) linspace(0,0,5)
...
      linspace(2.5,2.5,5) linspace(5,5,5)];
rng(400); zi=randi([-20 25],1,25);
%zi=zeros(1,25); % Flat plate for checking
xyzi=transpose([xi' yi' zi'])';
% Parameter ksi=s and eta=t
knotksi=[-1 -1 -1 -0.2 0.5 1 1 1 ones(1,p)]; % ones p-dummy
knoteta=[-1 -1 -1 -0.2 0.5 1 1 1 ones(1,q)]; % ones q-dummy
ns=size(knotksi,2)-p-1;
nt=size(knoteta,2)-q-1;
ws=ones(ns-p); ss=[-1 -0.5 0 0.5 1]; nss=size(ss,2);
wt=ones(nt-q); tt=[-1 -0.5 0 0.5 1]; ntt=size(tt,2);
Nurbske=zeros(ns-p,nt-q,nss,ntt);
%
for k=1:nss % Loop for creating T matrix
    Nurbsksi = Nurbs(p,knotksi,ws,ss(k));
for l=1:ntt
    Nurbseta = Nurbs(q,knoteta,wt,tt(l));
    for i=1:ns-p        % order of NURBS in ksi
    for j=1:nt-q        % order of NURBS in eta
        Nurbske(i,j,k,l) = Nurbsksi(i,p+2)*Nurbseta(j,q+2);
    end
    end
```

```
end
end
% [T] matrix
st=reshape(Nurbske,[(ns-p)*(nt-q),nss*ntt])';
% Control Points Px's, Py's and Pz's calculation
pxyz=st\xyzi;
px=reshape(pxyz(:,1),[(ns-p),(nt-q)]);
py=reshape(pxyz(:,2),[(ns-p),(nt-q)]);
pz=reshape(pxyz(:,3),[(ns-p),(nt-q)]);
% Gauss-Legendre Scheme
mg = 7; % eta
ng = 7;  % ksi
Gpwksi = Legendre(ng);
Gpweta = Legendre(mg);
Nurbsksi=zeros(ns-p,ng,mg);
DNurbsksi=zeros(ns-p,ng,mg);
Nurbseta=zeros(nt-q,ng,mg);
DNurbseta=zeros(nt-q,ng,mg);
for i = 1:mg  % number of gauss points of eta
    for j = 1:ng  % number of gauss points of ksi
        ksi = Gpwksi(j,1);
        eta = Gpweta(i,1);
        DNurbsBasiseta      = DNurbsLeibnitz(q,knoteta,wt,
eta); % eta
        DNurbsBasisksi      = DNurbsLeibnitz(p,knotksi,ws,
ksi); % ksi
        %
        Nurbseta(1:nt-q,i,j)  = DNurbsBasiseta(1:nt-q,q+5); %
0th-der
        DNurbseta(1:nt-q,i,j) = DNurbsBasiseta(1:nt-q,q+7); %
1st-der
        Nurbsksi(1:ns-p,i,j)  = DNurbsBasisksi(1:ns-p,p+5); %
0th-der
        DNurbsksi(1:ns-p,i,j) = DNurbsBasisksi(1:ns-p,p+7); %
1st-der
    end
end
%
Area = 0;
Tol = 1e-8;
for i = 1:mg      % eta gauss point
    for j = 1:ng  % ksi gauss point
       % Nurbs derivatives
       dxdksi = 0;      dxdeta = 0;
       dydksi = 0;      dydeta = 0;
       dzdksi = 0;      dzdeta = 0;
       for k = 1:nt-q  % eta basis functions 0 - m
       for l = 1:ns-p  % ksi basis function  0 - n
```

```
          dxdksi = dxdksi + DNurbsksi(l,j,i)*Nurbseta(k,j,i)*
px(l,k);
          dxdeta = dxdeta + Nurbsksi(l,j,i)*DNurbseta(k,j,i)*
px(l,k);
          dydksi = dydksi + DNurbsksi(l,j,i)*Nurbseta(k,j,i)*
py(l,k);
          dydeta = dydeta + Nurbsksi(l,j,i)*DNurbseta(k,j,i)*
py(l,k);
          dzdksi = dzdksi + DNurbsksi(l,j,i)*Nurbseta(k,j,i)*
pz(l,k);
          dzdeta = dzdeta + Nurbsksi(l,j,i)*DNurbseta(k,j,i)*
pz(l,k);
      end
      end
      Jac = det([dxdksi dydksi; dxdeta dydeta]);
      if abs(dxdksi) < Tol % x is not the function of ksi
           dzdx = dzdeta/dxdeta;
      elseif abs(dxdeta) < Tol % x is not the function of eta
           dzdx = dzdksi/dxdksi;
      elseif abs(dxdksi) < Tol && abs(dxdeta) < Tol
           dzdx = 0;
      else
           dzdx =dzdksi/dxdksi+dzdeta/dxdeta;
      end
      if abs(dydksi) < Tol % y is not the function of ksi
           dzdy = dzdeta/dydeta;
      elseif abs(dydeta) < Tol % x is not the function of eta
           dzdy = dzdksi/dydksi;
      elseif abs(dydksi) < Tol && abs(dydeta) < Tol
           dzdy = 0;
      else
           dzdy =dzdksi/dydksi+dzdeta/dydeta;
      end
      farea = sqrt(1+dzdx^2/Jac^2+dzdy^2/Jac^2);
      wi  = Gpweta(i,2);
      wj  = Gpwksi(j,2);
      Area = Area + farea * Jac * wi * wj;
    end
end
disp(vpa(sym(Area),17));
```

2.6 Curvature and Gradient of the NURBS Surface

The surface (see Figure 2.4) which is created randomly by using NURBS is considered for the curvature and gradient calculation example. The data to construct the surface are as follows: the polynomial order in x- and

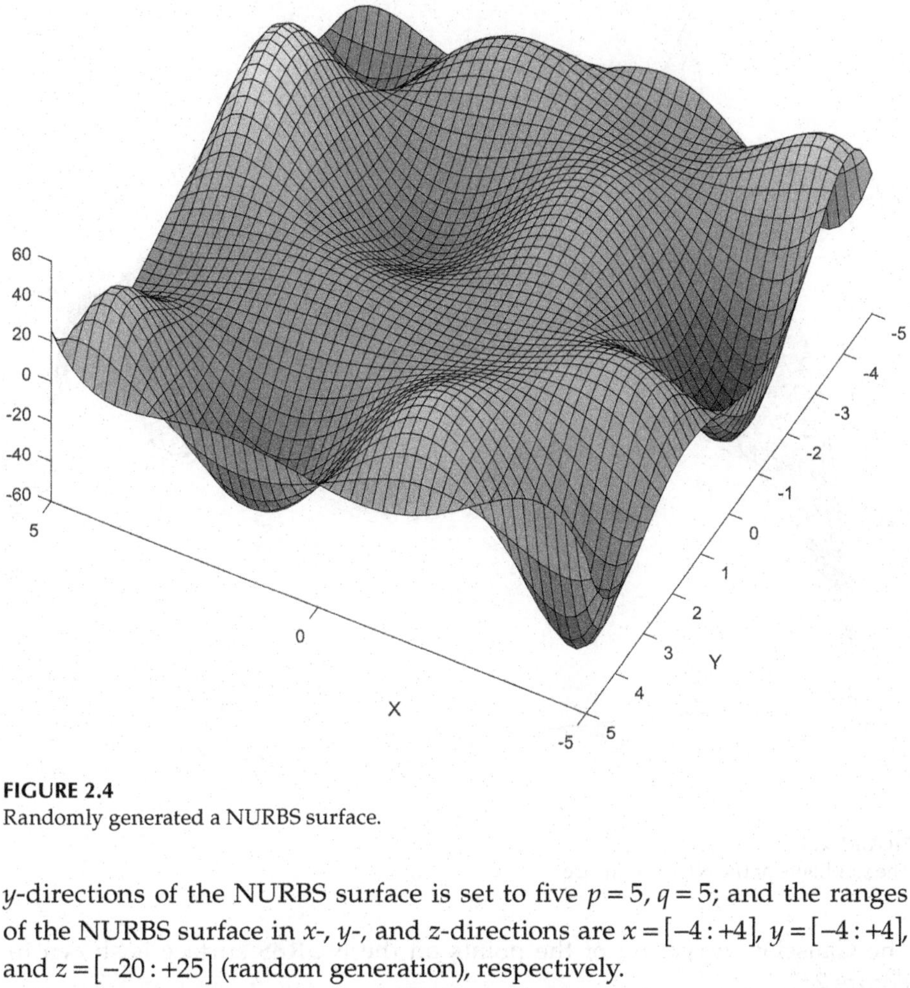

FIGURE 2.4
Randomly generated a NURBS surface.

y-directions of the NURBS surface is set to five $p=5, q=5$; and the ranges of the NURBS surface in x-, y-, and z-directions are $x=[-4:+4]$, $y=[-4:+4]$, and $z=[-20:+25]$ (random generation), respectively.

The Gaussian curvature of a NURBS surface (see Figure 2.5) at a point is the product of the principal curvatures, which is given by

$$\mathrm{K} = \kappa_x \cdot \kappa_y \tag{2.21}$$

where

$$\begin{aligned} \kappa_x &= \frac{\dfrac{dz}{d\xi}\dfrac{d^2x}{d\xi^2} - \dfrac{dx}{d\xi}\dfrac{d^2z}{d\xi^2}}{J^3} \\ \kappa_y &= \frac{\dfrac{dz}{d\eta}\dfrac{d^2y}{d\eta^2} - \dfrac{dy}{d\eta}\dfrac{d^2z}{d\eta^2}}{J^3} \end{aligned} \tag{2.22}$$

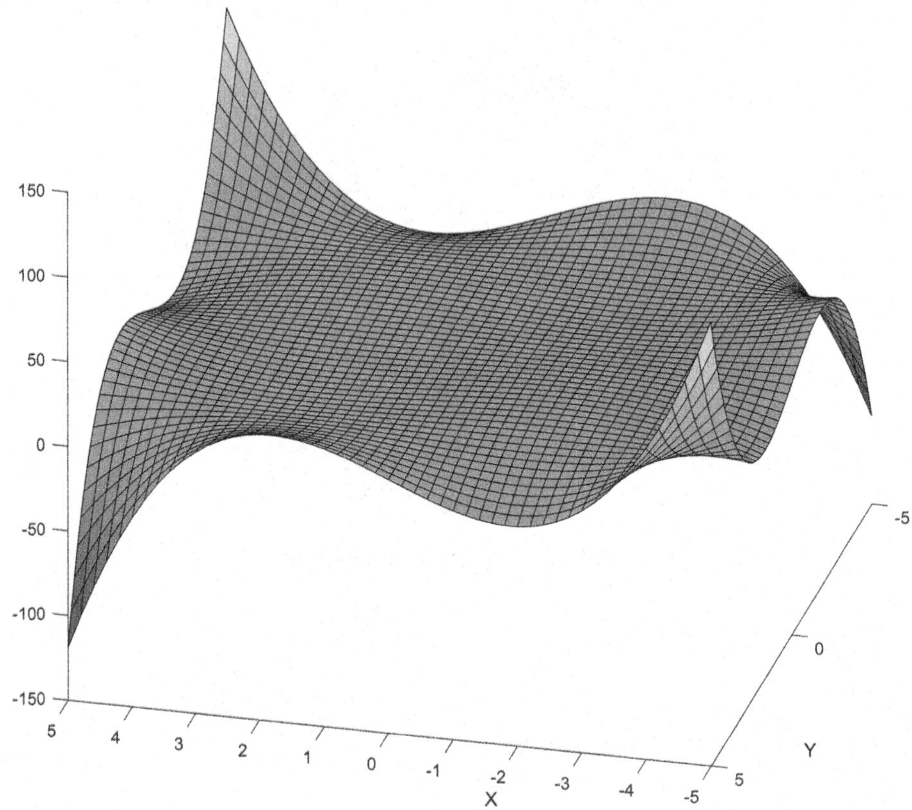

FIGURE 2.5
The gradients of the NURBS surface.

The Gaussian curvature of the points on the NURBS surface is shown in Figure 2.5.

The gradient of a point on the NURBS surface can be expressed as a scalar derivative function of the elevation $z(x,y)$ of the point:

$$\nabla z(x,y) = \frac{dz}{d\xi} + \frac{dz}{d\eta} \tag{2.23}$$

The Gaussian curvature of the points on the NURBS surface is shown in Figure 2.6.

MATLAB code NurbsSurfaceCurvGrad.m uses the functions DNurbsLeibnitz.m (Section 1.7.2) to calculate the NURBS basis functions and their derivatives of a random NURBS surface and draw the NURBS surface as shown in Figure 2.4, the gradient of the points on the NURBS surface as shown in Figure 2.5, and the Gaussian curvature of the points on the NURBS surface as shown in Figure 2.6.

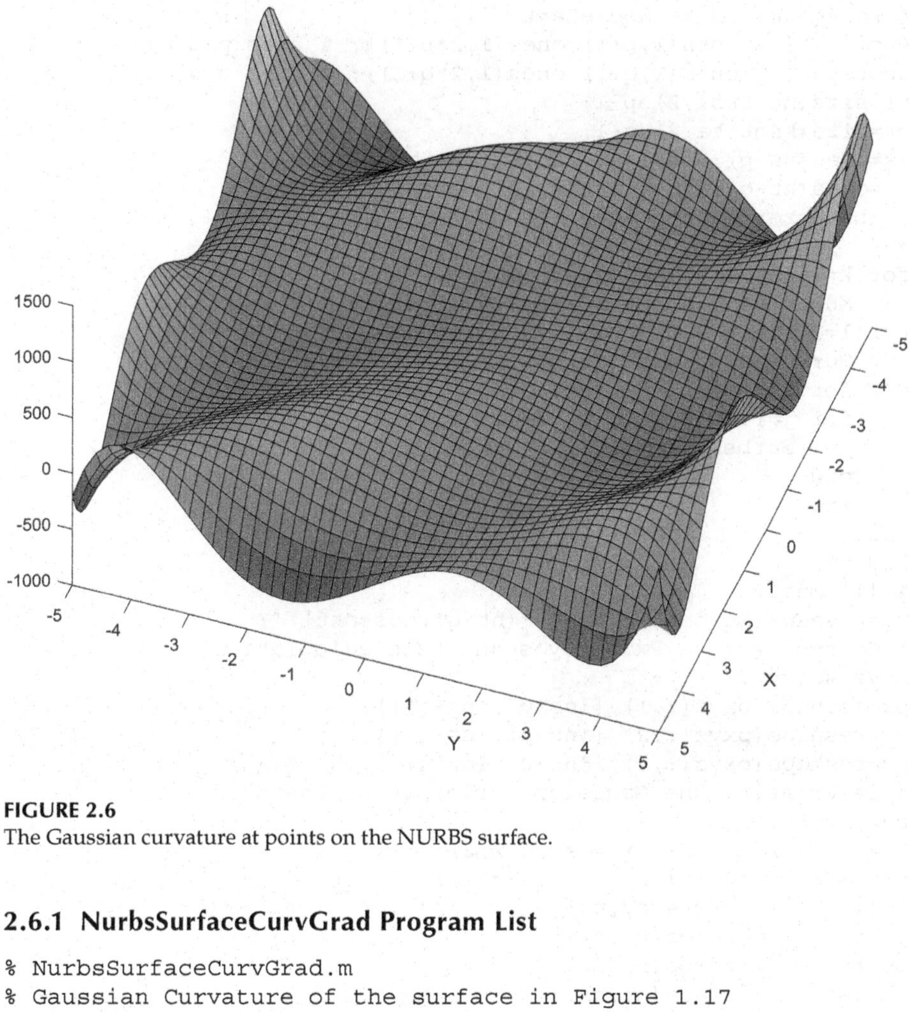

FIGURE 2.6
The Gaussian curvature at points on the NURBS surface.

2.6.1 NurbsSurfaceCurvGrad Program List

```
% NurbsSurfaceCurvGrad.m
% Gaussian Curvature of the surface in Figure 1.17
% Gauss point is stored in Gpw(n,1)
% Gauss weight is stored in Gpw(n,2)
clear variables;clc;
% NURBS surface parameters of Figure 1.17
p=5; m=0;
q=5; n=0;
numer = p+1; y0=-4; y1=4;
xi = repmat(linspace(y0,y1,numer),1,numer);
yi = [];
for iy = 0:numer-1
   yy = (y1-y0)*iy/(numer-1)+y0;
   yi = [yi linspace(yy,yy,numer)];
end
rng(400); zi=randi([-20 +25],1,numer^2);
xyzi=transpose([xi' yi' zi'])';
```

```
% Parameter ksi=s and eta=t
knotksi=[-1*ones(1,p+1) ones(1,2*p+1)]; % ones p-dummy
knoteta=[-1*ones(1,q+1) ones(1,2*q+1)]; % ones q-dummy
ns=size(knotksi,2)-p-1;
nt=size(knoteta,2)-q-1;
ws=ones(ns-p); ss=[-1:2/p:1]; nss=size(ss,2);
wt=ones(nt-q); tt=[-1:2/q:1]; ntt=size(tt,2);
Nurbske=zeros(ns-p,nt-q,nss,ntt);
%
for k=1:nss % Loop for creating T matrix
    Nurbsksi = Nurbs(p,knotksi,ws,ss(k));
for l=1:ntt
    Nurbseta = Nurbs(q,knoteta,wt,tt(l));
    for i=1:ns-p      % order of NURBS in ksi
    for j=1:nt-q      % order of NURBS in eta
        Nurbske(i,j,k,l) = Nurbsksi(i,p+2)*Nurbseta(j,q+2);
    end
    end
end
end
% [T] matrix
st=reshape(Nurbske,[(ns-p)*(nt-q),nss*ntt])';
% Control Points Px's, Py's and Pz's calculation
pxyz=st\xyzi;
px=reshape(pxyz(:,1),[(ns-p),(nt-q)]);
py=reshape(pxyz(:,2),[(ns-p),(nt-q)]);
pz=reshape(pxyz(:,3),[(ns-p),(nt-q)]);
% Calculating the Gaussian curvature of the surface
ndxy = 50;
x = -5:10/ndxy:5; nx = size(x,2);
y = -5:10/ndxy:5; ny = size(y,2);
xcor      = zeros(ny,nx);
ycor      = zeros(ny,nx);
surfz     = zeros(ny,nx);
curvature = zeros(ny,nx);
gradientz = zeros(ny,nx);
% NURBS basis functions
Nurbsksi=zeros(ns-p,nx,ny);
DNurbsksi=zeros(ns-p,nx,ny);
DDNurbsksi=zeros(ns-p,nx,ny);
Nurbseta=zeros(nt-q,nx,ny);
DNurbseta=zeros(nt-q,nx,ny);
DDNurbseta=zeros(nt-q,nx,ny);
for i = 1:ny  % number of data points of eta
    eta = y(i)/5;
    DNurbsBasiseta    = DNurbsLeibnitz(q,knoteta,wt,eta); % eta
    for j = 1:nx  % number of gauss points of ksi
        ksi = x(j)/5;
        DNurbsBasisksi      = DNurbsLeibnitz(p,knotksi,ws,
ksi); % ksi
```

```
        %
        Nurbseta(1:nt-q,i,j)  = DNurbsBasiseta(1:nt-q,q+5); %
0th-derv
        DNurbseta(1:nt-q,i,j) = DNurbsBasiseta(1:nt-q,q+7); %
1st-derv
        DDNurbseta(1:nt-q,i,j)= DNurbsBasiseta(1:nt-q,q+8); %
2nd-derv
        Nurbsksi(1:ns-p,i,j)  = DNurbsBasisksi(1:ns-p,p+5); %
0th-derv
        DNurbsksi(1:ns-p,i,j) = DNurbsBasisksi(1:ns-p,p+7); %
1st-derv
        DDNurbsksi(1:ns-p,i,j)= DNurbsBasisksi(1:ns-p,p+8); %
2nd-derv
    end
end
Tol = 1e-8;
for i = 1:ny      % y data
    for j = 1:nx  % x data
      % Nurbs derivatives
      dxdksi = 0;       dxdeta = 0;
      dydksi = 0;       dydeta = 0;
      dzdksi = 0;       dzdeta = 0;
      ddxdksi= 0;       ddydeta= 0;
      ddzdksi= 0;       ddzdeta= 0;
      z = 0;
      for k = 1:nt-q  % eta basis functions 0 - m
      for l = 1:ns-p  % ksi basis function  0 - n
          dxdksi = dxdksi + DNurbsksi(l,j,i)*Nurbseta(k,j,i)*
px(l,k);
          dxdeta = dxdeta + Nurbsksi(l,j,i)*DNurbseta(k,j,i)*
px(l,k);
          dydksi = dydksi + DNurbsksi(l,j,i)*Nurbseta(k,j,i)*
py(l,k);
          dydeta = dydeta + Nurbsksi(l,j,i)*DNurbseta(k,j,i)*
py(l,k);
          dzdksi = dzdksi + DNurbsksi(l,j,i)*Nurbseta(k,j,i)*
pz(l,k);
          dzdeta = dzdeta + Nurbsksi(l,j,i)*DNurbseta(k,j,i)*
pz(l,k);
          ddxdksi = ddxdksi + DDNurbsksi(l,j,i)*px(l,k);
          ddydeta = ddydeta + DDNurbseta(k,j,i)*py(l,k);
          ddzdksi = ddzdksi + DDNurbsksi(l,j,i)*pz(l,k);
          ddzdeta = ddzdeta + DDNurbseta(k,j,i)*pz(l,k);
          z = z + Nurbsksi(l,j,i)*Nurbseta(k,j,i)*pz(l,k);
      end
      end
      % Surface
      surfz(j,i) = z;
      % Jacobian operator
      Jac = det([dxdksi dydksi; dxdeta dydeta]);
```

```
        % Gaussian curvature of the surface
        curvx = -(dxdksi*ddzdksi-dzdksi*ddxdksi)/Jac^3;
        curvy = -(dydeta*ddzdeta-dzdeta*ddydeta)/Jac^3;
        curvature(j,i) = curvy*curvx;
        gradientz(j,i) = dzdksi+dzdeta;
    end
end
figure(1)
set(gcf,'position',[200,200,800,800])
surface(x,y,surfz);
xlabel('X'); ylabel('Y');
view(-152,62)
figure(2)
set(gcf,'position',[300,200,800,800])
surface(x,y,curvature);
xlabel('X'); ylabel('Y');
view(-166,28)
figure(3)
set(gcf,'position',[400,200,800,800])
surface(x,y,gradientz);
xlabel('X'); ylabel('Y');
view(-250,57)
```

References

Gradshteyn IS, Ryzhik IM (2000) *Table of Integrals, Series, and Products*, 6th edition. Academic Press, San Diego.

Hildebrand FB (1956) *Introduction to Numerical Analysis*. McGraw-Hill, New York.

Kaplan W (1984) *Advanced Calculus*, 3rd edition. Addison-Wesley, Reading.

Ralston A, Rabinowitz P (1978) *A First Course in Numerical Analysis*. McGraw-Hill, London.

3

Theory of Plate and Shell Elements

3.1 Theory of Plate and Shell

A plate structure is geometrically similar to a bidimensional beam problem, whereas in this case, it carries loads on its plane; these loads lead to the bending of the plate structure. A plate can be described as a structure where the thickness is considerably smaller than the dimensions on a plane. A shell is in the plate family of structure that has an even smaller thickness than a thin plate. The theories on the deformation of plates are approximated by reducing a three-dimensional into a two-dimensional problem. The classic thin plate theory was developed by Kirchhoff (1850) and is now associated with "Kirchhoff plate theory." Kirchhoff theory states that the section in the thickness direction remains straight and perpendicular to the midplane of the plate after the deformation; this assumption contrasts with thick plate theory introduced by Reissner (1945) and Mindlin (1951), which states that the section in the thickness direction remains straight but not necessarily perpendicular to the midplane of the plate after the deformation. In the thick plate theory, the shear deformation that causes the rotation of the section during the deformation is taken into consideration.

The objective of this section is to recall the theory behind the formulation of plates. The theory of plates is generally classified whether the effects of shear deformation are included in the formulations or neglected because the shear effect is negligibly small. In the following sections, we discuss the fundamental theories of plates.

3.2 Shell Meant by a Thin Plate

Shells are geometrically curved, so it is reasonable that shell elements look like curves. It makes sense that to model a shell by flat elements is to model a flat plate by curved elements. A curved shell element must model shell geometry and must combine the membrane and bending actions in its deformable behavior.

There are three choices in shell element formulation:

1. A flat element, made by combining a membrane element with a plate bending element
2. A curved element, formulated by shell theory
3. A solid or degenerate solid element, analogous to the plate elements

Choice 1 is easiest for both formulation and use, but accuracy is often average and sometimes unacceptable. Choice 2 is difficult, and the element may have many degrees of freedom (DOFs), some of which are higher displacement derivatives. Choice 3 occupies the consensus between Choices 1 and 2.

In Choice 3, the element is modeled by thin shells with 32 DOFs. The accuracy is good, but the unusual allocation of DOF requires special formulations around the edges of the element. Choice 2 has two drawbacks: shell theory is complicated, and theoreticians have proposed many different strain–displacement relations for the same geometry of shell because there is room for argument about how many terms can be discarded as negligible. Shallow shell theory uses simple relations; thus, the element cannot be used for the thick shell. Generally, the shell has a thin geometry. Hence, it can be modeled by a thin plate in which the transverse shear deformation effects are negligible. The thin shell is stiffer in stretching than in bending like a plate; hence, the thin plate theory considered in this chapter takes into consideration the effects of axial loadings at the boundary of the edge.

By combining the NURBS and condensation, this book attempts to eliminate all the drawbacks of Choice 3.

3.3 Thin Plate Formulation

As introduced previously, a plate is a structure with the thickness considerably smaller than the dimensions on the plane, whereas a midplane is taken as the reference plane ($z = 0$) for deriving the kinematic equations. The classical thin plate model is based on the kinematic assumptions made by Kirchhoff (1850) to formulate the governing equations.

The underlying assumption of Kirchhoff's thin plate theory is that the plane of the cross sections of both directions (x and y) remains perpendicular to the plate's midplane during the deformation. This implies that the rotations of the cross sections of the plate are equal to the slopes of x- and y-axes of the plate.

In Figure 3.1, the kinematic assumption of undeformed and deformed configurations of a plate which bent at both sides by moments along the y-axis with the transverse displacement w is the variable of vertical displacement. Point O is on the midplane ($z = 0$) of the plate that can be moved and deformed.

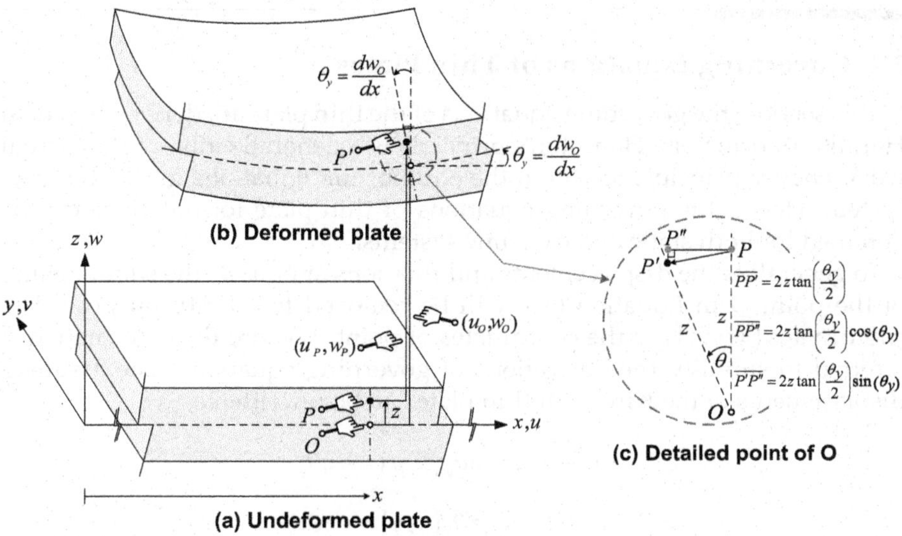

FIGURE 3.1
Conceptual kinematic assumptions of a flat Kirchhoff plate.

Under the small displacement assumption, we can apply the following approximations: $\sin(\theta_y) \approx \tan(\theta_y) \approx \theta_y$, $\cos(\theta_y) \approx 1$, and $\theta_y^2 \approx 0$.

Following the same principle, the kinematic assumptions can also be applied to the bending along the *x*-axis. We can see from Figure 3.1c that during the deformation of the plate, an arbitrary point *P* is deformed and rotated by the approximated displacement fields as given below.

$$
\begin{aligned}
u_P &= u_O - 2z\tan\left(\frac{\theta_y}{2}\right)\cos(\theta_y) \approx u_O - 2z\frac{\theta_y}{2}1 \approx u_O - z\theta_y \\
v_P &= v_O - 2z\tan\left(\frac{\theta_x}{2}\right)\cos(\theta_x) \approx v_O - 2z\frac{\theta_x}{2}1 \approx v_O - z\theta_x \\
w_P &= w_O + z - z - 2z\tan\left(\frac{\theta_y}{2}\right)\sin(\theta_y) \approx w_O - 2z\frac{\theta_y}{2}\theta_y \approx w_O \\
\theta_y &= \frac{dw_O}{dx}; \quad \theta_x = \frac{dw_O}{dy}
\end{aligned}
\tag{3.1}
$$

where u_O, v_O and u_P, v_P are the horizontal displacements of the point *O* and *P* in *x*- and *y*-axes, respectively. w_O, w_P are the vertical displacements of the point *O* and *P* in the *z*-direction. θ_x, θ_y are the rotations of the cross section of the plate in both *x*- and *y*-directions, which are equal to the slope of the plate at point *O* in both directions.

3.4 Governing Equations of Thin Plates

In this section, the governing equations of the thin plate are derived based on Hamilton's principle. Hamilton's principle is a generalization of the virtual work–energy principle to obtain the equilibrium equations in the dynamic system. Hence, the governing equations of thin plate formulations can be obtained for both static and dynamic systems.

To generalize the displacements and rotations at point P, the sub-indexing of the point O in Equation (3.3) will be removed for clarity purpose. The coordinates (x,y,z) are the coordinates of point P before the deformation is omitted to simplify the derivations of governing equations. The assumed displacements of the Kirchhoff thin plate can be rewritten as

$$
\begin{aligned}
u_P - u_O &= -z\theta_y(x,y) = -z\theta_y \\
v_P - v_O &= -z\theta_x(x,y) = -z\theta_x \\
w_P &= w(x,y) = w \\
\theta_y &= \frac{dw(x,y)}{dx} = \frac{dw}{dx} \\
\theta_x &= \frac{dw(x,y)}{dy} = \frac{dw}{dy}
\end{aligned}
\tag{3.2}
$$

From the theory of elasticity, we can obtain the strain components without shear deformations as

$$
\begin{aligned}
\varepsilon_{xx} &= -z\frac{d\theta_y}{dx} = -z\frac{d^2w}{dx^2} \\
\varepsilon_{yy} &= -z\frac{d\theta_x}{dy} = -z\frac{d^2w}{dy^2} \\
\varepsilon_{zz} &= \frac{dw_P}{dz} = 0 \\
\gamma_{xy} = \gamma_{yx} &= \frac{du}{dy} + \frac{dv}{dx} = -2z\frac{d^2w}{dxdy} \\
\gamma_{xz} = \gamma_{zx} &= -\theta_y + \frac{dw}{dx} = 0 \\
\gamma_{yz} = \gamma_{zy} &= -\theta_x + \frac{dw}{dy} = 0
\end{aligned}
\tag{3.3}
$$

For elastic plane stress problem, the general stress components in a plate take the form of

$$\sigma_{xx} = \frac{E_{xx}}{\left(1-\nu_{xy}^2\right)}\left(\varepsilon_{xx} + \nu_{xy}\varepsilon_{yy}\right)$$

$$\sigma_{yy} = \frac{E_{yy}}{\left(1-\nu_{yx}^2\right)}\left(\nu_{yx}\varepsilon_{xx} + \varepsilon_{yy}\right) \tag{3.4}$$

$$\tau_{xy} = \frac{E_{xx}}{2\left(1+\nu_{xy}\right)}\gamma_{xy} \quad \text{and} \quad \tau_{yx} = \frac{E_{yy}}{2\left(1+\nu_{yx}\right)}\gamma_{yx}$$

Under the assumption of isotropic and homogeneous material ($E_{xx} = E_{yy} = E_{xy} = E_{yx} = E$, $\nu_{xy} = \nu_{yx} = \nu$), the general stress components can be written as

$$\sigma_{xx} = \frac{E}{\left(1-\nu^2\right)}\left(\varepsilon_{xx} + \nu\varepsilon_{yy}\right)$$

$$\sigma_{yy} = \frac{E}{\left(1-\nu^2\right)}\left(\nu\varepsilon_{xx} + \varepsilon_{yy}\right) \tag{3.5}$$

$$\tau_{xy} = \frac{E}{2(1+\nu)}\gamma_{xy} \quad \text{and} \quad \tau_{yx} = \frac{E}{2(1+\nu)}\gamma_{yx}$$

where E and ν are the normal elastic modulus and Poisson's ratio of the plate material, respectively. Figure 3.2 shows the distribution of shear stress and normal stress due to bending of the plate.

In the illustration shown in Figure 3.3, the geometrical configurations for loadings and displacements of a plate are given.

The governing equations are then derived via Hamilton's principle as follows:

$$\delta H = \int_{t_1}^{t_2} \left(\delta S_E - \delta K_E - \delta W_E\right) dt = 0 \tag{3.6}$$

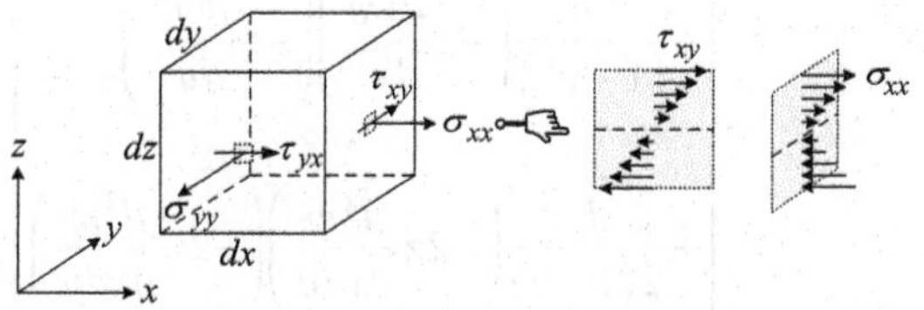

FIGURE 3.2
Definitions and distributions of stresses in the infinitesimal element of a plate.

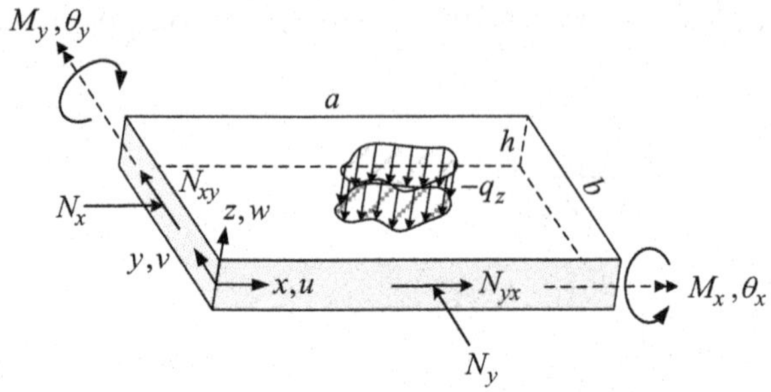

FIGURE 3.3
Geometrical configurations for loadings and displacements of a plate.

where

δH is the variation of total energy in plate; δS_E is the variation of plate strain energy; δK_E is the variation of plate kinetic energy; δW_E is the variation of external work applied to the plate.

The strain energy, S_E, of a uniform rectangular ($a \times b \times h$) plate is given by

$$S_E = \frac{1}{2}\int_0^b \int_0^a \int_{-h/2}^{h/2} \left\{ \sigma_{xx}\varepsilon_{xx} + \sigma_{yy}\varepsilon_{yy} + \tau_{xy}\gamma_{xy} \right\} dz\, dx\, dy \tag{3.7}$$

where a, b, h are the lengths of the plate in x- and y-directions and the uniform thickness of the plate, respectively.

Substituting Equations (3.3) and (3.5) into Equation (3.7) gives

$$S_E = \frac{1}{2}\int_0^b \int_0^a \int_{-h/2}^{+h/2} \left\{ \begin{bmatrix} \dfrac{E}{(1-\nu^2)}\left(-z\dfrac{d^2w}{dx^2}\right)\left(-z\dfrac{d^2w}{dx^2}\right) \\ +\dfrac{E}{(1-\nu^2)}\left(-z\dfrac{d^2w}{dy^2}\right)\left(-z\dfrac{d^2w}{dy^2}\right) \\ +\dfrac{2\nu E}{(1-\nu^2)}\left(-z\dfrac{d^2w}{dx^2}\right)\left(-z\dfrac{d^2w}{dy^2}\right) \\ +\dfrac{E}{2(1+\nu)}\left(-2z\dfrac{d^2w}{dxdy}\right)\left(-2z\dfrac{d^2w}{dxdy}\right) \end{bmatrix} \right\} dz\, dx\, dy \tag{3.8}$$

Applying the variational principle to the above equation, we can obtain

$$\delta S_E = \int_0^b \int_0^a \int_{-h/2}^{+h/2} \left\{ \begin{aligned} & \frac{E}{(1-\nu^2)} \left(z \frac{d^2 w}{dx^2} \right) \delta \left(z \frac{d^2 w}{dx^2} \right) \\ & + \frac{E}{(1-\nu^2)} \left(z \frac{d^2 w}{dy^2} \right) \delta \left(z \frac{d^2 w}{dy^2} \right) \\ & + \frac{2\nu E}{(1-\nu^2)} \left(-z \frac{d^2 w}{dx^2} \right) \delta \left(-z \frac{d^2 w}{dy^2} \right) \\ & + \frac{E}{2(1+\nu)} \left(2z \frac{d^2 w}{dxdy} \right) \delta \left(2z \frac{d^2 w}{dxdy} \right) \end{aligned} \right\} dz\,dx\,dy \tag{3.9}$$

For a rectangular plate with uniform thickness h,

$$\begin{aligned} \delta S_E = {} & D \int_0^b \int_0^a \left(\frac{d^2 w}{dx^2} \right) \delta \left(\frac{d^2 w}{dx^2} \right) dx\,dy + D \int_0^b \int_0^a \left(\frac{d^2 w}{dy^2} \right) \delta \left(\frac{d^2 w}{dy^2} \right) dx\,dy \\ & + D_{xxyy} \int_0^b \int_0^a \left(\frac{d^2 w}{dx^2} \right) \delta \left(\frac{d^2 w}{dy^2} \right) dx\,dy \\ & + D_{xy} \int_0^b \int_0^a \left(\frac{d^2 w}{dxdy} \right) \delta \left(\frac{d^2 w}{dxdy} \right) dx\,dy \end{aligned} \tag{3.10}$$

where

$$D = \frac{Eh^3}{12(1-\nu^2)},\ D_{xxyy} = \frac{\nu Eh^3}{6(1-\nu^2)},\ D_{xy} = \frac{Eh^3}{6(1+\nu)}$$

Applying the integration in part to the first term of the first integral term in Equation (3.10) results in

$$D \int_0^b \int_0^a \left(\frac{d^2 w}{dx^2} \right) \delta \left(\frac{d^2 w}{dx^2} \right) dx\,dy = M_y \big|_{BC}\, \delta\theta_y - D \int_0^b \int_0^a \left(\frac{d^3 w}{dx^3} \right) \delta \left(\frac{dw}{dx} \right) dx\,dy \tag{3.11}$$

where M_y is the rotational moment around the y-axis at plate ends, which is given by

$$M_y \big|_{BC} = D \left(\frac{d^2 w}{dx^2} \right) \bigg|_{BC} \tag{3.12}$$

where dw/dx is defined by θ_y.

By further applying the integration in part to the last integral term in Equation (3.11), we can have the following:

$$-D\int_0^b\int_0^a\left(\frac{d^3w}{dx^3}\right)\delta\left(\frac{dw}{dx}\right)dx\,dy=-Q_x\big|_{BC}\,\delta w+D\int_0^b\int_0^a\left(\frac{d^4w}{dx^4}\right)dx\,dy\,\delta w \tag{3.13}$$

where Q_x is the transversal nodal force at the plate ends which is given as

$$Q_x\big|_{BC}=D\left(\frac{d^3w}{dx^3}\right)\bigg|_{BC} \tag{3.14}$$

Summarizing from Equation (3.11) to Equation (3.14), the equations become

$$D\int_0^b\int_0^a\left(\frac{d^2w}{dx^2}\right)\delta\left(\frac{d^2w}{dx^2}\right)dx\,dy=M_y\big|_{BC}\,\delta\theta_y-Q_x\big|_{BC}\,\delta w+D\int_0^b\int_0^a\left(\frac{d^4w}{dx^4}\right)dx\,dy\,\delta w \tag{3.15}$$

Following the same procedure, by applying the integration in part to the second integral term of Equation (3.10) for the governing equations in the y-direction gives

$$D\int_0^b\int_0^a\left(\frac{d^2w}{dy^2}\right)\delta\left(\frac{d^2w}{dy^2}\right)dx\,dy=M_x\big|_{BC}\,\delta\theta_x-Q_y\big|_{BC}\,\delta w+D\int_0^b\int_0^a\left(\frac{d^4w}{dy^4}\right)dx\,dy\,\delta w \tag{3.16}$$

where M_x is the rotational moment around the y-axis at the plate ends, which is given by

$$M_x\big|_{BC}=D\left(\frac{d^2w}{dy^2}\right)\bigg|_{BC} \tag{3.17}$$

where dw/dy is defined by θ_x, and Q_y is the transversal nodal force at the plate ends, which is given as

$$Q_y\big|_{BC}=D\left(\frac{d^3w}{dy^3}\right)\bigg|_{BC} \tag{3.18}$$

Next, applying the integration in part to the third term of Equation (3.10) results in

$$D_{xxyy}\int_0^b\int_0^a\left(\frac{d^2w}{dx^2}\right)\delta\left(\frac{d^2w}{dy^2}\right)dx\,dy = M_{xy}\Big|_{BC}\,\delta\theta_x + M_{yx}\Big|_{BC}\,\delta\theta_y$$

$$+D_{xxyy}\int_0^b\int_0^a\left(\frac{d^4w}{dx^2dy^2}\right)dx\,dy\,\delta w \tag{3.19}$$

where M_{xy} and M_{yx} are the rotational moments about the midplane of the plate ends, which are given by

$$M_{xy}\Big|_{BC} = D_{xxyy}\left(\frac{d^2w}{dx^2}\right)\Bigg|_{BC}$$

$$M_{yx}\Big|_{BC} = D_{xxyy}\left(\frac{d^2w}{dy^2}\right)\Bigg|_{BC} \tag{3.20}$$

Finally, applying the integration in part to the fourth term of Equation (3.10) results in

$$D_{xy}\int_0^b\int_0^a\left(\frac{d^2w}{dxdy}\right)\delta\left(\frac{d^2w}{dxdy}\right)dx\,dy = M_{xy}\Big|_{BC}\,\delta\theta_x + M_{yx}\Big|_{BC}\,\delta\theta_y$$

$$+D_{xy}\int_0^b\int_0^a\left(\frac{d^4w}{dx^2dy^2}\right)dx\,dy\,\delta w \tag{3.21}$$

where M_{yx} and M_{xy} are the rotational moments about the midplane of the plate ends which are similar to Equation (3.20).

Summarizing the results of application of the variational principle to the strain energy, the governing equation of the plate can be expressed by using Equations (3.15), (3.16), (3.19), and (3.21) as

$$\delta S_E = D\int_0^b\int_0^a\left(\frac{d^4w}{dx^4}\right)dx\,dy\,\delta w + D\int_0^b\int_0^a\left(\frac{d^4w}{dy^4}\right)dx\,dy\,\delta w$$

$$+D_{xxyy}\int_0^b\int_0^a\left(\frac{d^4w}{dx^2dy^2}\right)dx\,dy\,\delta w + D_{xy}\int_0^b\int_0^a\left(\frac{d^4w}{dx^2dy^2}\right)dx\,dy\,\delta w \tag{3.22}$$

where the boundary conditions are given in Equations (3.12), (3.14), (3.17), (3.18), and (3.20).

The kinetic energy of the plate element is given as

$$K_E = \frac{1}{2}\int_0^b\int_0^a\int_{-h/2}^{+h/2}\rho\left(\frac{du}{dt}\right)^2 dz\,dx\,dy + \frac{1}{2}\int_0^b\int_0^a\int_{-h/2}^{+h/2}\rho\left(\frac{dv}{dt}\right)^2 dz\,dx\,dy$$

$$+\frac{1}{2}\int_0^b\int_0^a\int_{-h/2}^{+h/2}\rho\left(\frac{dw}{dt}\right)^2 dz\,dx\,dy \tag{3.23}$$

Substitution of Equation (3.2) gives

$$K_E = \frac{1}{2}\rho I\int_0^b\int_0^a\left(\frac{d\theta_y}{dt}\right)^2 dx\,dy + \frac{1}{2}\rho I\int_0^b\int_0^a\left(\frac{d\theta_x}{dt}\right)^2 dx\,dy$$

$$+\frac{1}{2}\rho I_{xy}\int_0^b\int_0^a\left(\frac{dw}{dt}\right)^2 dx\,dy \tag{3.24}$$

where

$$I = \int_{-h/2}^{+h/2} z^2\,dz = \frac{h^3}{12}$$

$$I_{xy} = \int_{-h/2}^{+h/2} 1\,dz = h$$

Applying the variational principle to Equation (3.24) gives

$$\delta K_E = \rho I\int_0^b\int_0^a\left(\frac{d\theta_y}{dt}\right)\delta\left(\frac{d\theta_y}{dt}\right)dx\,dy + \rho I\int_0^b\int_0^a\left(\frac{d\theta_x}{dt}\right)\delta\left(\frac{d\theta_x}{dt}\right)dx\,dy$$

$$+\rho I_{xy}\int_0^b\int_0^a\left(\frac{dw}{dt}\right)\delta\left(\frac{dw}{dt}\right)dx\,dy \tag{3.25}$$

Applying the integration in part to the first term of Equation (3.25) yields

$$\rho I\int_0^b\int_0^a\left(\frac{d\theta_y}{dt}\right)\delta\left(\frac{d\theta_y}{dt}\right)dx\,dy = \dot{M}_y\Big|_{BC}\delta\theta_y - \rho I\int_0^b\int_0^a\left(\frac{d^2\theta_y}{dt^2}\right)dx\,dy\,\delta\theta_y \tag{3.26}$$

where $\dot{M}_y$ is the time-rate rotational moment about the midplane of the plate ends, which is given by

$$\dot{M}_y\Big|_{BC} = \rho I\left(\frac{d\theta_y}{dt}\right)\Bigg|_{BC} \tag{3.27}$$

with $\dot{M}_y = \dfrac{d(M_y)}{dt}$.

Similarly, applying the integration in part to the second term of Equation (3.25) results in

$$\rho I\int_0^b\int_0^a\left(\frac{d\theta_x}{dt}\right)\delta\left(\frac{d\theta_x}{dt}\right)dx\,dy=\dot{M}_x\Big|_{BC}\delta\theta_x-\rho I\int_0^b\int_0^a\left(\frac{d^2\theta_x}{dt^2}\right)dx\,dy\,\delta\theta_x \tag{3.28}$$

where $\dot{M}_x$ is the time-rate rotational moment about the mid-plane of the plate ends, which is given by

$$\dot{M}_x\Big|_{BC}=\rho I\left(\frac{d\theta_x}{dt}\right)\Bigg|_{BC} \tag{3.29}$$

Applying the integration in part to the third term of Equation (3.25) yields

$$\rho I_{xy}\int_0^b\int_0^a\left(\frac{dw}{dt}\right)\delta\left(\frac{dw}{dt}\right)dx\,dy=\dot{Q}_z\Big|_{BC}\delta w-\rho I_{xy}\int_0^b\int_0^a\left(\frac{d^2w}{dt^2}\right)dx\,dy\,\delta w \tag{3.30}$$

where

$$\dot{Q}_z\Big|_{BC}=\rho I_{xy}\left(\frac{dw}{dt}\right)\Bigg|_{BC} \tag{3.31}$$

Summarizing the results of application of the variational principle to the kinetic energy, the governing equation of the plate can be expressed by using Equations (3.26), (3.28), and (3.30) as

$$\delta K_E=+\rho I\int_0^b\int_0^a\left(\frac{d^2\theta_y}{dt^2}\right)dx\,dy\,\delta\theta_y+\rho I\int_0^b\int_0^a\left(\frac{d^2\theta_x}{dt^2}\right)dx\,dy\,\delta\theta_x$$
$$+\rho I_{xy}\int_0^b\int_0^a\left(\frac{d^2w}{dt^2}\right)dx\,dy\,\delta w \tag{3.32}$$

where the boundary conditions are given in Equations (3.27), (3.29), and (3.31).

The external work applied to the plate element due to distributed load on the top surface of the plate, $q(x,y)$, and in-plane compressive and shear forces in normal and tangential directions on the edges of the plate, N_x, N_y, N_{xy}, as shown in Figure 3.3 can be expressed as

$$W_E=\frac{1}{2}\int_0^b\int_0^a N_x\left(\frac{dw}{dx}\right)^2dx\,dy+\frac{1}{2}\int_0^b\int_0^a N_y\left(\frac{dw}{dy}\right)^2dx\,dy$$
$$+\frac{1}{2}\int_0^b\int_0^a N_{xy}\left(\frac{d^2w}{dxdy}\right)dx\,dy+\frac{1}{2}\int_0^b\int_0^a N_{yx}\left(\frac{d^2w}{dxdy}\right)dx\,dy+\int_0^b\int_0^a q_z w\,dx\,dy \tag{3.33}$$

Substituting Equation (3.3) into the above equation results in

$$W_E = \frac{1}{2}\int_0^b\int_0^a N_x\left(\frac{dw}{dx}\right)^2 dx\,dy + \frac{1}{2}\int_0^b\int_0^a N_y\left(\frac{dw}{dy}\right)^2 dx\,dy$$
$$+\frac{1}{2}\int_0^b\int_0^a N_{xy}\left(\frac{dw}{dx}\right)\left(\frac{dw}{dy}\right)dx\,dy + \frac{1}{2}\int_0^b\int_0^a N_{yx}\left(\frac{dw}{dy}\right)\left(\frac{dw}{dx}\right)dx\,dy$$
$$+\frac{1}{2}\int_0^b\int_0^a qw\,dx\,dy \tag{3.34}$$

Applying the variational principle to the above equation results in

$$\delta W_E = \int_0^b\int_0^a N_x\left(\frac{dw}{dx}\right)\delta\left(\frac{dw}{dx}\right)dx\,dy + \int_0^b\int_0^a N_y\left(\frac{dw}{dy}\right)\delta\left(\frac{dw}{dy}\right)dx\,dy$$
$$+\int_0^b\int_0^a N_{xy}\left(\frac{dw}{dx}\right)\delta\left(\frac{dw}{dy}\right)dx\,dy + \int_0^b\int_0^a N_{yx}\left(\frac{dw}{dy}\right)\delta\left(\frac{dw}{dx}\right)dx\,dy$$
$$+\int_0^b\int_0^a q_z\,\delta w\,dx\,dy \tag{3.35}$$

Next, applying the integration in part to the first term of Equation (3.34) yields

$$\int_0^b\int_0^a N_x\left(\frac{dw}{dx}\right)\delta\left(\frac{dw}{dx}\right)dx\,dy = p_y\big|_{BC}\,\delta w - \int_0^b\int_0^a N_x\left(\frac{d^2w}{dx^2}\right)dx\,dy\,\delta w \tag{3.36}$$

where p_y is the distributed normal force on the midplane of the plate ends, which is given by

$$p_y\big|_{BC} = N_x\left(\frac{dw}{dx}\right)\bigg|_{BC} \tag{3.37}$$

Similarly, applying the integration in part to the second term of Equation (3.34) yields

$$\int_0^b\int_0^a N_y\left(\frac{dw}{dy}\right)\delta\left(\frac{dw}{dy}\right)dx\,dy = p_x\big|_{BC}\,\delta w - \int_0^b\int_0^a N_y\left(\frac{d^2w}{dy^2}\right)dx\,dy\,\delta w \tag{3.38}$$

where p_x is the distributed normal force on the midplane of the plate ends, which is given by

$$p_x\big|_{BC} = N_y\left(\frac{dw}{dy}\right)\bigg|_{BC} \tag{3.39}$$

Next, applying the integration in part to the third term of Equation (3.34) yields

$$\int_0^b\int_0^a N_{xy}\left(\frac{dw}{dx}\right)\delta\left(\frac{dw}{dy}\right)dx\,dy = p_{xy}\big|_{BC}\,\delta w - \int_0^b\int_0^a N_{xy}\left(\frac{d^2w}{dxdy}\right)dx\,dy\,\delta w \tag{3.40}$$

where p_{xy} is the distributed shearing force on the midplane of the plate ends, which is given by

$$p_{xy}\big|_{BC} = N_{xy}\left(\frac{dw}{dx}\right)\bigg|_{BC} \tag{3.41}$$

Similarly, applying the integration in part to the fourth term of Equation (3.34) yields

$$\int_0^b\int_0^a N_{yx}\left(\frac{dw}{dy}\right)\delta\left(\frac{dw}{dx}\right)dx\,dy = p_{yx}\big|_{BC}\,\delta w - \int_0^b\int_0^a N_{yx}\left(\frac{d^2w}{dydx}\right)dx\,dy\,\delta w \tag{3.42}$$

where p_{yx} is the distributed shearing force on the midplane of the plate ends, which is given by

$$p_{yx}\big|_{BC} = N_{yx}\left(\frac{dw}{dy}\right)\bigg|_{BC} \tag{3.43}$$

Finally, summarizing the results of application of the variational principle to the external work, the governing equation of the plate can be expressed by using Equations (3.36), (3.38), (3.40), and (3.42) as

$$\begin{aligned}\delta W_E = &-\int_0^b\int_0^a N_x\left(\frac{d^2w}{dx^2}\right)dx\,dy\,\delta w - \int_0^b\int_0^a N_y\left(\frac{d^2w}{dy^2}\right)dx\,dy\,\delta w \\ &-\int_0^b\int_0^a N_{xy}\left(\frac{d^2w}{dxdy}\right)dx\,dy\,\delta w - \int_0^b\int_0^a N_{yx}\left(\frac{d^2w}{dydx}\right)dx\,dy\,\delta w \\ &+\int_0^b\int_0^a q_z\,dx\,dy\,\delta w\end{aligned} \tag{3.44}$$

where the boundary conditions are given in Equations (3.37), (3.39), (3.41), and (3.43).

Finally, Hamilton's principle in Equation (3.6) for isotropic and elastic material plate element can be written as follows:

$$\delta H=\int_{t_1}^{t_2}\left(\begin{array}{l} D\int_0^b\int_0^a\left(\dfrac{d^4w}{dx^4}\right)dx\,dy\,\delta w+D\int_0^b\int_0^a\left(\dfrac{d^4w}{dy^4}\right)dx\,dy\,\delta w \\ +D_{xxyy}\int_0^b\int_0^a\left(\dfrac{d^4w}{dx^2dy^2}\right)dx\,dy\,\delta w+D_{xy}\int_0^b\int_0^a\left(\dfrac{d^4w}{dx^2dy^2}\right)dy\,dx\,\delta w \\ +\rho I\int_0^b\int_0^a\left(\dfrac{d^2\theta_y}{dt^2}\right)dx\,dy\,\delta\theta_y+\rho I\int_0^b\int_0^a\left(\dfrac{d^2\theta_x}{dt^2}\right)dx\,dy\,\delta\theta_x \\ +\rho I_{xy}\int_0^b\int_0^a\left(\dfrac{d^2w}{dt^2}\right)dx\,dy\,\delta w \\ +\int_0^b\int_0^a N_x\left(\dfrac{d^2w}{dx^2}\right)dx\,dy\,\delta w+\int_0^b\int_0^a N_y\left(\dfrac{d^2w}{dy^2}\right)dx\,dy\,\delta w \\ +\int_0^b\int_0^a N_{xy}\left(\dfrac{d^2w}{dxdy}\right)dx\,dy\,\delta w+\int_0^b\int_0^a N_{yx}\left(\dfrac{d^2w}{dydx}\right)dx\,dy\,\delta w \\ -\int_0^b\int_0^a q_z\,dx\,dy\,\delta w \end{array}\right)dt=0 \tag{3.45}$$

The static governing equations can be obtained at an instantaneous zero time frame. The equations inside the double integral can be written as

$$\begin{aligned} &D\left(\frac{d^4w}{dx^4}\right)\delta w+D\left(\frac{d^4w}{dy^4}\right)\delta w+D_{xxyy}\left(\frac{d^4w}{dx^2dy^2}\right)\delta w+D_{xy}\left(\frac{d^4w}{dx^2dy^2}\right)\delta w \\ &+\rho I\left(\frac{d^2\theta_y}{dt^2}\right)\delta\theta_y+\rho I\left(\frac{d^2\theta_x}{dt^2}\right)\delta\theta_x+\rho I_{xy}\left(\frac{d^2w}{dt^2}\right)\delta w \\ &+N_x\left(\frac{d^2w}{dx^2}\right)\delta w+N_y\left(\frac{d^2w}{dy^2}\right)\delta w+N_{xy}\left(\frac{d^2w}{dxdy}\right)\delta w \\ &+N_{yx}\left(\frac{d^2w}{dydx}\right)\delta w-q_z\delta w=0 \end{aligned} \tag{3.46}$$

Knowing that $D_{xxyy}+D_{xy}=2D$ and eliminating all the variations, the above equation can be written as

$$
D\left(\frac{d^4w}{dx^4}\right)+D\left(\frac{d^4w}{dy^4}\right)+2D\left(\frac{d^4w}{dx^2dy^2}\right)
$$
$$
+\rho I\left(\frac{d^2\theta_y}{dt^2}\right)+\rho I\left(\frac{d^2\theta_x}{dt^2}\right)+\rho I_{xy}\left(\frac{d^2w}{dt^2}\right)
$$
$$
=q_z-N_x\left(\frac{d^2w}{dx^2}\right)-N_y\left(\frac{d^2w}{dy^2}\right)-N_{xy}\left(\frac{d^2w}{dxdy}\right)-N_{yx}\left(\frac{d^2w}{dydx}\right) \tag{3.47}
$$

The above equation is the governing differential equation of the well-known classical plate theory (CPT) of an isotropic rectangular plate. It is worth noting that finding the exact solution for the above governing equation is quite difficult. Except for a few simple problem of rectangular plates.

To summarize, by differentiating the above governing equations, the differential equations of motion that correspond to the generalized displacements can be grouped as

$$
\delta w: D\left(\frac{d^4w}{dx^4}\right)+D\left(\frac{d^4w}{dy^4}\right)+2D\left(\frac{d^4w}{dx^2dy^2}\right)+\rho I_{xy}\left(\frac{d^2w}{dt^2}\right)
$$
$$
=q_z-N_x\left(\frac{d^2w}{dx^2}\right)-N_y\left(\frac{d^2w}{dy^2}\right)-N_{xy}\left(\frac{d^2w}{dxdy}\right)-N_{yx}\left(\frac{d^2w}{dydx}\right)
$$
$$
\delta\theta_x: \rho I\left(\frac{d^2\theta_x}{dt^2}\right) \tag{3.48}
$$
$$
\delta\theta_y: \rho I\left(\frac{d^2\theta_y}{dt^2}\right)
$$

The essential and natural boundary conditions of the governing equations are given by

$$
\delta w: Q=-D\left(\frac{d^3w}{dx^3}\right)-D\left(\frac{d^3w}{dy^3}\right)-\rho I_{xy}\left(\frac{dw}{dt}\right)
$$
$$
\delta\theta_x: M_x=D\left(\frac{d^2w}{dy^2}\right)+2D_{xxyy}\left(\frac{d^2w}{dx^2}\right)-\rho I\left(\frac{d\theta_x}{dt}\right) \tag{3.49}
$$
$$
\delta\theta_y: M_y=D\left(\frac{d^2w}{dx^2}\right)+2D_{xxyy}\left(\frac{d^2w}{dy^2}\right)-\rho I\left(\frac{d\theta_y}{dt}\right)
$$

3.5 Navier Solutions for Rectangular Plates

In this section, we discuss the Navier solutions to solve the bending and natural vibration problems of isotropic rectangular plates. The Navier solutions can be used for simply supported rectangular plates under the following conditions:

1. It is applicable only for simply supported plates.
2. It solves bending solutions under arbitrary transverse loading $q_z(x,y)$.
3. It solves natural frequency problems of plates.
4. It solves buckling problems under in-plane biaxial compressive loadings defined by $\hat{N}_y = \lambda\hat{N}_x = \lambda N_0$, where λ is the ratio between the applied axial compressive loadings.

In case of a simply supported boundary condition of a rectangular plate (see Figure 3.4), the following initial displacement conditions at the supports are as follows:

$$w(-a/2,y,t)=0;\quad w(+a/2,y,t)=0;\quad w(x,-b/2,t)=0;$$
$$w(x,+b/2,t)=0 \tag{3.50}$$

Moreover, the initial moments are as follows:

$$M_x(-a/2,y,t)=0;\quad M_x(+a/2,y,t)=0;\quad M_y(x,-b/2,t)=0;$$
$$M_y(x,+b/2,t)=0 \tag{3.51}$$

In Navier's method of solution, the displacement w is expanded into double sine series with unknown coefficients $W_{mn}(t)$ as the function of time:

$$w(x,y,t)=\sum_{n=1}^{\infty}\sum_{m=1}^{\infty}W_{mn}(t)\sin(\alpha_m x)\sin(\beta_n y) \tag{3.52}$$

where

$$\alpha_m=\frac{m\pi}{a};\quad \beta_n=\frac{n\pi}{b}$$

The coefficients $W_{mn}(t)$ should be determined in such a way that Equation (3.47) is satisfied everywhere in the plate. Equation (3.52) is adopted by the

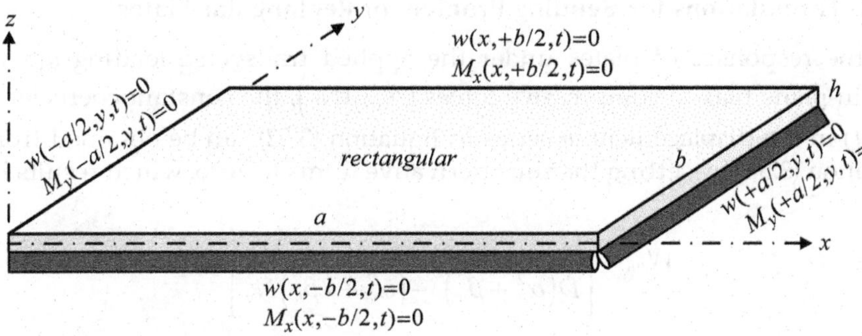

FIGURE 3.4
Simply supported boundary conditions of a rectangular plate.

fact that the double sine series can satisfy the simply supported boundary conditions in Equations (3.50) and (3.51). The external loading q can also be expanded in the double sine series as

$$q(x,y,t)=\sum_{n=1}^{\infty}\sum_{m=1}^{\infty}q_{mn}(t)\sin(\alpha_m x)\sin(\beta_n y) \tag{3.53}$$

where

$$q_{mn}(x,y,t)=\frac{1}{ab}\int_{-b}^{+b}\int_{-a}^{+a}q(x,y,t)\sin(\alpha_m x)\sin(\beta_n y)\,dx\,dy \tag{3.54}$$

Substituting Equations (3.51) and (3.52) into Equation (3.46) results in

$$0=\sum_{n=1}^{\infty}\sum_{m=1}^{\infty}\begin{Bmatrix}D\left(\alpha_m^4+2\alpha_m^2\beta_n^2+\beta_n^4\right)W_{mn}-\left(\alpha_m^2+\lambda\beta_n^2\right)N_o\\-q_{mn}+\left[I_0+I_2\left(\alpha_m^2+\beta_n^2\right)\right]\ddot{W}_{mn}\end{Bmatrix}\sin(\alpha_m x)\sin(\beta_n y) \tag{3.55}$$

For any location (x, y), it follows that

$$\left[I_0+I_2\left(\alpha_m^2+\beta_n^2\right)\right]\ddot{W}_{mn}+\left[D\left(\alpha_m^2+\beta_n^2\right)^2-\left(\alpha_m^2+\lambda\beta_n^2\right)N_o\right]W_{mn}=q_{mn} \tag{3.56}$$

for every pair of m and n.

Equation (3.56) can also be used for solving static bending, natural vibration, and buckling of simply supported rectangular plates.

The coefficients $q_{mn}(x,y,t)$ for various types of loadings can be calculated by using Equation (3.54).

3.5.1 Formulations for Bending Problem of Rectangular Plates

For the responses of plates under the applied transverse loading $q(x,y)$ and in-plane biaxial compressive forces $(\hat{N}_0, \lambda\hat{N}_0)$, the constant coefficients $W_{mn}(t)$ in the displacement w series in Equation (3.52) can be obtained from Equation (3.56) by setting the time derivative terms to zero, which results in

$$W_{mn} = \frac{q_{mn}}{\left[D\left(\alpha_m^2 + \beta_n^2\right)^2 - \left(\alpha_m^2 + \lambda\beta_n^2\right)N_o\right]} \tag{3.57}$$

Moreover, the static solution of displacement in Equation (3.51) becomes

$$w(x,y) = \sum_{n=1}^{\infty}\sum_{m=1}^{\infty} W_{mn}\sin(\alpha_m x)\sin(\beta_n y) \tag{3.58}$$

It is worth noting that the compressive force N_0 has the effect of increasing or reducing the displacement, depending on whether the in-plane force is tensile or not. When the in-plane compressive loads are zero, we can have the following:

$$W_{mn} = \frac{q_{mn}}{D\left(\alpha_m^2 + \beta_n^2\right)^2} = \frac{b^4}{D\pi^4}\frac{q_{mn}}{\left(m^2s^2 + n^2\right)^2} \tag{3.59}$$

where s is the plate aspect ratio, defined by $s = b/a$, and the solution of displacement equation becomes

$$w(x,y) = \frac{b^4}{D\pi^4}\sum_{n=1}^{\infty}\sum_{m=1}^{\infty}\frac{q_{mn}}{\left(m^2s^2 + n^2\right)^2}\sin(\alpha_m x)\sin(\beta_n y) \tag{3.60}$$

The bending moments can be calculated from

$$\begin{aligned}
M_x &= D\sum_{n=1}^{\infty}\sum_{m=1}^{\infty}\frac{\pi^2}{b^2}\left(m^2s^2 + n^2\right)W_{mn}\sin(\alpha_m x)\sin(\beta_n y) \\
M_y &= D\sum_{n=1}^{\infty}\sum_{m=1}^{\infty}\frac{\pi^2}{b^2}\left(\nu m^2s^2 + n^2\right)W_{mn}\sin(\alpha_m x)\sin(\beta_n y) \\
M_{xy} &= -(1-\nu)\frac{s\pi^2}{b^2}D\sum_{n=1}^{\infty}\sum_{m=1}^{\infty} mnW_{mn}\cos(\alpha_m x)\cos(\beta_n y)
\end{aligned} \tag{3.61}$$

The shear forces can be computed as follows:

$$Q_x = \frac{\partial M_x}{\partial x} + \frac{\partial M_{xy}}{\partial y} = D\sum_{n=1}^{\infty}\sum_{m=1}^{\infty} S_x W_{mn}\cos(\alpha_m x)\sin(\beta_n y)$$
$$Q_y = \frac{\partial M_{yx}}{\partial x} + \frac{\partial M_y}{\partial y} = D\sum_{n=1}^{\infty}\sum_{m=1}^{\infty} S_y W_{mn}\sin(\alpha_m x)\cos(\beta_n y) \tag{3.62}$$

where

$$S_x = D\left(\alpha_m^3 + \nu\alpha_m\beta_n^2\right),\ S_y = D\left(\nu\alpha_m^3 + \alpha_m\beta_n^2\right)$$

The effective shear forces (i.e., reaction forces) defined by $\hat{Q}_x$ and $\hat{Q}_y$ along the simply supported edges of $x = \pm a/2$ and $y = \pm b/2$, respectively, can be obtained from

$$\hat{Q}_x(a,y) = Q_x + \frac{\partial M_{xy}}{\partial y} = \frac{\partial M_x}{\partial x} + 2\frac{\partial M_{xy}}{\partial y} = \sum_{n=1}^{\infty}\sum_{m=1}^{\infty}(-1)^m \hat{S}_x W_{mn}\sin(\beta_n y)$$
$$\hat{Q}_y(x,b) = Q_y + \frac{\partial M_{xy}}{\partial y} = 2\frac{\partial M_{xy}}{\partial y} + \frac{\partial M_y}{\partial y} = \sum_{n=1}^{\infty}\sum_{m=1}^{\infty}(-1)^n \hat{S}_y W_{mn}\sin(\alpha_m y) \tag{3.63}$$

where

$$\hat{S}_x = D\left[\alpha_m^3 + (2-\nu)\alpha_m\beta_n^2\right],\ \hat{S}_y = D\left[\beta_n^3 + (2-\nu)\beta_m\alpha_n^2\right]$$

Therefore, the distribution of the reaction forces along the edges of the plate follows a sinusoidal function.

Besides the reaction forces given in Equation (3.63) along the edges, the plate also raises concentrated forces at the corners due to twisting moment M_{xy} per unit length. The concentrated forces at the corners are given by

$$F_{corner} = -2M_{xy} = 2(1-\nu)D\frac{dw^2}{dxdy}$$
$$= 2(1-\nu)D\sum_{n=1}^{\infty}\sum_{m=1}^{\infty}\left(\frac{m\pi}{a}\right)\left(\frac{n\pi}{b}\right)(-1)^{m+n}W_{mn} \tag{3.64}$$

In the pure bending case considered herewith, the stresses in the simply supported rectangular plate are given by

$$\sigma_{xx} = \frac{E}{(1-\nu^2)}\left(-z\frac{d^2w}{dx^2} - \nu z\frac{d^2w}{dy^2}\right) = z\sum_{n=1}^{\infty}\sum_{m=1}^{\infty} W_{mn}R_x \sin(\alpha_m x)\sin(\beta_n y)$$

$$\sigma_{yy} = \frac{E}{(1-\nu^2)}\left(-\nu z\frac{d^2w}{dx^2} - z\frac{d^2w}{dy^2}\right) = z\sum_{n=1}^{\infty}\sum_{m=1}^{\infty} W_{mn}R_y \sin(\alpha_m x)\sin(\beta_n y) \quad (3.65)$$

$$\tau_{xy} = \frac{E}{2(1+\nu)}\left(-2z\frac{d^2w}{dxdy}\right) = -z\sum_{n=1}^{\infty}\sum_{m=1}^{\infty} W_{mn}R_{xy} \cos(\alpha_m x)\cos(\beta_n y)$$

where

$$R_x = \frac{\pi^2 E}{b^2(1-\nu^2)}\left(m^2s^2 + \nu n^2\right)$$

$$R_y = \frac{\pi^2 E}{b^2(1-\nu^2)}\left(\nu m^2s^2 + n^2\right)$$

$$R_{xy} = mns\frac{\pi^2 E}{b^2(1+\nu)}$$

The maximum stresses occur at the bottom and top middle points on the surface of the plate at $(x,y,z) = (a/2, b/2, \pm h/2)$.

In the classical plate theory, the transverse stresses are vanishing when computed from the constitutive equations because the transverse shear strains are assumed to be zero. However, they can be computed by using the 3D stress equilibrium equations for any $-h/2 \le z \le +h/2$ as

$$\tau_{xz} = -\int_{-h/2}^{z}\left(\frac{d\sigma_x}{dx} + \frac{d\tau_{xy}}{dy}\right)dz + C_1(x,y)$$

$$\tau_{yz} = -\int_{-h/2}^{z}\left(\frac{d\tau_{yx}}{dx} + \frac{d\sigma_y}{dy}\right)dz + C_2(x,y) \quad (3.66)$$

where the constants C_i can be determined from the boundary conditions of $\sigma_{xz}(x,y,-h/2) = 0$ and $\sigma_{yz}(x,y,-h/2) = 0$; we can obtain the values of $C_i = 0$ and the following equations:

$$\tau_{xz} = \frac{h^2}{16}\left[1-\left(\frac{2z}{h}\right)^2\right]\sum_{n=1}^{\infty}\sum_{m=1}^{\infty} S_{mn}^{xz}\cos\left(\frac{m\pi x}{2a}\right)\sin\left(\frac{n\pi y}{2b}\right)$$

$$\tau_{yz} = \frac{h^2}{16}\left[1-\left(\frac{2z}{h}\right)^2\right]\sum_{n=1}^{\infty}\sum_{m=1}^{\infty} S_{mn}^{yz}\sin\left(\frac{m\pi x}{2a}\right)\cos\left(\frac{n\pi y}{2b}\right) \quad (3.67)$$

where

$$S_{mn}^{xz} = S_{13}W_{mn} \text{ and } S_{mn}^{yz} = S_{23}W_{mn}$$

where S_{ij} are defined by

$$S_{13} = \frac{E}{(1-\nu^2)}\left(\alpha_m^3 + \alpha_m\beta_n^2\right)$$

$$S_{23} = \frac{E}{(1-\nu^2)}\left(\beta_n^3 + \alpha_m^2\beta_n\right)$$

It is worth noting that the stresses of σ_{xz} and σ_{yz} are zero. The transverse shear stresses σ_{xz} and σ_{yz} have the maximum values at $(0,b/2,0)$ and $(a/2,0,0)$, respectively.

3.5.2 Solution for Bending of a Rectangular Plate Example

An isotropic rectangular plate subjected to uniformly distributed load is considered. From Equation (3.59) and Table 3.1, the displacement of the plate can be given by

$$w(x,y) = \frac{16q_0b^4}{D\pi^6}\sum_{n=1,3,\ldots}^{\infty}\sum_{m=1,3,\ldots}^{\infty}\frac{1}{mn\left(m^2s^2+n^2\right)^2}\sin\left(\frac{m\pi x}{a}\right)\sin\left(\frac{n\pi y}{b}\right)$$

From the above equation, the maximum values of $w_{\max}$ and $M_{x\max}$ are given by

$$w_{\max}(a/2,b/2) = \frac{16q_0b^4}{D\pi^6}\sum_{n=1,3,\ldots}^{\infty}\sum_{m=1,3,\ldots}^{\infty}\frac{(-1)^{(m+n)/2-1}}{mn\left(m^2s^2+n^2\right)^2}$$

$$M_{\max}(a/2,b/2) = \frac{16q_0b^2}{D\pi^4}\sum_{n=1,3,\ldots}^{\infty}\sum_{m=1,3,\ldots}^{\infty}\frac{(-1)^{(m+n)/2-1}\times\left(m^2s^2+\nu n^2\right)^2}{mn\left(m^2s^2+n^2\right)^2}$$

In case of a square plate $(a=b)$, we can have

$$w_{\max} = \frac{16q_0a^4}{D\pi^6}\sum_{n=1,3,\ldots}^{\infty}\sum_{m=1,3,\ldots}^{\infty}\frac{(-1)^{(m+n)/2-1}}{mn\left(m^2s^2+n^2\right)^2}$$

$$M_{\max} = \frac{16q_0a^2}{\pi^4}\sum_{n=1,3,\ldots}^{\infty}\sum_{m=1,3,\ldots}^{\infty}\frac{(-1)^{(m+n)/2-1}\times\left(m^2s^2+\nu n^2\right)^2}{mn\left(m^2s^2+n^2\right)^2}$$

TABLE 3.1

Coefficients in the Double Sine Series Expansion of Loads in Navier's Method

Loading Type	q_{mn} Coefficient
Uniform 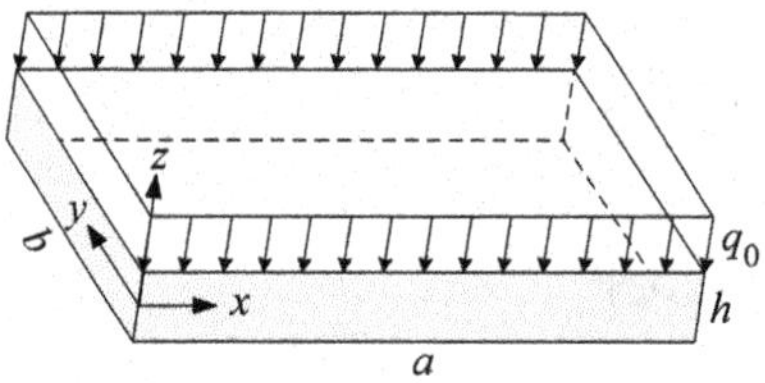	$q(x,y)=q_0$ $q_{mn}=\dfrac{16q_0}{\pi^2 mn}$ $(m,n=1,3,5,\ldots)$
Triangular prism 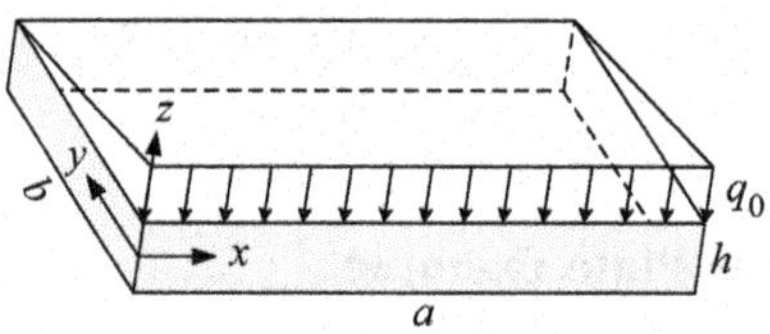	$q(x,y)=q_0\dfrac{y}{b}$ $q_{mn}=\dfrac{8q_0}{\pi^2 mn}(-1)^{n+1}$ $(m=1,3,5,\ldots;\ \ n=1,2,3,\ldots)$
Concentrated 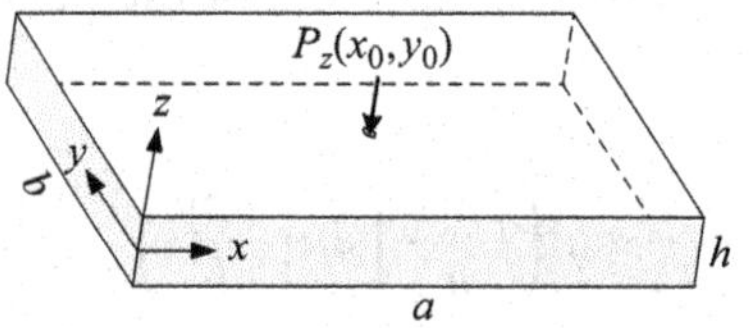	$P_z(x_0,y_0)=P_0$ $q_{mn}=\dfrac{4P_0}{ab}\sin\left(\dfrac{m\pi x_0}{a}\right)\sin\left(\dfrac{n\pi y_0}{b}\right)$ $(m,n=1,2,3,\ldots)$
Line 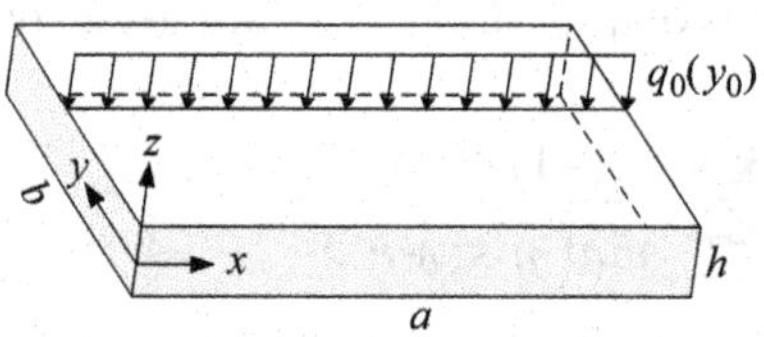	$q(x,y)=q_0\delta(y-y_0)$ $q_{mn}=\dfrac{8q_0}{\pi^2 bm}\sin\left(\dfrac{n\pi y_0}{b}\right)$ $(m=1,3,5,\ldots;\ \ n=1,2,3,\ldots)$

Assuming $\nu=0.3$, the solutions of the first term of the above series are as follows:

$$w_{\max}(m=1,n=1)=\frac{4q_0a^4}{D\pi^6}=0.00416\frac{q_0a^4}{D},$$

$$M_{\max}=0.05338q_0a^2,\text{ and }\sigma_{x\text{-max}}=0.3203\frac{q_0a^2}{h^2}.$$

The displacement is about 2.4% in error compared to the solution obtained with $m=n=1,3,\ldots,9$. Hence, it shows that the series of $w_{\max}$ converges rapidly. However, the bending moment $w_{\max}$ does not converge so fast. We need the $m=n=1,3,\ldots,29$ to converge to

$$w_{\max} = 0.004062\frac{q_0 a^4}{D}, \ M_{\max} = 0.04789 q_0 a^2, \text{ and } \sigma_{x-\max} = 0.2816\frac{q_0 a^2}{h^2}.$$

For an isotropic rectangular plate under a concentrated loading $P_z(x_0, y_0)$, the displacement w can be computed by

$$w(x,y) = \frac{4P_z b^2 s}{D\pi^4} \sum_{n=1,3,\ldots}^{\infty} \sum_{m=1,3,\ldots}^{\infty} \frac{\sin\left(\frac{m\pi x_0}{a}\right)\sin\left(\frac{n\pi y_0}{b}\right)}{\left(m^2 s^2 + n^2\right)^2} \sin\left(\frac{m\pi x}{a}\right)\sin\left(\frac{n\pi y}{b}\right)$$

The displacement w at the center of the plate is given by

$$w(a/2, b/2) = \frac{4P_z b^2 s}{D\pi^4} \sum_{n=1,3,\ldots}^{\infty} \sum_{m=1,3,\ldots}^{\infty} \frac{1}{\left(m^2 s^2 + n^2\right)^2}$$

For a square plate, the displacement w at the center of the plate becomes

$$w_{\max} = w(a/2, b/2) = \frac{4P_z a^2}{D\pi^4} \sum_{n=1,3,\ldots}^{\infty} \sum_{m=1,3,\ldots}^{\infty} \frac{1}{\left(m^2 + n^2\right)^2}$$

Assuming $\nu = 0.3$, the first term of the series in the above equation can be given by

$$w_{\max} = 0.01027\frac{P_z a^2}{D} = 0.1121\frac{P_z a^2}{Eh^3}, \ M_{\max} = 0.1317 P_z a, \text{ and } \sigma_{x-\max} = 0.7903\frac{P_z a}{h^2}.$$

Taking the first four terms of the series results in

$$w_{\max} = 0.01121\frac{P_z a^2}{D} = 0.1225\frac{P_z a^2}{Eh^3}, \ M_{\max} = 0.199 P_z a, \text{ and } \sigma_{x-\max} = 1.194\frac{P_z a}{h^2}.$$

The maximum displacement is about 3.4% in error compared to the solution obtained by using $m = n = 1, 3, \ldots, 19$ of $w_{\max} = 0.01160\frac{P_z a^2}{D}$.

Figure 3.5 shows the transverse displacement w of a flat rectangular plate subjected to the uniformly distributed loading on the surface of the plate.

Table 3.2 tabulates the nondimensional maximum transverse displacement $\bar{w}$ and stresses of square and rectangular plates under various types of loadings (see Table 3.1). The maximum value of 15 of m and n is used in the analysis. The plate is discretized into a 30×30 mesh. The loading data and material properties are given as follows: $Q_0 = q_0 = -1.0$, $x_0 = a/2$, $y_0 = b/2$, $E = 1.0$, $a = 1.0$, $b = 1.0, 3.0$, $h = 1.0$, and $\nu = 0.3$.

The transverse displacement w and stresses are nondimensional by using the following:

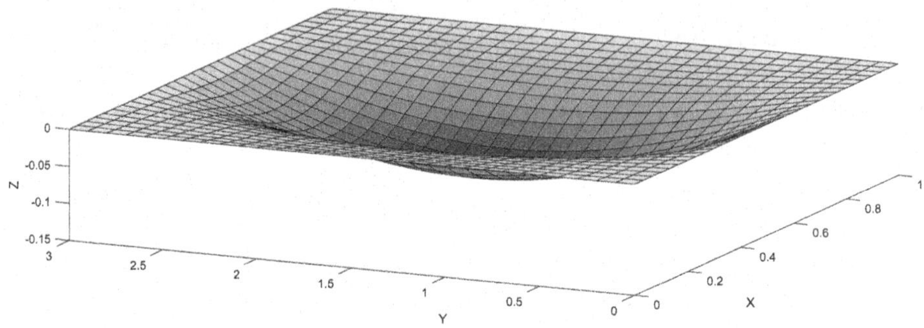

FIGURE 3.5
Bending solution of a rectangular plate.

$$\overline{w} = w(a/2, b/2)\left(\frac{q_0}{Eh^2a^4}\right),$$

$$\overline{\sigma}_{xx} = \sigma_{xx}(a/2, b/2, h/2)\left(\frac{q_0}{h^2a^2}\right), \overline{\sigma}_{yy} = \sigma_{yy}(a/2, b/2, h/2)\left(\frac{q_0}{h^2a^2}\right),$$

$$\overline{\tau}_{xy} = \tau_{xy}(a, b, h/2)\left(\frac{q_0}{h^2a^2}\right), \overline{\tau}_{xz} = \tau_{xz}(0, b/2, 0)\left(\frac{q_0}{2ha}\right),$$

$$\overline{\tau}_{yz} = \tau_{yz}(a/2, 0, 0)\left(\frac{q_0}{2ha}\right).$$

TABLE 3.2
Transverse Displacements and Stresses of a Rectangular/Square Isotropic Plate

Loading Type	$\overline{w}$	$\overline{\sigma}_{xx}$	$\overline{\sigma}_{yy}$	$\overline{\tau}_{xy}$	$\overline{\tau}_{xz}$	$\overline{\tau}_{yz}$
Square plates $(s = b/a = 1)$						
Uniform	0.0444	0.2872	0.2872	0.1945	0.4869	0.4869
Triangular	0.0224	0.1617	0.1474	0.1161	0.1375	0.2737
Concentrated	0.1264	2.0453	2.0453	0.3661	5.1951	5.1951
Line	0.0735	0.5290	0.6887	0.2475	1.3735	0.7309
Rectangular plates $(s = b/a = 3)$						
Uniform	0.1336	0.7128	0.2693	0.2821	0.7207	0.4996
Triangular	0.0669	0.3671	0.1378	0.1598	0.2470	0.2720
Concentrated	0.1833	1.9626	1.4330	0.2228	3.1221	2.2528
Line	0.1088	0.6533	0.5005	0.1771	1.0255	0.4590

Note: The number in parentheses shows the maximum values of m and n used to obtain the results.

3.5.3 BendingPlateNavier Program List

```
% BendingPlateNavier.m
% Solution for bending of a rectangular plate example
clear variables; clc;
% Geometrical data
h = 1.0;
a = 1;
b = 3;        % b=1 or b=3
s = b/a;
% Series
mmax = 15;
nmax = 15;
z3 = +h/2;  % top plate
z2 = 0;     % mid-plane
z1 = -h/2;  % bottom plate
z  = z2;    % select from above
% Loading selection index
id = 4;     % selection : 1,2,3,4
%%
nx = 31;
xi=linspace(0,a,nx);
ny = 31;
yi=linspace(0,b,ny);
wxy = zeros(nx,ny);  % Transverse Displacement w
sxx = zeros(nx,ny);  % Normal Stress sigma-xx
syy = zeros(nx,ny);  % Normal Stress sigma-yy
txy = zeros(nx,ny);  % Shear Stress tau-xy
txz = zeros(nx,ny);  % Shear Stress tau-xz
tyz = zeros(nx,ny);  % Shear Stress tau-yz
% Plate's material
E = 1;
nu = 0.3;
D = E*h^3/12/(1-nu^2);
q0 = -1.0;
Q00 = -1.0;
Q0x = a/2;
Q0y = b/2;
wcoef = b^4/D/pi^4;
%
switch id % Loading types
  case 1  % Uniform distributed loading type
    for i = 1:nx
    for j = 1:ny
      for n = 1:2:nmax  % odd number
      for m = 1:2:mmax  % odd number
        qmn = 16*q0/pi^2/m/n;
        wdenom = (m^2*s^2+n^2)^2;
        Wmn = wcoef*qmn/wdenom;
```

```
          wxy(i,j) = wxy(i,j)+Wmn*sin(m*pi/a*xi(i))
*sin(n*pi/b*yi(j));
          Rx = pi^2*E/b^2/(1-nu^2)*(m^2*s^2+nu*n^2);
          sxx(i,j) = sxx(i,j)-z*Wmn*Rx*sin(m*pi/a*xi(i))
*sin(n*pi/b*yi(j));
          Ry = pi^2*E/b^2/(1-nu^2)*(nu*m^2*s^2+n^2);
          syy(i,j) = syy(i,j)-z*Wmn*Ry*sin(m*pi/a*xi(i))
*sin(n*pi/b*yi(j));
          Rxy = pi^2*E/b^2/(1+nu)*m*s*n;
          txy(i,j) = txy(i,j)+z*Wmn*Rxy*cos(m*pi/a*xi(i))
*cos(n*pi/b*yi(j));
          S13 = E/(1-nu^2)*(m^3*pi^3/a^3+m*n^2*s*pi^3/b^3);
          Sxzc = h^2/16*(1-4*z^2/h^2)*S13;
          txz(i,j) = txz(i,j)+Wmn*Sxzc*cos(m*pi/a*xi(i))
*sin(n*pi/b*yi(j));
          S23 = E/(1-nu^2)*(n^3*pi^3/b^3+m^2*n*s^2*pi^3/b^3);
          Syzc = h^2/16*(1-4*z^2/h^2)*S23;
          tyz(i,j) = tyz(i,j)+Wmn*Syzc*sin(m*pi/a*xi(i))
*cos(n*pi/b*yi(j));
        end
        end
      end
      end
    case 2  % Triangular prism distributed loading type
      for i = 1:nx
      for j = 1:ny
        mmax = floor(nmax/2);
        for n = 1:2:nmax  % odd number
        for m = 1:1:mmax  % odd number
          qmn = 8*q0/pi^2/m/n*(-1)^(n+1);
          wdenom = (m^2*s^2+n^2)^2;
          Wmn = wcoef*qmn/wdenom;
          wxy(i,j) = wxy(i,j)+Wmn*sin(m*pi/a*xi(i))
*sin(n*pi/b*yi(j));
          Rx = pi^2*E/b^2/(1-nu^2)*(m^2*s^2+nu*n^2);
          sxx(i,j) = sxx(i,j)-z*Wmn*Rx*sin(m*pi/a*xi(i))
*sin(n*pi/b*yi(j));
          Ry = pi^2*E/b^2/(1-nu^2)*(nu*m^2*s^2+n^2);
          syy(i,j) = syy(i,j)-z*Wmn*Ry*sin(m*pi/a*xi(i))
*sin(n*pi/b*yi(j));
          Rxy = pi^2*E/b^2/(1+nu)*m*s*n;
          txy(i,j) = txy(i,j)+z*Wmn*Rxy*cos(m*pi/a*xi(i))
*cos(n*pi/b*yi(j));
          S13 = E/(1-nu^2)*(m^3*pi^3/a^3+m*n^2*s*pi^3/b^3);
          Sxzc = h^2/16*(1-4*z^2/h^2)*S13;
          txz(i,j) =
txz(i,j)+Wmn*Sxzc*cos(m*pi/a*xi(i))*sin(n*pi/b*yi(j));
          S23 = E/(1-nu^2)*(n^3*pi^3/b^3+m^2*n*s^2*pi^3/b^3);
```

```
        Syzc = h^2/16*(1-4*z^2/h^2)*S23;
        tyz(i,j) = tyz(i,j)+Wmn*Syzc*sin(m*pi/a*xi(i))
*cos(n*pi/b*yi(j));
      end
      end
    end
    end
  case 3  % Concentrated Force loading type
    for i = 1:nx
    for j = 1:ny
      for n = 1:1:nmax  % odd number
      for m = 1:1:mmax  % odd number
        qmn = 4*Q00/a/b*sin(m*pi/a*Q0x)*sin(n*pi/b*Q0y);
        wdenom = (m^2*s^2+n^2)^2;
        Wmn = wcoef*qmn/wdenom;
        wxy(i,j) = wxy(i,j)+Wmn*sin(m*pi/a*xi(i))
*sin(n*pi/b*yi(j));
        Rx = pi^2*E/b^2/(1-nu^2)*(m^2*s^2+nu*n^2);
        sxx(i,j) = sxx(i,j)-z*Wmn*Rx*sin(m*pi/a*xi(i))
*sin(n*pi/b*yi(j));
        Ry = pi^2*E/b^2/(1-nu^2)*(nu*m^2*s^2+n^2);
        syy(i,j) =
syy(i,j)-z*Wmn*Ry*sin(m*pi/a*xi(i))*sin(n*pi/b*yi(j));
        Rxy = pi^2*E/b^2/(1+nu)*m*s*n;
        txy(i,j) = txy(i,j)+z*Wmn*Rxy*cos(m*pi/a*xi(i))
*cos(n*pi/b*yi(j));
        S13 = E/(1-nu^2)*(m^3*pi^3/a^3+m*n^2*s*pi^3/b^3);
        Sxzc = h^2/16*(1-4*z^2/h^2)*S13;
        txz(i,j) = txz(i,j)+Wmn*Sxzc*cos(m*pi/a*xi(i))
*sin(n*pi/b*yi(j));
        S23 = E/(1-nu^2)*(n^3*pi^3/b^3+m^2*n*s^2*pi^3/b^3);
        Syzc = h^2/16*(1-4*z^2/h^2)*S23;
        tyz(i,j) = tyz(i,j)+Wmn*Syzc*sin(m*pi/a*xi(i))
*cos(n*pi/b*yi(j));
      end
      end
    end
    end
  case 4  % Line loading parallel with x-axis type
    for i = 1:nx
    for j = 1:ny
      nmax = floor(mmax/2);
      for n = 1:1:nmax  % odd number
      for m = 1:2:mmax  % odd number
        qmn = 8*q0/pi/b/m*sin(n*pi/b*Q0y);
        wdenom = (m^2*s^2+n^2)^2;
        Wmn = wcoef*qmn/wdenom;
        wxy(i,j) =
wxy(i,j)+Wmn*sin(m*pi/a*xi(i))*sin(n*pi/b*yi(j));
```

```
        Rx = pi^2*E/b^2/(1-nu^2)*(m^2*s^2+nu*n^2);
        sxx(i,j) = sxx(i,j)-z*Wmn*Rx*sin(m*pi/a*xi(i))
*sin(n*pi/b*yi(j));
        Ry = pi^2*E/b^2/(1-nu^2)*(nu*m^2*s^2+n^2);
        syy(i,j) = syy(i,j)-z*Wmn*Ry*sin(m*pi/a*xi(i))
*sin(n*pi/b*yi(j));
        Rxy = pi^2*E/b^2/(1+nu)*m*s*n;
        txy(i,j) = txy(i,j)+z*Wmn*Rxy*cos(m*pi/a*xi(i))
*cos(n*pi/b*yi(j));
        S13 = E/(1-nu^2)*(m^3*pi^3/a^3+m*n^2*s*pi^3/b^3);
        Sxzc = h^2/16*(1-4*z^2/h^2)*S13;
        txz(i,j) = txz(i,j)+Wmn*Sxzc*cos(m*pi/a*xi(i))
*sin(n*pi/b*yi(j));
        S23 = E/(1-nu^2)*(n^3*pi^3/b^3+m^2*n*s^2*pi^3/b^3);
        Syzc = h^2/16*(1-4*z^2/h^2)*S23;
        tyz(i,j) = tyz(i,j)+Wmn*Syzc*sin(m*pi/a*xi(i))
*cos(n*pi/b*yi(j));
      end
      end
    end
    end
  otherwise
        disp('Wrong loading index [1 to 4]')
        disp('press any key.....')
        pause;
        return;
end
% Normalization
wN = abs(q0/E/h^2/a^4);
sN = abs(q0/h^2/a^2);
sN2 = abs(q0/2/a/h);
text=['Minimum value of normalized w = ',
num2str(min(min(wxy))/wN)];
disp(text);
text=['Normal stress Sxx = ',num2str(max(max(sxx))/sN)];
disp(text);
text=['Normal stress Syy = ',num2str(max(max(syy))/sN)];
disp(text);
text=['Shear stress Txy = ',num2str(max(max(txy))/sN)];
disp(text);
text=['Shear stress Txz = ',num2str(max(max(txz))/sN2)];
disp(text);
text=['Shear stress Tyz = ',num2str(max(max(tyz))/sN2)];
disp(text);
figure(1)
set(gcf,'position',[400,400,1200,400])
for l=1:1
    hold on;
    surface(repmat(xi,nx,1),repmat(yi',1,ny),wxy)
```

```
    xlabel('X'); ylabel('Y'); zlabel('Z');
    azm = -65;
    view(azm,50)
end
hold off;
```

3.5.4 Formulations for Free Vibration Problem of Rectangular Plates

In the free vibration problem of a plate, the displacement w of the plate is assumed to be periodic in time as follows:

$$w(x,y,t)=\sum_{n=1}^{\infty}\sum_{m=1}^{\infty}W_{mn}e^{i\omega_{mn}t}\sin\left(\frac{m\pi x}{a}\right)\sin\left(\frac{n\pi y}{b}\right) \tag{3.68}$$

with

$$W_{mn}(t)=W_{mn}e^{i\omega_{mn}t} \tag{3.69}$$

where W_{mn} is the amplitude of the free vibration and ω_{mn} is the natural frequency of free vibration which is associated with the mth and n^{th} mode shapes of vibration:

$$\sin\left(\frac{m\pi x}{a}\right)\sin\left(\frac{n\pi y}{b}\right) \tag{3.70}$$

Substituting (3.68) into (3.55) results in

$$D\left(\alpha_m^2+\beta_n^2\right)^2-\left(\alpha_m^2+\lambda\beta_n^2\right)N_0-\omega_{mn}^2\left[I_0+I_2\left(\alpha_m^2+\beta_n^2\right)\right]=0 \tag{3.71}$$

Solving for ω_{mn}^2 as

$$\omega_{mn}^2=\frac{\pi^2}{\tilde{I}_0 b^2}\left[\frac{\pi^2 D}{b^2}\left(m^2s^2+n^2\right)^2-N_o\left(m^2s^2+\lambda n^2\right)\right];\quad s=\frac{b}{a} \tag{3.72}$$

where

$$\tilde{I}_0=I_0+I_2\left[\left(\frac{m\pi}{a}\right)^2+\left(\frac{n\pi}{b}\right)^2\right]$$

In case where m and n have different values, there is a unique natural frequency ω_{mn}, and its corresponding mode shape can be obtained by using Equation (3.70). For thin plates, where the ratio of $h/b < 10$, the rotary inertia I_2 term can be neglected.

The smallest value of ω_{mn} is called the *fundamental frequency* of the plate. In case where the rotary inertia I_2 is neglected, the natural frequency of a rectangular isotropic plate without in-plane forces N_0 yields

$$\omega_{mn} = \frac{\pi^2}{b^2}\sqrt{\frac{D}{\rho h}}\left(m^2 s^2 + n^2\right); \quad s = \frac{b}{a} \tag{3.73}$$

In a more specific case of a square isotropic plate, the natural frequency is given by

$$\omega_{mn} = \frac{\pi^2}{a^2}\sqrt{\frac{D}{\rho h}}\left(m^2 + n^2\right) \tag{3.74}$$

The fundamental frequency of the square isotropic plate of the first mode $(m = 1; n = 1)$ can be obtained by

$$\omega_{11} = \frac{2\pi^2}{a^2}\sqrt{\frac{D}{\rho h}} \tag{3.75}$$

3.5.5 Solution for Free Vibration of a Rectangular Plate Example

Table 3.3 tabulates the nondimensional frequencies $\bar{\omega}_{mn}$ of isotropic $(\nu = 0.25)$ plates for $(m,n) = (1,1),(1,2),(2,1)$.

3.5.6 Formulations for Buckling Problem of Rectangular Plates

In the buckling problem of a rectangular plate subjected to in-plane biaxial compressive loadings at all edges (see Figure 3.6), Equation (3.55) becomes

$$\left[D\left(\alpha_m^2 + \beta_n^2\right)^2 - \left(\alpha_m^2 + \lambda\beta_n^2\right)N_0\right]W_{mn} = 0 \tag{3.76}$$

TABLE 3.3

Nondimensional Frequencies $\bar{\omega}_{mn}$ of an Isotropic $(\nu = 0.3)$ Plate

	$\bar{\omega}_{11}$			$\bar{\omega}_{12}$			$\bar{\omega}_{21}$		
b/a	*w/o*	0.01 (h/b)	0.1 (h/b)	*w/o*	0.01 (h/b)	0.1 (h/b)	*w/o*	0.01 (h/b)	0.1 (h/b)
0.5	1.250	1.250	1.249	4.250	4.249	4.243	2.000	2.000	1.998
1.0	2.000	2.000	1.998	5.000	4.999	4.990	5.000	4.999	4.990
1.5	3.250	3.250	3.246	6.250	6.248	6.234	10.000	9.996	9.959
2.0	5.000	4.999	4.990	8.000	7.997	7.974	17.000	16.988	16.882
2.5	7.250	7.248	7.228	10.250	10.246	10.207	26.000	25.972	25.726
3.0	10.000	9.996	9.959	13.000	12.993	12.931	37.000	36.944	36.450

Note: *w/o* denotes the results without rotary inertia for $h/b = 0.01$.

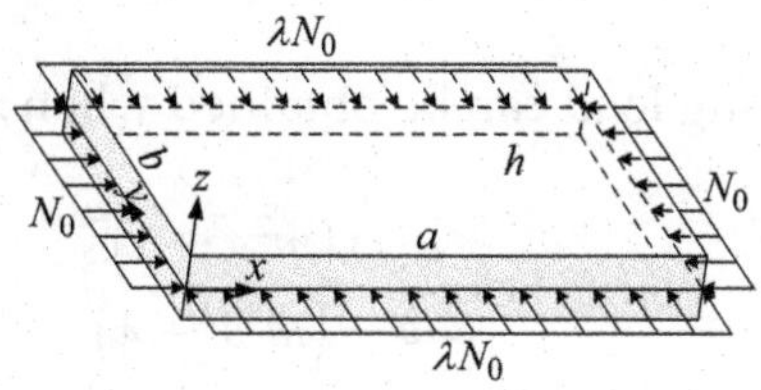

FIGURE 3.6
Buckling loads of a rectangular plate.

Solving the loading term N_0, leads to,

$$N_0(m,n)=\frac{\pi^2 D}{b^2}\frac{\left(m^2s^2+n^2\right)^2}{\left(m^2s^2+\lambda n^2\right)} \tag{3.77}$$

where $s(=b/a)$ is the aspect ratio of the plate. For each combination of m and n, there is a corresponding unique value of buckling load N_0. The smallest value of N_0 is called the *critical buckling load* (N_{cr}) of a plate.

3.5.6.1 Case 1: Biaxial Compression of a Plate

For a rectangular isotropic plate under biaxial compression ($\lambda = 1$), we can obtain the buckling load from

$$N_0(m,n)=\frac{\pi^2 D}{b^2}\left(m^2s^2+n^2\right) \tag{3.78}$$

The critical buckling load can be obtained at $m = n = 1$, which is given by

$$N_{cr}=\frac{\pi^2 D}{b^2}\left(1+s^2\right);\quad s=\frac{b}{a} \tag{3.79}$$

Moreover, in case the plate is square $(a = b)$, the above equation reduces to

$$N_{cr}=\frac{2\pi^2 D}{a^2}=\frac{2\pi^2 D}{b^2} \tag{3.80}$$

3.5.6.2 Case 2: Biaxial Compression and Tension of a Plate

Suppose the edges $x = 0, a$ of a square plate are subjected to compressive load $\hat{N}_x = N_0$ and the edges $y = 0, b$ are subjected to tension load $\hat{N}_y = -\lambda N_0$, then Equation (3.78) becomes

$$N_0(m,n)=\frac{\pi^2 D}{a^2}\frac{\left(m^2+n^2\right)^2}{\left(m^2+\lambda n^2\right)} \tag{3.81}$$

when $\lambda n^2 < m^2$.

The minimum buckling load can be obtained when $n = 1$ as

$$N_0(m,1) = \frac{\pi^2 D}{b^2} \frac{\left(m^2 s^2 + 1\right)^2}{\left(m^2 s^2 - \lambda\right)} \tag{3.82}$$

Theoretically, the minimum value of $N_0(m,1)$ can be obtained when $m^2 s^2 = 1 + 2\lambda$. Hence, for a square plate with $\lambda = 0.5$, we get

$$N_0(1,1) = \frac{8\pi^2 D}{a^2}; \quad N_0(2,1) = \frac{7.1429\pi^2 D}{a^2} = N_{cr} \tag{3.83}$$

3.5.6.3 Case 3: Uniaxial Compression of a Rectangular Plate

Suppose a rectangular plate is subjected to uniform compressive load N_0 at the edges $x = 0, a$, that is, $\lambda = 0$, then the buckling load can be given by

$$N_0(m,n) = \frac{\pi^2 a^2 D}{m^2} \left(\frac{m^2}{a^2} + \frac{n^2}{b^2} \right)^2 \tag{3.84}$$

and

$$N_0(m,1) = \frac{\pi^2 D}{a^2} \left(m + \frac{1}{m} \frac{a^2}{b^2} \right)^2. \tag{3.85}$$

In case of a given aspect ratio s, there are two different modes of m_1 and m_2 that have the same buckling load when $\sqrt{m_1 m_2} = a/b$. Especially, when the intersection of curves m and $m+1$ arises for the aspect ratios of $a/b = \sqrt{2}, \sqrt{6}, \sqrt{12}, \sqrt{20}, \ldots, \sqrt{m^2 + m}$.

Hence, there is a mode change at those aspect ratios from m half-waves to $m+1$ half-waves.

By letting $m = 1$ in Equation (3.84), we can obtain the following:

$$N_{cr} = \frac{\pi^2 D}{b^2} \left(\frac{a}{b} + \frac{b}{a} \right)^2 \tag{3.86}$$

In case of a square plate, the above equation becomes

$$N_{cr} = \frac{4\pi^2 D}{b^2} \tag{3.87}$$

TABLE 3.4

The Nondimensional Buckling Load of Simply Supported Rectangular Plates

$\lambda = 0$				$\lambda = 1$			
a/b	$\bar{N}$	a/b	$\bar{N}$	a/b	$\bar{N}$	a/b	$\bar{N}$
0.5	6.250	2.0	$4.000^{(2,1)}$	0.5	5.000	2.0	1.250
1.0	4.000	2.5	$4.134^{(3,1)}$	1.0	2.000	2.5	1.160
1.5	$4.340^{(2,1)}$	3.0	$4.000^{(3,1)}$	1.5	1.444	3.0	1.111

Note: The numbers in parentheses show the values of m and n used to obtain the buckling load, while $(m,n)=(1,1)$ for other cases.

3.5.7 Solution for Buckling Problem of a Rectangular Plate Example

Table 8.6 tabulates the effect of plate aspect ratio and loading ratio on the critical buckling loads $\bar{N} = N_{cr}b^2/(\pi^2 D)$ of rectangular isotropic $(\nu = 0.3)$ plates under uniform axial compression $(\lambda = 0)$ and biaxial compression $(\lambda = 1)$. In all of the cases, the critical buckling mode is considered for $(m,n)=(1,1)$ (Table 3.4).

References

Kirchhoff G (1850) Uber das Gleichgewicht und die Bewegung einer elastischen Scheibe. *J Reine Angewandte Mathematik* 40:51–88.

Mindlin RD (1951) Influence of rotatory inertia and shear on flexural motions of isotropic, elastic plates. *J Appl Mech (ASME)* 18:31–38.

Reissner E (1945) The effect of transverse shear deformation on the bending of elastic plates. *J Appl Mech (ASME)* 12(2):69–77.

4

Theories of Thick Plate and Shell Elements

4.1 Various Theories of Plate and Shell

In general, the behavior of plate and shell under mechanical, electrical, and thermal loadings can be predicted using either three-dimensional (3D) elasticity theory or equivalent single-layer (ESL) theories. The ESL models can be obtained from the 3D elasticity theory by making appropriate assumptions about the kinematics of deformation through the thickness of plate and shell (Reddy 2004). The ESL theories take into consideration both shear and normal deformation effects depending on the level of assumptions. The simplest ESL model is the classical plate theory (CPT) discussed in Chapter 3, also known as Kirchhoff theory (Kirchhoff 1850), which neglects both shear and normal deformation effects. The next theory in the hierarchy of ESL models is the first-order shear deformation theory (FSDT) developed by Mindlin (1951). The FSDT takes into account the shear deformation effect by means of a linear variation of in-plane displacements through the thickness by using a shear correction factor. The shear correction factor is difficult to be determined because it depends on the geometric parameters, loading, and boundary conditions. To avoid the use of the shear correction factor, higher order shear deformation theories (HSDTs) were proposed. The HSDT was developed by expanding the displacement and rotational components in power series through the thickness coordinate. In principle, the theories developed by using HSDT are accurate by increasing the number of parameters in the series. Among the HSDTs, the third-order shear deformation theory (TSDT) developed by Reddy (1984) is the most widely used theory due to its simplicity and accuracy. This chapter describes the existing theories for the modeling and analysis of plate and shell with the main emphasis on the ESL models such as the CPT, FSDT, TSDT, HSDT, simplified theories, and mixed theories. In addition, the 3D elasticity solutions and a unified formulation will also be introduced.

4.2 Classical Plate Theory

The CPT (given in Chapter 3) is based on the Kirchhoff–Love hypothesis, which stated that the straight lines remain straight and perpendicular to the midplane after deformation. These assumptions imply the vanishing of the shear and normal strains, and consequently, neglecting the shear and normal deformation effects. The CPT is the simplest ESL model, and it is only accurate for thin plates and shells where the shear and normal deformation effects are negligible. The governing equations derived from the CPT can be analytically solved using a Levy-type solution approach or numerically solved using Navier solution. The semi-analytical differential quadrature method (DQM) and the modal superposition approach were mostly used to calculate the natural frequency and transient response of rectangular plates of the Levy-type solution.

4.3 First-Order Shear Deformation Theory

The FSDT developed by Mindlin (1951) takes into account the shear deformation effect employing a linear variation of the in-plane displacements through the thickness of the plate and shell. It should be noted that the theory developed by Reissner (1945, 1947) also takes into account the shear deformation effects in the analysis of plates. The Reissner theory was based on the assumption of a linear bending stress distribution and a parabolic shear stress distribution; its formulation will certainly lead to the variation of displacement which is not necessarily linear across the thickness of the plate. Thus, FSDT is a particular case of the theory developed by Reissner by neglecting the normal stress in the formulation.

4.4 Third-Order Shear Deformation Theory

The TSDT developed by Reddy (1984) for laminated composite plates takes into account the transverse shear deformation effect, which satisfies the zero-traction boundary conditions on the top and bottom surfaces of a plate. A shear correction factor is henceforth not necessary. It is worth noting that the displacement field of Reddy theory is identical to that of Levinson theory (1980). However, the governing equations of both theories are different. This is because Levinson used the equilibrium equations of the FSDT which are variationally inconsistent with those derived from the variational approach by Reddy. Reddy (2000) presented both analytical and finite element formulations based on the TSDT to analyze functionally graded plates. The formulations account for the thermomechanical coupling, time dependency, and von Karman-type geometric nonlinearity.

4.5 Higher Order Shear Deformation Theory

The HSDT takes into account the higher order variations of the in-plane displacements or both in-plane and transverse displacements (i.e., quasi-3D theory) through the thickness, and consequently, considers the effects of shear deformation or both shear and normal deformations. The HSDT can be developed using polynomial shape functions or non-polynomial shape functions.

4.5.1 Polynomial Function-Based Models

The displacement field of the quasi-3D theory of Lo et al. (1977a, 1977b) has 11 unknowns and accounts for a cubic variation of the in-plane displacements and a quadratic variation of the transverse displacement through the thickness.

4.5.2 Non-polynomial Function-Based Models

The non-polynomial function was first introduced by Levy (1877) with a sinusoidal function to develop a refined theory for thick isotropic plates. The sinusoidal function was later adopted by Stein (1986) and Touratier (1991) to develop a five-parameter sinusoidal shear deformation theory (SSDT) for isotropic and laminated composite plate theories. The SSDT was extensively used to study the thermal bending of composite plates (Ferreira et al. 2005; Zenkour 2004a, 2004b), buckling of composite plates (Zenkour 2004c), bending of functionally graded sandwich plates (Zenkour 2005a; Zenkour and Alghamdi 2010a), buckling and vibration of functionally graded sandwich plates (Zenkour 2005b; Zenkour and Sobhy 2010b), vibration of Functionally Graded plates (Zenkour 2005c), bending of functionally graded plates (Zenkour 2006), thermal bending of functionally graded plates resting on an elastic foundation (Zenkour 2009), and thermal buckling of functionally graded plates resting on an elastic foundation (Zenkour and Sobhy 2011) and nanoplates (Thai et al. 2014).

4.6 Simplified Theories

If we expand the displacements in power series of the HSDT and quasi-3D theories in the thickness direction, computationally the cost becomes expensive because additional series in the thickness direction will increase the unknown parameters to the theories. Hence, there is a need to simplify the existing HSDT and quasi-3D theories by developing simple theories with fewer unknown parameters. Senthilnathan et al. (1987) simplified the TSDT by splitting the transverse displacement into the bending and shear components and making additional assumptions to the TSDT. They reduced the number of the unknown parameters to only one. The idea of splitting the transverse displacement into the bending and shear parts was first proposed by Huffington (1963) and later modified by Krishna Murty (1987). The uncoupling of the displacement terms into the bending and shear parts not only can reduce the number of unknown parameters but also show the roles and contributions of both parts in the total displacements. Shimpi (2002) developed the refined plate theory (RPT) for isotropic plates by splitting the displacements into the bending and shear components. The RPT has only two unknown parameters compared with the FSDT and TSDT which have three unknown parameters, but RPT has adequate accuracy in predicting the global responses of isotropic plates (Shimpi 2002; Thai and Choi 2013; Thai et al. 2013) and orthotropic plates (Thai and Kim 2012a; Shimpi and Patel 2006; Kim et al. 2009a; Thai and Kim 2011; Thai and Kim 2012b). Kim et al. (2009b) extended the RPT concept to the laminated composite plates and modified the RPT by taking into account the extension component of the transverse displacement. The conclusion was that the RPT plate element could accurately predict the static bending, buckling, and free vibration behaviors of laminated composite plates (Thai and Kim 2010; Thai and Choi 2014; Tran et al. 2014). Based on trigonometric functions, Mantari and Soares (2014, 2015) proposed quasi-3D theories with four unknown parameters for functionally graded plates by combining the shear and stretching parts with the transverse displacement.

4.7 Mixed Theories

In the mixed theories of plate element, all the ESL models mentioned previously are developed based on the principle of virtual displacements (PVDs) where the displacement components are regarded as the primary variables, and the stress components are calculated from the displacement components using the strain–displacement and constitutive relationships. In the variational approach, the Reissner mixed variational theorem (RMVT), was proposed by Reissner (1984, 1986) by assuming two independent parameters for displacements and transverse stresses. The advantage of the RMVT over the PVD is that the compatibility of displacements and the equilibrium between two adjacent layers can be automatically satisfied. The RMVT is considered as a powerful tool for the analysis of multilayered plates. Murakami (1986) was the first to apply the RMVT to develop a mixed laminate theory using a first-order zigzag displacement model. Based on the RMVT, Demasi (2009a–e) developed a variety of mixed laminate theories in a series of five papers including layer-wise theories, HSDT, and zigzag models for plate elements. The meshless collocation method and the element-free Galerkin method were developed by Wu et al. (2011) and Wu and Chiu (2011) using the differential reproducing kernel interpolation (Wang et al. 2010) for the RMVT.

4.8 3D Elasticity

The development of exact solutions of 3D elasticity theory is useful in assessing the accuracy and validity of the ESL plate element.

Cheng and Batra (2000) derived the exact solutions for 3D bending analysis of functionally graded clamped elliptic plates under thermal loadings using an asymptotic expansion method. Reddy and Cheng (2001) also adopted the asymptotic expansion method to derive the exact solutions for 3D bending analysis of functionally graded simply supported plates under thermal loadings. Instead of using the asymptotic expansion method, Vel and Batra (2002) adopted the power series method to derived the exact solutions for the 3D bending analysis of functionally graded simply supported plates subjected to thermal and mechanical loadings.

The exact 3D solutions for the free vibration of functionally graded rectangular plates under general boundary conditions were also provided by Jin et al. (2014) using the Rayleigh–Ritz method.

4.9 Unified Formulation

The unified formulation proposed by Carrera (1995, 1997, 1998a–c, 1999a–e, 2000a–d, 2001, 2003) for multilayered composite structures is a hierarchical formulation of the procedure to implement various plate/shell theories in a unified manner by referring to a few fundamental nuclei concepts. All theories can be easily developed in the framework of the Carrera Unified Formulation (CUF) by expanding the displacement variables in the thickness direction by using Taylor's expansions of N-order with N being an unrestricted parameter. More detailed information and applications of the CUF can be found in the books authored by Carrera et al. (2011a, 2011b, 2014).

4.10 Shell Meant by a Thick Plate

Shells are geometrically curved, so it is reasonable that shell elements look like curves. It makes sense that to model a shell by flat elements is to model a flat plate by curved elements. A curved surface shell element must model shell geometry and must combine the membrane and bending actions in its deformable behavior.

There are three choices in shell element formulation:

1. A flat element, made by combining a membrane element with a plate bending element
2. A curved element, formulated by shell theory
3. A solid or degenerate solid element, analogous to the plate elements

Choice 1 is easiest for both formulation and use, but accuracy is often average and sometimes unacceptable. Choice 2 is difficult, and the element may have many degrees of freedom (DOFs), some of which are higher displacement derivatives. Choice 3 occupies the consensus between Choices 1 and 2.

In Choice 3, the element is modeled by thin shells with 32 DOFs. The accuracy is good, but the unusual allocation of DOF requires special formulations around the edges of the element. Choice 2 has two drawbacks: shell theory is complicated, and theoreticians have proposed many different strain–displacement relations for the same geometry of shell because there is room for argument about how many terms can be discarded as negligible. Shallow shell theory uses simple relations; thus, the element cannot be used for the thick shell. Generally, the shell has a thin geometry. Hence, it can be modeled by a thin plate in which the transverse shear deformation effects are negligible. The thin shell is stiffer in stretching than in bending like a plate; hence, the thin plate theory considered in this chapter takes into consideration the effects of axial loadings at the boundary of the edge.

By combining the NURBS and condensation, this book is an attempt to eliminate all the drawbacks of Choice 3.

4.11 Thick Plate and Shell Formulation

In Chapter 3, we have discussed the theory of classic thin plate theory developed by Kirchhoff (1850) known as CPT. In CPT, the sections at the edges of the plate remains straight and perpendicular to the midplane of the plate after deformation; this assumption contrasts with the thick plate theory introduced by Reissner (1945) and FSDT by Mindlin (1951), which states that the section in the thickness direction remains straight but is not necessarily perpendicular to the midplane of the plate after the deformation. In the thick plate theory, the shear deformation that causes the rotation of the section during the deformation is taken into consideration.

The thick plate model FSDT is based on the kinematic assumptions made by Mindlin (1951) to formulate the governing equations.

The underlying assumption of Mindlin thick plate theory is that the plane of the cross sections of both directions (x and y) is not necessarily perpendicular to the plate's midplane during the deformation. This implies that the rotations of the cross sections of the plate are not equal to the slopes of x- and y-axes of the plate due to shear strain involvement.

In Figure 4.1, the kinematic assumption (undeformed and deformed) of a plate bent at both sides by moments along the y-axis with the transverse displacement w is the variable of vertical displacement. Point O is on the midplane ($z = 0$) of the plate that can be moved and deformed.

Under the small displacement assumption, we can apply the following approximations: $\sin(\theta_y) \approx \tan(\theta_y) \approx \theta_y$, $\cos(\theta_y) \approx 1$, and $\theta_y^2 \approx 0$.

Following the same principle, the kinematic assumptions can also be applied to the bending along the x-axis. We can see from Figure 4.1c that during the deformation of the plate, an arbitrary point P is deformed and rotated by the approximated displacement fields as given below.

$$u_P = u_O - 2z\tan\left(\frac{\theta_y}{2}\right)\cos(\theta_y) \approx u_O - 2z\frac{\theta_y}{2}1 \approx u_O - z\theta_y$$

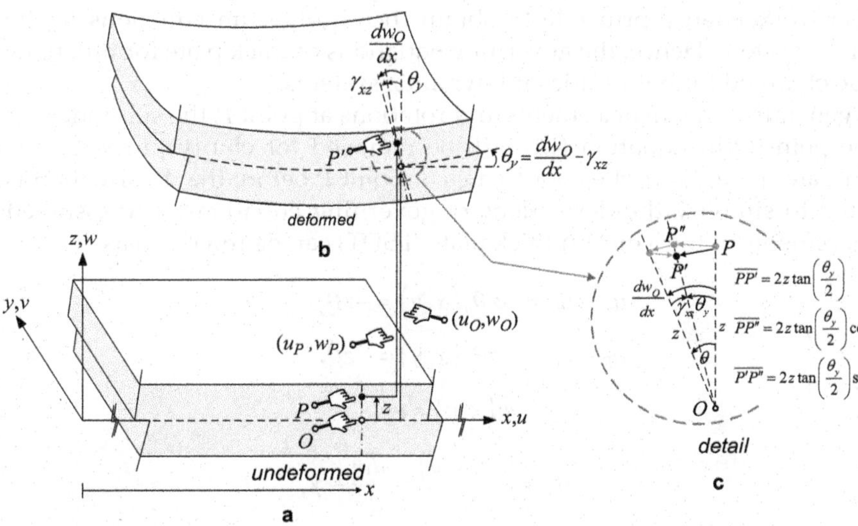

FIGURE 4.1
Conceptual kinematic assumptions of a flat Mindlin plate.

$$v_P = v_O - 2z\tan\left(\frac{\theta_x}{2}\right)\cos(\theta_x) \approx v_O - 2z\frac{\theta_x}{2}1 \approx v_O - z\theta_x$$

$$w_P = w_O + z - z - 2z\tan\left(\frac{\theta_y}{2}\right)\sin(\theta_y) \approx w_O - 2z\frac{\theta_y}{2}\theta_y \approx w_O \tag{4.1}$$

$$\theta_y = \frac{dw_O}{dx} - \gamma_{xz};\quad \theta_x = \frac{dw_O}{dy} - \gamma_{yz}$$

where u_O, v_O and u_P, v_P are the horizontal displacements of the point O and P in x- and y-axes, respectively. w_O, w_P are the vertical displacements of the point O and P in the z-direction. θ_x, θ_y are the rotations of the cross section of the plate in both x- and y-directions, which are not equal to the slope $\left(\frac{dw_O}{dx}, \frac{dw_O}{dy}\right)$ of the plate at point O in both directions. The shear strains γ_{xz} and γ_{yz} can be imagined as a *shearing relieving action* to the section after the plate is bent, usually need to be considered in the thick place element.

4.12 Governing Equations of Thick Plates

In this section, the governing equations of the thick plate are derived based on Hamilton's principle. Hamilton's principle is a generalization of the virtual work–energy principle to obtain the equilibrium equations in the dynamic system. Hence, the governing equations of thick plate formulations can be obtained for both static and dynamic systems.

To generalize the displacements and rotations at point P, the sub-indexing of the point O in Equation (4.1) will be removed for clarity purpose. The coordinates (x, y, z) are the coordinates of point P before the deformation is omitted to simplify the derivations of governing equations. The assumed displacements of the Mindlin thick plate (FSDT) can be rewritten as

$$u_P - u_O = -z\theta_y(x,y) = -z\theta_y$$

$$v_P - v_O = -z\theta_x(x,y) = -z\theta_x$$

$$w_P = w(x,y) = w$$

$$\theta_y = \frac{dw(x,y)}{dx} = \frac{dw}{dx} - \gamma_{xz}$$

$$\theta_x = \frac{dw(x,y)}{dy} = \frac{dw}{dy} - \gamma_{yz} \tag{4.2}$$

From the theory of elasticity, we can obtain the strain components considering shear deformations as

$$
\begin{aligned}
\varepsilon_{xx} &= -z\frac{d\theta_y}{dx} \\
\varepsilon_{yy} &= -z\frac{d\theta_x}{dy} \\
\varepsilon_{zz} &= \frac{dw_P}{dz} = 0 \\
\gamma_{xy} = \gamma_{yx} &= \frac{du}{dy} + \frac{dv}{dx} = -z\frac{d\theta_y}{dy} - z\frac{d\theta_x}{dx} \\
\gamma_{xz} = \gamma_{zx} &= \frac{dw}{dx} - \theta_y \\
\gamma_{yz} = \gamma_{zy} &= \frac{dw}{dy} - \theta_x
\end{aligned}
\tag{4.3}
$$

For elastic plane stress problem, the general stress components in a plate take the form of,

$$
\begin{aligned}
\sigma_{xx} &= \frac{E_{xx}}{\left(1-\nu_{xy}^2\right)}\left(\varepsilon_{xx} + \nu_{xy}\varepsilon_{yy}\right) \\
\sigma_{yy} &= \frac{E_{yy}}{\left(1-\nu_{yx}^2\right)}\left(\nu_{yx}\varepsilon_{xx} + \varepsilon_{yy}\right) \\
\tau_{xy} &= \frac{E_{xx}}{2\left(1+\nu_{xy}\right)}\gamma_{xy}, \quad \tau_{xz} = \frac{\kappa_{xz}E_{xx}}{2\left(1+\nu_{xz}\right)}\gamma_{xz}, \quad \text{and} \quad \tau_{yz} = \frac{\kappa_{yz}E_{yy}}{2\left(1+\nu_{yz}\right)}\gamma_{yz}
\end{aligned}
\tag{4.4}
$$

where κ_{xz} and κ_{yz} are the *shear correction factors.*

Under the assumption of isotropic and homogeneous material ($E_{xx} = E_{yy} = E_{xy} = E_{yx} = E$, $\nu_{xy} = \nu_{yx} = \nu_{xz} = \nu_{yz} = \nu$, $\kappa_{xz} = \kappa_{yz} = \kappa$), the general stress components can be written as

$$
\begin{aligned}
\sigma_{xx} &= \frac{E}{\left(1-\nu^2\right)}\left(\varepsilon_{xx} + \nu\varepsilon_{yy}\right) \\
\sigma_{yy} &= \frac{E}{\left(1-\nu^2\right)}\left(\nu\varepsilon_{xx} + \varepsilon_{yy}\right) \\
\tau_{xy} &= \frac{E}{2(1+\nu)}\gamma_{xy}, \quad \tau_{xz} = \frac{\kappa E}{2(1+\nu)}\gamma_{xz}, \quad \text{and} \quad \tau_{yz} = \frac{\kappa E}{2(1+\nu)}\gamma_{yz}
\end{aligned}
\tag{4.5}
$$

where E and ν are the normal elastic modulus and Poisson's ratio of the plate material, respectively.

Figure 4.2 shows the distribution of shear stress and normal stress due to bending of the plate.

In the illustration shown in Figure 4.3, the geometrical configurations for loadings and displacements of a plate are given as,

The governing equations are then derived via Hamilton's principle as follows:

$$\delta H = \int_{t_1}^{t_2} (\delta S_E - \delta K_E - \delta W_E)dt = 0 \tag{4.6}$$

where

δH is the variation of total energy in plate; δS_E is the variation of plate strain energy; δK_E is the variation of plate kinetic energy; δW_E is the variation of external work applied to the plate.

The strain energy, S_E, of a uniform rectangular ($a \times b \times h$) plate is given by

$$S_E = \frac{1}{2}\int_0^b \int_0^a \int_{-h/2}^{h/2} \{\sigma_{xx}\varepsilon_{xx} + \sigma_{yy}\varepsilon_{yy} + \tau_{xy}\gamma_{xy} + \tau_{xz}\gamma_{xz} + \tau_{yz}\gamma_{yz}\} dz\,dx\,dy \tag{4.7}$$

where a, b, h are the lengths of the plate in x- and y-directions and the uniform thickness of the plate, respectively.

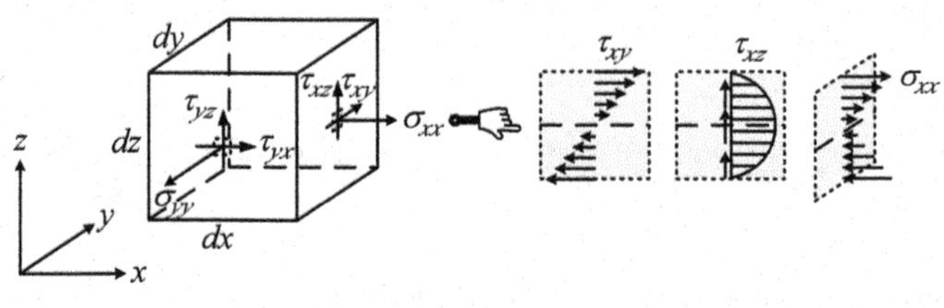

FIGURE 4.2
Definitions and distributions of stresses in the infinitesimal element of a plate.

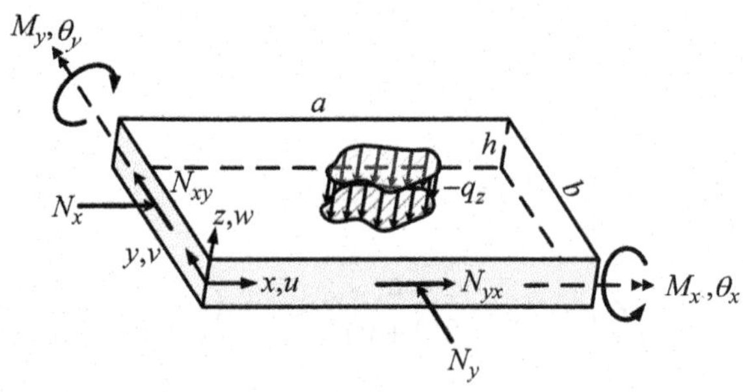

FIGURE 4.3
Geometrical configurations for loadings and displacements of a plate.

Substituting Equations (4.3) and (4.5) into Equation (4.7) gives

$$S_E = \frac{1}{2}\int_0^b \int_0^a \int_{-h/2}^{+h/2} \left\{ \begin{aligned} & \frac{E}{(1-\nu^2)}\left(-z\frac{d\theta_y}{dx}\right)\left(-z\frac{d\theta_y}{dx}\right) \\ & + \frac{E}{(1-\nu^2)}\left(-z\frac{d\theta_x}{dy}\right)\left(-z\frac{d\theta_x}{dy}\right) \\ & + \frac{2\nu E}{(1-\nu^2)}\left(-z\frac{d\theta_y}{dx}\right)\left(-z\frac{d\theta_x}{dy}\right) \\ & + \frac{E}{2(1+\nu)}\left(-z\frac{d\theta_x}{dx} - z\frac{d\theta_y}{dy}\right)\left(-z\frac{d\theta_x}{dx} - z\frac{d\theta_y}{dy}\right) \\ & + \frac{\kappa E}{2(1+\nu)}\left(\frac{dw}{dx} - \theta_y\right)\left(\frac{dw}{dx} - \theta_y\right) \\ & + \frac{\kappa E}{2(1+\nu)}\left(\frac{dw}{dy} - \theta_x\right)\left(\frac{dw}{dy} - \theta_x\right) \end{aligned} \right\} dx\,dy\,dz \tag{4.8}$$

Applying the variational principle to the above equation, we can obtain

$$\delta S_E = \frac{1}{2}\int_0^b \int_0^a \int_{-h/2}^{+h/2} \left\{ \begin{aligned} & \frac{E}{(1-\nu^2)}\left(-z\frac{d\theta_y}{dx}\right)\delta\left(-z\frac{d\theta_y}{dx}\right) \\ & + \frac{E}{(1-\nu^2)}\left(-z\frac{d\theta_x}{dy}\right)\delta\left(-z\frac{d\theta_x}{dy}\right) \\ & + \frac{2\nu E}{(1-\nu^2)}\left(-z\frac{d\theta_y}{dx}\right)\delta\left(-z\frac{d\theta_x}{dy}\right) \\ & + \frac{2\nu E}{(1-\nu^2)}\left(-z\frac{d\theta_x}{dy}\right)\delta\left(-z\frac{d\theta_y}{dx}\right) \\ & + \frac{E}{2(1+\nu)}\left(-z\frac{d\theta_x}{dx} - z\frac{d\theta_y}{dy}\right)\delta\left(-z\frac{d\theta_x}{dx} - z\frac{d\theta_y}{dy}\right) \\ & + \frac{\kappa E}{2(1+\nu)}\left(\frac{dw}{dx} - \theta_y\right)\delta\left(\frac{dw}{dx} - \theta_y\right) \\ & + \frac{\kappa E}{2(1+\nu)}\left(\frac{dw}{dy} - \theta_x\right)\delta\left(\frac{dw}{dy} - \theta_x\right) \end{aligned} \right\} dx\,dy\,dz \tag{4.9}$$

For a rectangular plate with uniform thickness h,

$$\begin{aligned}
\delta S_E &= D\int_0^b\int_0^a\left(\frac{d\theta_y}{dx}\right)\delta\left(\frac{d\theta_y}{dx}\right)dx\,dy + D\int_0^b\int_0^a\left(\frac{d\theta_x}{dy}\right)\delta\left(\frac{d\theta_x}{dy}\right)dx\,dy \\
&+ D_{xxyy}\int_0^b\int_0^a\left(\frac{d\theta_y}{dx}\right)\delta\left(\frac{d\theta_x}{dy}\right)dx\,dy + D_{xxyy}\int_0^b\int_0^a\left(\frac{d\theta_x}{dy}\right)\delta\left(\frac{d\theta_y}{dx}\right)dx\,dy \\
&+ D_{xy}\int_0^b\int_0^a\left(\frac{d\theta_x}{dx}+\frac{d\theta_y}{dy}\right)\delta\left(\frac{d\theta_x}{dx}+\frac{d\theta_y}{dy}\right)dx\,dy \\
&+ D_{xz}\int_0^b\int_0^a\left(\frac{dw}{dx}-\theta_y\right)\delta\left(\frac{dw}{dx}-\theta_y\right)dx\,dy \\
&+ D_{yz}\int_0^b\int_0^a\left(\frac{dw}{dy}-\theta_x\right)\delta\left(\frac{dw}{dy}-\theta_x\right)dx\,dy
\end{aligned} \tag{4.10}$$

where

$$D=\frac{Eh^3}{12\left(1-v^2\right)},\ D_{xxyy}=\frac{vEh^3}{6\left(1-v^2\right)},\ D_{xy}=\frac{Eh^3}{24(1+v)},\ D_{xz}=D_{yz}=\kappa Gh,\ G=\frac{E}{2(1+v)}$$

Applying the integration in part to the first term of the first integral term in Equation (4.10) results in

$$D\int_0^b\int_0^a\left(\frac{d\theta_y}{dx}\right)\delta\left(\frac{d\theta_y}{dx}\right)dx\,dy = M_y\big|_{BC}\,\delta\theta_y - D\int_0^b\int_0^a\left(\frac{d^2\theta_y}{dx^2}\right)dx\,dy\,\delta\theta_y \tag{4.11}$$

where M_y is the rotational moment around the y-axis at plate ends, which is given by

$$M_y\big|_{BC} = D\left(\frac{d\theta_y}{dx}\right)\bigg|_{BC} \tag{4.12}$$

Following the same procedure, by applying the integration in part to the second integral term of Equation (4.10) for the governing equations in the y-direction gives

$$D\int_0^b\int_0^a\left(\frac{d\theta_x}{dy}\right)\delta\left(\frac{d\theta_x}{dy}\right)dx\,dy = M_x\big|_{BC}\,\delta\theta_x - D\int_0^b\int_0^a\left(\frac{d^2\theta_x}{dy^2}\right)dx\,dy\,\delta\theta_x \tag{4.13}$$

where M_x is the rotational moment around the y-axis at plate ends, which is given by

$$M_x\big|_{BC} = D\left(\frac{d\theta_x}{dy}\right)\Bigg|_{BC} \tag{4.14}$$

Next, applying the integration in part to the third and fourth terms of Equation (4.10) results in

$$\begin{aligned} &D_{xxyy}\int_0^b\int_0^a\left(\frac{d\theta_y}{dx}\right)\delta\left(\frac{d\theta_x}{dy}\right)dx\,dy + D_{xxyy}\int_0^b\int_0^a\left(\frac{d\theta_x}{dy}\right)\delta\left(\frac{d\theta_y}{dx}\right)dx\,dy \\ &= M_{yx}\big|_{BC}\,\delta\theta_y - D_{xxyy}\int_0^b\int_0^a\left(\frac{d^2\theta_y}{dxdy}\right)dx\,dy\,\delta\theta_y + M_{xy}\big|_{BC}\,\delta\theta_x \\ &- D_{xxyy}\int_0^b\int_0^a\left(\frac{d^2\theta_x}{dxdy}\right)dx\,dy\,\delta\theta_x \end{aligned} \tag{4.15}$$

where M_{yx} and M_{xy} are the rotational moments about the midplane of the plate ends, which are given by

$$\begin{aligned} M_{yx}\big|_{BC} &= D_{xxyy}\left(\frac{d\theta_y}{dx}\right)\Bigg|_{BC} \\ M_{xy}\big|_{BC} &= D_{xxyy}\left(\frac{d\theta_x}{dy}\right)\Bigg|_{BC} \end{aligned} \tag{4.16}$$

Next, applying the integration in part to the fifth term of Equation (4.10) results in

$$\begin{aligned} &D_{xy}\int_0^b\int_0^a\left(\frac{d\theta_x}{dx}+\frac{d\theta_y}{dy}\right)\delta\left(\frac{d\theta_x}{dx}+\frac{d\theta_y}{dy}\right)dx\,dy \\ &= D_{xy}\int_0^b\int_0^a\left(\frac{d\theta_x}{dx}+\frac{d\theta_y}{dy}\right)\delta\left(\frac{d\theta_x}{dx}\right)dx\,dy \\ &+ D_{xy}\int_0^b\int_0^a\left(\frac{d\theta_x}{dx}+\frac{d\theta_y}{dy}\right)\delta\left(\frac{d\theta_y}{dy}\right)dx\,dy \\ &= M_{xy}\big|_{BC}\,\delta\theta_y - D_{xy}\int_0^b\int_0^a\left(\frac{d^2\theta_y}{dy^2}+\frac{d^2\theta_y}{dxdy}\right)dx\,dy\,\delta\theta_y \\ &+ M_{yx}\big|_{BC}\,\delta\theta_x - D_{xy}\int_0^b\int_0^a\left(\frac{d^2\theta_x}{dx^2}+\frac{d^2\theta_x}{dxdy}\right)dx\,dy\,\delta\theta_x \end{aligned} \tag{4.17}$$

where M_{yx} and M_{xy} are the rotational moments about the midplane of the plate ends, which are given by

$$\begin{aligned} M_{xx}\big|_{BC} &= D_{xy}\left(\frac{d\theta_x}{dx}\right)\bigg|_{BC} \\ M_{yy}\big|_{BC} &= D_{xy}\left(\frac{d\theta_y}{dy}\right)\bigg|_{BC} \end{aligned} \tag{4.18}$$

Next, applying the integration in part to the sixth term of Equation (4.10) results in

$$\begin{aligned} & D_{xz}\int_0^b\int_0^a\left(\frac{dw}{dx}-\theta_y\right)\delta\left(\frac{dw}{dx}-\theta_y\right)dx\,dy \\ &= D_{xz}\int_0^b\int_0^a\left(\frac{dw}{dx}-\theta_y\right)\delta\left(\frac{dw}{dx}\right)dx\,dy - D_{xz}\int_0^b\int_0^a\left(\frac{dw}{dx}-\theta_y\right)\delta\theta_y\,dx\,dy \\ &= Q_{xz}\big|_{BC}\,\delta w - D_{xz}\int_0^b\int_0^a\left(\frac{d^2w}{dx^2}-\frac{d\theta_y}{dx}\right)dx\,dy\,\delta w - D_{xz}\int_0^b\int_0^a\left(\frac{dw}{dx}-\theta_y\right)dx\,dy\,\delta\theta_y \end{aligned} \tag{4.19}$$

where Q_{xz} is the transverse load at the x-boundaries of the plate ends, which is given by

$$Q_{xz}\big|_{BC} = D_{xz}\left(\frac{dw}{dx}-\theta_y\right)\bigg|_{BC} = D_{xz}\gamma_{xz}\big|_{BC} \tag{4.20}$$

Finally, applying the integration in part to the seventh term of Equation (4.10) results in

$$\begin{aligned} & D_{yz}\int_0^b\int_0^a\left(\frac{dw}{dy}-\theta_x\right)\delta\left(\frac{dw}{dy}-\theta_x\right)dx\,dy \\ &= D_{yz}\int_0^b\int_0^a\left(\frac{dw}{dy}-\theta_x\right)\delta\left(\frac{dw}{dy}\right)dx\,dy - D_{yz}\int_0^b\int_0^a\left(\frac{dw}{dy}-\theta_x\right)\delta\theta_x\,dx\,dy \\ &= Q_{yz}\big|_{BC}\,\delta w - D_{yz}\int_0^b\int_0^a\left(\frac{d^2w}{dy^2}-\frac{d\theta_x}{dy}\right)dx\,dy\,\delta w - D_{yz}\int_0^b\int_0^a\left(\frac{dw}{dy}-\theta_x\right)dx\,dy\,\delta\theta_x \end{aligned} \tag{4.21}$$

where Q_{yz} is the transverse load at the x-boundaries of the plate ends, which is given by

$$Q_{yz}\big|_{BC} = D_{yz}\left(\frac{dw}{dy} - \theta_x\right)\Bigg|_{BC} = D_{yz}\gamma_{yz}\big|_{BC} \tag{4.22}$$

Summarizing the results of application of the variational principle to the strain energy, the governing equation of the plate can be expressed by using Equations (4.11), (4.13), (4.15), (4.17), (4.19), and (4.21) as

$$\begin{aligned}
\delta S_E = &-D\int_0^b\int_0^a\left(\frac{d^2\theta_y}{dx^2}\right)dx\,dy\,\delta\theta_y - D\int_0^b\int_0^a\left(\frac{d^2\theta_x}{dy^2}\right)dx\,dy\,\delta\theta_x\\
&-D_{xxyy}\int_0^b\int_0^a\left(\frac{d^2\theta_y}{dxdy}\right)dx\,dy\delta\theta_x - D_{xxyy}\int_0^b\int_0^a\left(\frac{d^2\theta_x}{dxdy}\right)dx\,dy\,\delta\theta_y\\
&-D_{xy}\int_0^b\int_0^a\left(\frac{d^2\theta_y}{dy^2}+\frac{d^2\theta_y}{dxdy}\right)dx\,dy\,\delta\theta_y - D_{xy}\int_0^b\int_0^a\left(\frac{d^2\theta_x}{dx^2}+\frac{d^2\theta_x}{dxdy}\right)dx\,dy\,\delta\theta_x\\
&-D_{xz}\int_0^b\int_0^a\left(\frac{d^2w}{dx^2}-\frac{d\theta_y}{dx}\right)dx\,dy\,\delta w - D_{xz}\int_0^b\int_0^a\left(\frac{dw}{dx}-\theta_y\right)dx\,dy\,\delta\theta_y\\
&-D_{yz}\int_0^b\int_0^a\left(\frac{d^2w}{dy^2}-\frac{d\theta_x}{dy}\right)dx\,dy\,\delta w - D_{yz}\int_0^b\int_0^a\left(\frac{dw}{dy}-\theta_x\right)dx\,dy\,\delta\theta_x
\end{aligned} \tag{4.23}$$

where the boundary conditions are given in Equations (4.12), (4.14), (4.16), (4.18), (4.20), and (4.22).

The kinetic energy of the plate element is given as

$$\begin{aligned}
K_E = &\frac{1}{2}\int_0^b\int_0^a\int_{-h/2}^{+h/2}\rho\left(\frac{du}{dt}\right)^2 dz\,dx\,dy + \frac{1}{2}\int_0^b\int_0^a\int_{-h/2}^{+h/2}\rho\left(\frac{dv}{dt}\right)^2 dz\,dx\,dy\\
&+\frac{1}{2}\int_0^b\int_0^a\int_{-h/2}^{+h/2}\rho\left(\frac{dw}{dt}\right)^2 dz\,dx\,dy
\end{aligned} \tag{4.24}$$

Substitution of Equation (4.2) gives

$$K_E = \frac{1}{2}\rho I \int_0^b \int_0^a \left(\frac{d\theta_y}{dt}\right)^2 dx\,dy + \frac{1}{2}\rho I \int_0^b \int_0^a \left(\frac{d\theta_x}{dt}\right)^2 dx\,dy + \frac{1}{2}\rho I_{xy} \int_0^b \int_0^a \left(\frac{dw}{dt}\right)^2 dx\,dy \tag{4.25}$$

where

$$I = \int_{-h/2}^{+h/2} z^2 dz = \frac{h^3}{12}$$

$$I_{xy} = \int_{-h/2}^{+h/2} 1dz = h$$

Applying the variational principle to Equation (4.25) gives

$$\delta K_E = \rho I \int_0^b \int_0^a \left(\frac{d\theta_y}{dt}\right) \delta \left(\frac{d\theta_y}{dt}\right) dx\,dy + \rho I \int_0^b \int_0^a \left(\frac{d\theta_x}{dt}\right) \delta \left(\frac{d\theta_x}{dt}\right) dx\,dy$$
$$+ \rho I_{xy} \int_0^b \int_0^a \left(\frac{dw}{dt}\right) \delta \left(\frac{dw}{dt}\right) dx\,dy \tag{4.26}$$

Applying the integration in part to the first term of Equation (4.26) yields

$$\rho I \int_0^b \int_0^a \left(\frac{d\theta_y}{dt}\right) \delta \left(\frac{d\theta_y}{dt}\right) dx\,dy = \dot{M}_y\Big|_{BC} \delta\theta_y - \rho I \int_0^b \int_0^a \left(\frac{d^2\theta_y}{dt^2}\right) dx\,dy \delta\theta_y \tag{4.27}$$

where $\dot{M}_y$ is the time-rate rotational moment about the midplane of the plate ends, which is given by

$$\dot{M}_y\Big|_{BC} = \rho I \left(\frac{d\theta_y}{dt}\right)\Big|_{BC} \tag{4.28}$$

with $\dot{M}_y = \dfrac{d(M_y)}{dt}$

Similarly, applying the integration in part to the second term of Equation (4.26) results in

$$\rho I \int_0^b \int_0^a \left(\frac{d\theta_x}{dt}\right) \delta \left(\frac{d\theta_x}{dt}\right) dx\,dy = \dot{M}_x\Big|_{BC} \delta\theta_x - \rho I \int_0^b \int_0^a \left(\frac{d^2\theta_x}{dt^2}\right) dx\,dy\, \delta\theta_x \tag{4.29}$$

where $\dot{M}_x$ is the time-rate rotational moment about the midplane of the plate ends, which is given by

$$\dot{M}_x\Big|_{BC} = \rho I\left(\frac{d\theta_x}{dt}\right)\Bigg|_{BC} \tag{4.30}$$

Applying the integration in part to the third term of Equation (4.26) yields

$$\rho I_{xy}\int_0^b\int_0^a\left(\frac{dw}{dt}\right)\delta\left(\frac{dw}{dt}\right)dx\,dy = \dot{Q}_z\Big|_{BC}\delta w - \rho I_{xy}\int_0^b\int_0^a\left(\frac{d^2w}{dt^2}\right)dx\,dy\,\delta w \tag{4.31}$$

where

$$\dot{Q}_z\Big|_{BC} = \rho I_{xy}\left(\frac{dw}{dt}\right)\Bigg|_{BC} \tag{4.32}$$

Summarizing the results of application of the variational principle to the kinetic energy, the governing equation of the plate can be expressed by using Equations (4.27), (4.29), and (4.31) as

$$\delta K_E = -\rho I\int_0^b\int_0^a\left(\frac{d^2\theta_y}{dt^2}\right)dx\,dy\,\delta\theta_y - \rho I\int_0^b\int_0^a\left(\frac{d^2\theta_x}{dt^2}\right)dx\,dy\,\delta\theta_x$$

$$-\rho I_{xy}\int_0^b\int_0^a\left(\frac{d^2w}{dt^2}\right)dx\,dy\,\delta w \tag{4.33}$$

where the boundary conditions are given in Equations (4.28), (4.30), and (4.32).

The external work applied to the plate element due to distributed load on the top surface of the plate, $q(x,y)$, and in-plane compressive and shear forces in normal and tangential directions on the edges of the plate, N_x, N_y, N_{xy}, as shown in Figure 4.3 can be expressed as

$$W_E = \frac{1}{2}\int_0^b\int_0^a N_x\left(\frac{dw}{dx}\right)^2 dx\,dy + \frac{1}{2}\int_0^b\int_0^a N_y\left(\frac{dw}{dy}\right)^2 dx\,dy$$

$$+\frac{1}{2}\int_0^b\int_0^a N_{xy}\left(\frac{d^2w}{dxdy}\right)dx\,dy + \frac{1}{2}\int_0^b\int_0^a N_{yx}\left(\frac{d^2w}{dxdy}\right)dxdy + \int_0^b\int_0^a q_z w\,dx\,dy \tag{4.34}$$

Substituting Equation (4.3) into the above equation results in

$$W_E = \frac{1}{2}\int_0^b\int_0^a N_x\left(\frac{dw}{dx}\right)^2 dx\,dy + \frac{1}{2}\int_0^b\int_0^a N_y\left(\frac{dw}{dy}\right)^2 dx\,dy$$
$$+\frac{1}{2}\int_0^b\int_0^a N_{xy}\left(\frac{dw}{dx}\right)\left(\frac{dw}{dy}\right)dx\,dy$$
$$+\frac{1}{2}\int_0^b\int_0^a N_{yx}\left(\frac{dw}{dy}\right)\left(\frac{dw}{dx}\right)dx\,dy + \frac{1}{2}\int_0^b\int_0^a qw\,dxdy \tag{4.35}$$

Applying the variational principle to the above equation results in

$$\delta W_E = \int_0^b\int_0^a N_x\left(\frac{dw}{dx}\right)\delta\left(\frac{dw}{dx}\right)dx\,dy + \int_0^b\int_0^a N_y\left(\frac{dw}{dy}\right)\delta\left(\frac{dw}{dy}\right)dx\,dy$$
$$+\int_0^b\int_0^a N_{xy}\left(\frac{dw}{dx}\right)\delta\left(\frac{dw}{dy}\right)dx\,dy + \int_0^b\int_0^a N_{yx}\left(\frac{dw}{dy}\right)\delta\left(\frac{dw}{dx}\right)dx\,dy$$
$$+\int_0^b\int_0^a q_z\,\delta w\,dx\,dy \tag{4.36}$$

Next, applying the integration in part to the first term of Equation (4.36) yields

$$\int_0^b\int_0^a N_x\left(\frac{dw}{dx}\right)\delta\left(\frac{dw}{dx}\right)dx\,dy = p_y\big|_{BC}\,\delta w - \int_0^b\int_0^a N_x\left(\frac{d^2w}{dx^2}\right)dx\,dy\,\delta w \tag{4.37}$$

where p_y is the distributed normal force on the midplane of the plate ends, which is given by

$$p_y\big|_{BC} = N_x\left(\frac{dw}{dx}\right)\bigg|_{BC} \tag{4.38}$$

Similarly, applying the integration in part to the second term of Equation (4.36) yields

$$\int_0^b\int_0^a N_y\left(\frac{dw}{dy}\right)\delta\left(\frac{dw}{dy}\right)dx\,dy = p_x\big|_{BC}\,\delta w - \int_0^b\int_0^a N_y\left(\frac{d^2w}{dy^2}\right)dx\,dy\,\delta w \tag{4.39}$$

where p_x is the distributed normal force on the midplane of the plate ends, which is given by

$$p_x\big|_{BC} = N_y\left(\frac{dw}{dy}\right)\bigg|_{BC} \tag{4.40}$$

Next, applying the integration in part to the third term of Equation (4.36) yields

$$\int_0^b\int_0^a N_{xy}\left(\frac{dw}{dx}\right)\delta\left(\frac{dw}{dy}\right)dx\,dy = p_{xy}\big|_{BC}\,\delta w - \int_0^b\int_0^a N_{xy}\left(\frac{d^2w}{dxdy}\right)dx\,dy\,\delta w \tag{4.41}$$

where p_{xy} is the distributed shearing force on the midplane of the plate ends, which is given by

$$p_{xy}\big|_{BC} = N_{xy}\left(\frac{dw}{dx}\right)\bigg|_{BC} \tag{4.42}$$

Similarly, applying the integration in part to the fourth term of Equation (4.36) yields

$$\int_0^b\int_0^a N_{yx}\left(\frac{dw}{dy}\right)\delta\left(\frac{dw}{dx}\right)dx\,dy = p_{yx}\big|_{BC}\,\delta w - \int_0^b\int_0^a N_{yx}\left(\frac{d^2w}{dydx}\right)dx\,dy\,\delta w \tag{4.43}$$

where p_{yx} is the distributed shearing force on the midplane of the plate ends, which is given by

$$p_{yx}\big|_{BC} = N_{yx}\left(\frac{dw}{dy}\right)\bigg|_{BC} \tag{4.44}$$

Finally, summarizing the results of application of the variational principle to the external work, the governing equation of the plate can be expressed by using Equations (4.37), (4.39), (4.41), and (4.43) as

$$\begin{aligned}\delta W_E = &-\int_0^b\int_0^a N_x\left(\frac{d^2w}{dx^2}\right)dx\,dy\,\delta w - \int_0^b\int_0^a N_y\left(\frac{d^2w}{dy^2}\right)dx\,dy\,\delta w\\ &-\int_0^b\int_0^a N_{xy}\left(\frac{d^2w}{dxdy}\right)dx\,dy\,\delta w - \int_0^b\int_0^a N_{yx}\left(\frac{d^2w}{dydx}\right)dx\,dy\,\delta w\\ &+\int_0^b\int_0^a q_z\,dx\,dy\,\delta w\end{aligned} \tag{4.45}$$

where the boundary conditions are given in Equations (4.38), (4.40), (4.42), and (4.44).

In summary, Hamilton's principle in Equation (4.6) for an isotropic and elastic material plate element can be written as follows:

$$
\delta H = \int_{t_1}^{t_2} \left(\begin{array}{l}
-D\int_0^b \int_0^a \left(\frac{d^2\theta_y}{dx^2} \right) dx\,dy\,\delta\theta_y - D\int_0^b \int_0^a \left(\frac{d^2\theta_x}{dy^2} \right) dx\,dy\,\delta\theta_x \\
-D_{xxyy}\int_0^b \int_0^a \left(\frac{d^2\theta_y}{dxdy} \right) dx\,dy\,\delta\theta_x - D_{xxyy}\int_0^b \int_0^a \left(\frac{d^2\theta_x}{dxdy} \right) dx\,dy\,\delta\theta_y \\
-D_{xy}\int_0^b \int_0^a \left(\frac{d^2\theta_y}{dy^2} + \frac{d^2\theta_y}{dxdy} \right) dx\,dy\,\delta\theta_y \\
-D_{xy}\int_0^b \int_0^a \left(\frac{d^2\theta_x}{dx^2} + \frac{d^2\theta_x}{dxdy} \right) dx\,dy\,\delta\theta_x \\
-D_{xz}\int_0^b \int_0^a \left(\frac{d^2w}{dx^2} - \frac{d\theta_y}{dx} \right) dx\,dy\,\delta w \\
-D_{xz}\int_0^b \int_0^a \left(\frac{dw}{dx} - \theta_y \right) dx\,dy\,\delta\theta_y \\
-D_{yz}\int_0^b \int_0^a \left(\frac{d^2w}{dy^2} - \frac{d\theta_x}{dy} \right) dx\,dy\,\delta w \\
-D_{yz}\int_0^b \int_0^a \left(\frac{dw}{dy} - \theta_x \right) dx\,dy\,\delta\theta_x \\
+\rho I\int_0^b \int_0^a \left(\frac{d^2\theta_y}{dt^2} \right) dx\,dy\,\delta\theta_y + \rho I\int_0^b \int_0^a \left(\frac{d^2\theta_x}{dt^2} \right) dx\,dy\,\delta\theta_x \\
+\rho I_{xy}\int_0^b \int_0^a \left(\frac{d^2w}{dt^2} \right) dx\,dy\,\delta w + \int_0^b \int_0^a N_x \left(\frac{d^2w}{dx^2} \right) dx\,dy\,\delta w \\
+\int_0^b \int_0^a N_y \left(\frac{d^2w}{dy^2} \right) dx\,dy\,\delta w + \int_0^b \int_0^a N_{xy} \left(\frac{d^2w}{dxdy} \right) dx\,dy\,\delta w \\
+\int_0^b \int_0^a N_{yx} \left(\frac{d^2w}{dydx} \right) dx\,dy\,\delta w - \int_0^b \int_0^a q_z\,dx\,dy\,\delta w
\end{array} \right) dt = 0 \tag{4.46}
$$

The static governing equations can be obtained at an instantaneous zero time frame. The equations inside the double integral can be written as

$$-D\left(\frac{d^2\theta_y}{dx^2}\right)\delta\theta_y - D\left(\frac{d^2\theta_x}{dy^2}\right)\delta\theta_x - D_{xxyy}\left(\frac{d^2\theta_y}{dxdy}\right)\delta\theta_x - D_{xxyy}\left(\frac{d^2\theta_x}{dxdy}\right)\delta\theta_y$$

$$-D_{xy}\left(\frac{d^2\theta_y}{dy^2}+\frac{d^2\theta_y}{dxdy}\right)\delta\theta_y - D_{xy}\left(\frac{d^2\theta_x}{dx^2}+\frac{d^2\theta_x}{dxdy}\right)\delta\theta_x$$

$$-D_{xz}\left(\frac{d^2w}{dx^2}-\frac{d\theta_y}{dx}\right)\delta w - D_{xz}\left(\frac{dw}{dx}-\theta_y\right)\delta\theta_y$$

$$-D_{yz}\left(\frac{d^2w}{dy^2}-\frac{d\theta_x}{dy}\right)\delta w - D_{yz}\left(\frac{dw}{dy}-\theta_x\right)\delta\theta_x$$

$$+\rho I\left(\frac{d^2\theta_y}{dt^2}\right)\delta\theta_y + \rho I\left(\frac{d^2\theta_x}{dt^2}\right)\delta\theta_x + \rho I_{xy}\left(\frac{d^2w}{dt^2}\right)\delta w$$

$$+N_x\left(\frac{d^2w}{dx^2}\right)\delta w + N_y\left(\frac{d^2w}{dy^2}\right)\delta w + N_{xy}\left(\frac{d^2w}{dxdy}\right)\delta w + N_{yx}\left(\frac{d^2w}{dydx}\right)\delta w - q_z\delta w = 0 \tag{4.47}$$

The above equation is the governing differential equation of the well-known FSDT of an isotropic rectangular plate. It is worth noting that finding the exact solution for the above governing equation is quite difficult. Except for a few simple problem of rectangular plates.

To summarize, by differentiating the above governing equations, the differential equations of motion that correspond to the generalized displacements can be grouped as

$$\delta w: -D_{xz}\left(\frac{d^2w}{dx^2}-\frac{d\theta_y}{dx}\right) - D_{yz}\left(\frac{d^2w}{dy^2}-\frac{d\theta_x}{dy}\right) + \rho I_{xy}\left(\frac{d^2w}{dt^2}\right) =$$

$$q_z - N_x\left(\frac{d^2w}{dx^2}\right) - N_y\left(\frac{d^2w}{dy^2}\right) - N_{xy}\left(\frac{d^2w}{dxdy}\right) - N_{yx}\left(\frac{d^2w}{dydx}\right)$$

$$\delta\theta_x: -D\left(\frac{d^2\theta_x}{dy^2}\right) - D_{xxyy}\left(\frac{d^2\theta_y}{dxdy}\right) - D_{xy}\left(\frac{d^2\theta_x}{dx^2}+\frac{d^2\theta_x}{dxdy}\right)$$

$$- D_{yz}\left(\frac{dw}{dy}-\theta_x\right) + \rho I\left(\frac{d^2\theta_x}{dt^2}\right) = 0$$

$$\delta\theta_y: -D\left(\frac{d^2\theta_y}{dx^2}\right) - D_{xxyy}\left(\frac{d^2\theta_x}{dxdy}\right) - D_{xy}\left(\frac{d^2\theta_y}{dy^2}+\frac{d^2\theta_y}{dxdy}\right)$$

$$- D_{xz}\left(\frac{dw}{dx}-\theta_y\right) + \rho I\left(\frac{d^2\theta_y}{dt^2}\right) = 0 \tag{4.48}$$

The essential and natural boundary conditions of the governing equations are given by

$$\delta w : Q = D_{xz}\left(\frac{dw}{dx} - \theta_y\right) + D_{yz}\left(\frac{dw}{dy} - \theta_x\right) - \rho I_{xy}\left(\frac{dw}{dt}\right) + N_x\left(\frac{dw}{dx}\right)$$

$$+ N_y\left(\frac{dw}{dy}\right) + N_{xy}\left(\frac{dw}{dx}\right) + N_{yx}\left(\frac{dw}{dy}\right)$$

$$\delta\theta_x : M_x = D\left(\frac{d\theta_x}{dy}\right) + D_{xxyy}\left(\frac{d\theta_x}{dy}\right) + D_{xy}\left(\frac{d\theta_x}{dx}\right) - \rho I\left(\frac{d\theta_x}{dt}\right)$$

$$\delta\theta_y : M_y = D\left(\frac{d\theta_y}{dx}\right) + D_{xxyy}\left(\frac{d\theta_y}{dx}\right) + D_{xy}\left(\frac{d\theta_y}{dy}\right) - \rho I\left(\frac{d\theta_y}{dt}\right) \quad (4.49)$$

4.13 Navier Solutions for FSDT Rectangular Plates

Similar to the CPT solutions described in Chapter 3, we discuss the Navier solutions to solve the bending and natural vibration problems of isotropic FSDT Mindlin rectangular plate. The Navier solutions can be used for solving the simply supported rectangular plates under the following conditions:

1. It is applicable only for simply supported plates.
2. It solves bending solutions under arbitrary transverse loading $q_z(x,y)$.
3. It solves natural frequency problems of plates.
4. It solves buckling problems under in-plane biaxial compressive loadings defined by $\hat{N}_y = \lambda \hat{N}_x = \lambda N_0$, where λ is the ratio between the applied axial compressive loadings.

In case of a simply supported boundary condition of a rectangular plate (see Figure 4.4), the following initial displacement conditions at the supports are as follows:

$$w(-a/2,y,t) = 0; \quad w(+a/2,y,t) = 0; \quad w(x,-b/2,t) = 0; \quad w(x,+b/2,t) = 0 \quad (4.50)$$

The initial rotations are as follows:

$$\theta_x(-a/2,y,t) = 0; \quad \theta_x(+a/2,y,t) = 0; \quad \theta_y(x,-b/2,t) = 0; \quad \theta_y(x,+b/2,t) = 0 \quad (4.51)$$

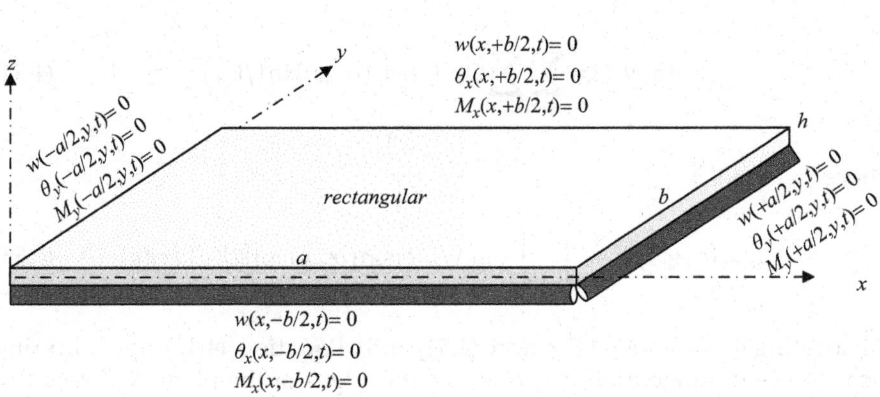

FIGURE 4.4
Simply supported boundary conditions of a rectangular plate.

The initial moments are as follows:

$$M_x(-a/2,y,t)=0; \quad M_x(+a/2,y,t)=0;$$
$$M_y(x,-b/2,t)=0; \quad M_y(x,+b/2,t)=0 \tag{4.52}$$

In Navier's method of solution, the displacement w and rotations θ_x, θ_y are expanded into double sine series with unknown coefficients $W_{mn}(t)$, $X_{mn}(t)$, and $Y_{mn}(t)$ as the function of time:

$$w(x,y,t)=\sum_{n=1}^{\infty}\sum_{m=1}^{\infty}W_{mn}(t)\sin(\alpha_m x)\sin(\beta_n y)$$
$$\theta_x(x,y,t)=\sum_{n=1}^{\infty}\sum_{m=1}^{\infty}X_{mn}(t)\sin(\alpha_m x)\cos(\beta_n y) \tag{4.53}$$
$$\theta_y(x,y,t)=\sum_{n=1}^{\infty}\sum_{m=1}^{\infty}Y_{mn}(t)\cos(\alpha_m x)\sin(\beta_n y)$$

where

$$\alpha_m=\frac{m\pi}{a}; \quad \beta_n=\frac{n\pi}{b}$$

The coefficients $W_{mn}(t)$, $X_{mn}(t)$, and $Y_{mn}(t)$ should be determined in such a way that Equation (4.47) is satisfied everywhere in the plate. Equation (4.53) is adopted from the fact that the double sine and cosine series can satisfy the simply supported boundary conditions in Equations (4.50), (4.51), and (4.52). The external loading q can also be expanded in the double sine series as

$$q(x,y,t)=\sum_{n=1}^{\infty}\sum_{m=1}^{\infty}Q_{mn}(t)\sin(\alpha_m x)\sin(\beta_n y) \tag{4.54}$$

where

$$Q_{mn}(x,y,t)=\frac{1}{ab}\int_{-b}^{+b}\int_{-a}^{+a}q(x,y,t)\sin(\alpha_m x)\sin(\beta_n y)dx\,dy \tag{4.55}$$

Substituting Equations (4.53) and (4.54) into Equation (4.47) and omitting shear forces in tangential directions on the edges of the plate, N_{xy}, result in the matrix form as

$$\begin{bmatrix} s_{11}-\bar{s}_{11} & s_{12} & s_{13} \\ s_{21} & s_{22} & s_{23} \\ s_{31} & s_{32} & s_{33} \end{bmatrix}\begin{Bmatrix} W_{mn} \\ X_{mn} \\ Y_{mn} \end{Bmatrix}+\begin{bmatrix} m_{11} & 0 & 0 \\ 0 & m_{22} & 0 \\ 0 & 0 & m_{33} \end{bmatrix}\begin{Bmatrix} \ddot{W}_{mn} \\ \ddot{X}_{mn} \\ \ddot{Y}_{mn} \end{Bmatrix}$$
$$=\begin{Bmatrix} Q_{mn} \\ 0 \\ 0 \end{Bmatrix} \tag{4.56}$$

where

$$s_{11}=\left(\alpha_m^2+\beta_n^2\right)D_{xz} \qquad s_{12}=-\beta_n D_{xz} \qquad s_{13}=-\alpha_m D_{xz}$$

$$s_{21}=-\beta_n D_{xz} \qquad s_{22}=D_{xz}+\beta_n^2 D+\alpha_m^2 D_{xy} \qquad s_{23}=\alpha_m\beta_n\left(D_{xxyy}+D_{xy}\right)$$

$$s_{31}=-\alpha_m D_{xz} \qquad s_{32}=\alpha_m\beta_n\left(D_{xxyy}+D_{xy}\right) \qquad s_{33}=D_{xz}+\alpha_m^2 D+\beta_n^2 D_{xy}$$

$$m_{11}=\rho I_{xy} \qquad m_{22}=\rho I \qquad m_{33}=\rho I$$

and

$$\bar{s}_{11}=\alpha_m^2 N_x+\beta_n^2 N_y$$

Equation (4.56) can be used for solving static bending, natural vibration, and buckling of simply supported rectangular plates.

The coefficients $Q_{mn}(x,y,t)$ for various types of loadings can be obtained from Table 3.1.

4.13.1 Formulations for Bending Problem of Square FSDT Plates

By setting the time derivative terms to zero, Equation (4.56) can be written as

$$\begin{bmatrix} s_{11}-\bar{s}_{11} & s_{12} & s_{13} \\ s_{21} & s_{22} & s_{23} \\ s_{31} & s_{32} & s_{33} \end{bmatrix} \begin{Bmatrix} W_{mn} \\ X_{mn} \\ Y_{mn} \end{Bmatrix} = \begin{Bmatrix} Q_{mn} \\ 0 \\ 0 \end{Bmatrix} \tag{4.57}$$

By inversing the coefficient matrix and multiplying the loading vector, the solution of Equation (4.57) for each $m,n=1,2,3,\ldots$ will give the values of W_{mn}, X_{mn}, Y_{mn}.

The bending moments can then be calculated as

$$M_x = D\sum_{n=1}^{\infty}\sum_{m=1}^{\infty}(\alpha_m X_{mn} + \nu\beta_n Y_{mn})\sin(\alpha_m x)\sin(\beta_n y)$$

$$M_y = D\sum_{n=1}^{\infty}\sum_{m=1}^{\infty}(\nu\alpha_m X_{mn} + \beta_n Y_{mn})\sin(\alpha_m x)\sin(\beta_n y)$$

$$M_{xy} = -(1-\nu)\frac{D}{2}\sum_{n=1}^{\infty}\sum_{m=1}^{\infty}(\beta_n X_{mn} + \alpha_m Y_{mn})\cos(\alpha_m x)\cos(\beta_n y) \tag{4.58}$$

The shear forces can be computed as follows:

$$Q_x = \frac{\partial M_x}{\partial x} + \frac{\partial M_{xy}}{\partial y} = D_{xz}\sum_{n=1}^{\infty}\sum_{m=1}^{\infty}(\alpha_m W_{mn} + X_{mn})\cos(\alpha_m x)\sin(\beta_n y)$$

$$Q_y = \frac{\partial M_{yx}}{\partial x} + \frac{\partial M_y}{\partial y} = D_{xz}\sum_{n=1}^{\infty}\sum_{m=1}^{\infty}(\beta_m W_{mn} + Y_{mn})\sin(\alpha_m x)\cos(\beta_n y) \tag{4.59}$$

In the pure bending case considered herewith, the stresses in the simply supported rectangular plate are given by

$$\sigma_{xx} = z\sum_{n=1}^{\infty}\sum_{m=1}^{\infty} R_x \sin(\alpha_m x)\sin(\beta_n y)$$

$$\sigma_{yy} = z\sum_{n=1}^{\infty}\sum_{m=1}^{\infty} R_y \sin(\alpha_m x)\sin(\beta_n y)$$

$$\tau_{xy} = -z\sum_{n=1}^{\infty}\sum_{m=1}^{\infty} R_{xy} \cos(\alpha_m x)\cos(\beta_n y)$$

$$
\begin{aligned}
\tau_{xz} &= -h\sum_{n=1}^{\infty}\sum_{m=1}^{\infty} R_{xz}\sin(\alpha_m x)\cos(\beta_n y) \\
\tau_{yz} &= -h\sum_{n=1}^{\infty}\sum_{m=1}^{\infty} R_{yz}\cos(\alpha_m x)\sin(\beta_n y)
\end{aligned}
\tag{4.60}
$$

where

$$R_x = \frac{E}{(1-v^2)}(\alpha_m Y_{mn} + v\beta_n X_{mn})$$

$$R_y = \frac{E}{(1-v^2)}(v\alpha_m Y_{mn} + \beta_n X_{mn})$$

$$R_{xy} = \frac{E}{2(1+v)}(\beta_n Y_{mn} + \alpha_m X_{mn})$$

$$R_{xz} = G(\alpha_m W_{mn} - Y_{mn})$$

$$R_{yz} = G(\beta_n W_{mn} - X_{mn})$$

4.13.2 Solution for Bending of a Square Plate Example

An isotropic square plate subjected to uniformly distributed load is considered. The solutions can be obtained by solving the coefficients in Equation (4.57) by using Table 3.1 and substituting back the coefficients of series into Equation (4.53) to obtain the displacements and Equation (4.61) to compute the stresses in the plate.

Figure 4.5 shows the deformed square plate subjected to the uniformly distributed loading on the surface of the plate.

Table 4.1 tabulates the maximum non-dimensionalized transverse displacement $\bar{w}$ and stresses of a square plate under the uniformly distributed loading (see Table 3.1). The sum of series m and n of 19 are used in the analysis. The loading data and material properties are given as follows: $Q_0 = q_0 = -1.0,\ x_0 = a/2,\ y_0 = b/2,\ E = 1.0,\ a = 1.0,\ b = 1.0,\ b/h = 5,10,20,50,$ $\kappa = 5/6$, and $v = 0.25$.

The transverse displacement w and stresses are non-dimensionalized by using the following:

$$\bar{w} = w\left(\frac{Eh^3}{q_0 b^4}\right), \bar{\sigma}_{xx} = \sigma_{xx}(a/2,b/2,h/2)\left(\frac{h^2}{q_0 b^2}\right),$$

$$\bar{\sigma}_{yy} = \sigma_{yy}(a/2,b/2,h/2)\left(\frac{h^2}{q_0 b^2}\right)$$

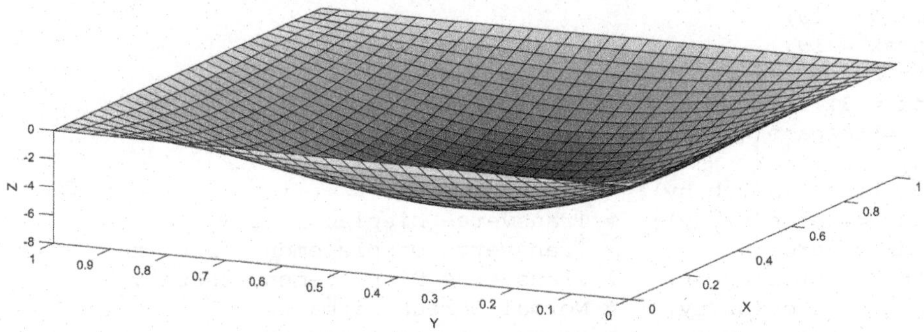

FIGURE 4.5
Deformation of a square plate.

TABLE 4.1
Transverse Displacements and Stresses of a Square Isotropic Plate

b/h	$\bar{w}$	$\bar{\sigma}_{xx}$	$\bar{\sigma}_{yy}$	$\bar{\tau}_{xy}$	$\bar{\tau}_{xz}$	$\bar{\tau}_{yz}$
Square Plates $(s = b/a = 1)$						
5	0.0513	0.2908	0.2908	0.1682	0.3861	0.3861
10	0.0440 [0.4259]	0.2903 [0.2762]	0.2903 [0.2762]	0.1689 [0.2085]	0.3864 [0.3927]	0.3864 [0.3927]
20	0.0421 [0.4111]	0.2901 [0.2762]	0.2901 [0.2762]	0.1692 [0.2085]	0.3865 [0.3927]	0.3865 [0.3927]
50	0.0416 [0.4070]	0.2900 [0.2762]	0.2900 [0.2762]	0.1693 [0.2085]	0.3865 [0.3927]	0.3865 [0.3927]
100	0.0415 [0.4060]	0.2900 [0.2762]	0.2900 [0.2762]	0.1693 [0.2085]	0.3865 [0.3927]	0.3865 [0.3927]

Values in the brackets [ ] are taken from Reddy (2002).

$$\bar{\tau}_{xy} = \tau_{xy}(a,b,h/2)\left(\frac{h^2}{q_0 b^2}\right), \bar{\tau}_{xz} = \tau_{xz}(0,b/2,0)\left(\frac{2h}{q_0 b}\right), \bar{\tau}_{yz} = \tau_{yz}(a/2,0,0)\left(\frac{2h}{q_0 b}\right).$$

4.13.3 BendingPlateNavier Program List

```
% BendingPlateNavier.m R7
% Solution for bending of a square FSDT plate example
clear variables; clc;
% Geometrical data
bh = 100;     % 5, 10, 20 or 50
a = 1;
b = 1;
s = b/a;
h = b/bh;
z3 = +h/2;  % top plate
z2 = 0;     % mid-plane
z1 = -h/2;  % bottom plate
%%
z = z3;     % select from above
% Series
```

```
mmax = 19;
nmax = 19;
%%
nx = 31;
xi=linspace(0,a,nx);
ny = 31;
yi=linspace(0,b,ny);
wxy = zeros(nx,ny);  % Transverse Displacement w
thx = zeros(nx,ny);  % Transverse Displacement theta-x
thy = zeros(nx,ny);  % Transverse Displacement theta-y
sxx = zeros(nx,ny);  % Normal Stress sigma-xx
syy = zeros(nx,ny);  % Normal Stress sigma-yy
txy = zeros(nx,ny);  % Shear Stress tau-xy
txz = zeros(nx,ny);  % Shear Stress tau-xz
tyz = zeros(nx,ny);  % Shear Stress tau-yz
% Plate's material
E     = 1.0;
nu    = 0.25;
Io    = h^3/12;
Ixy   = h;
kappa = 5/6;
Dd    = E*h^3/12/(1-nu^2);    % D Isotropic material
Dxxyy = nu*E*h^3/6/(1-nu^2); % Dxxyy
Dxy   = E*h^3/24/(1+nu);      % Dxy = Dyx
Dxz   = kappa*E*h/2/(1+nu);   % Dxz = Dyz
% distributed loading
q0 = -1.0;
%
smat=zeros(3,3);
qvec=zeros(3,1);
% Distributed loading
for i = 1:nx                 % x divisions
  for j = 1:ny               % y divisions
    for n = 1:1:nmax     % y series of solutions
      for m = 1:1:mmax  % x series of solutions
        % stiffness matrix equivalent
        alpham = m*pi/a;
        betan  = n*pi/b;
        smat(1,1) = (alpham^2+betan^2)*Dxz;
        smat(1,2) = -betan*Dxz;
        smat(2,1) = smat(1,2);
        smat(2,2) = Dxz+betan^2*Dd+alpham^2*Dxy;
        smat(1,3) = -alpham*Dxz;
        smat(3,1) = smat(1,3);
        smat(2,3) = alpham*betan*(Dxxyy+Dxy);
        smat(3,2) = smat(2,3);
        smat(3,3) = Dxz+alpham^2*Dd+betan^2*Dxy;
        % Loading vector
        qvec(1,1) = 16*q0/pi^2/m/n; % qmn
        coef=smat\qvec;
```

```
        Wmn = coef(1,1);
        Xmn = coef(2,1);
        Ymn = coef(3,1);
        % Displacement w and rotations theta-x,y
        wxy(i,j) = wxy(i,j)+Wmn*sin(alpham*xi(1,i))*sin(betan*
yi(1,j));
        thx(i,j) = thx(i,j)+Xmn*sin(alpham*xi(1,i))*cos(betan*
yi(1,j));
        thy(i,j) = thy(i,j)+Ymn*cos(alpham*xi(1,i))*sin(betan*
yi(1,j));
        % Stresses
        Rx = E/(1-nu^2)*(alpham*Ymn+nu*betan*Xmn);
        sxx(i,j) = sxx(i,j)-z*Rx*sin(alpham*xi(1,i))*sin(betan*
yi(1,j));
        Ry = E/(1-nu^2)*(nu*alpham*Ymn+betan*Xmn);
        syy(i,j) = syy(i,j)-z*Ry*sin(alpham*xi(1,i))*sin(betan*
yi(1,j));
        Rxy = E/2/(1+nu)*(betan*Ymn+alpham*Xmn);
        txy(i,j) = txy(i,j)+z*Rxy*cos(alpham*xi(1,i))*cos(betan*
yi(1,j));
        Rxz = Dxz/h*(alpham*Wmn-Ymn);
        txz(i,j) = txz(i,j)+Rxz*cos(alpham*xi(1,i))*sin(betan*
yi(1,j));
        Ryz = Dxz/h*(betan*Wmn-Xmn);
        tyz(i,j) = tyz(i,j)+Ryz*sin(alpham*xi(1,i))*cos(betan*
yi(1,j));
      end
    end
  end
end
% Normalization
wN = abs(E*h^3/q0/b^4);
sN = abs(h^2/b^2/q0);
sN2 = abs(2*h/b/q0);
text=['Minimum value of normalized w = ',
num2str(min(min(wxy))*wN)];
disp(text);
text=['Normal stress Sxx = ',num2str(max(max(sxx))*sN)];
disp(text);
text=['Normal stress Syy = ',num2str(max(max(syy))*sN)];
disp(text);
text=['Shear stress Txy = ',num2str(max(max(txy))*sN)];
disp(text);
text=['Shear stress Txz = ',num2str(max(max(txz))*sN2)];
disp(text);
text=['Shear stress Tyz = ',num2str(max(max(tyz))*sN2)];
disp(text);
% figure(1)
% set(gcf,'position',[400,400,1200,400])
% for l=1:1
```

```
%       hold;
%       surface(repmat(xi,nx,1),repmat(yi',1,ny),wxy)
%       xlabel('X'); ylabel('Y'); zlabel('Z');
%       azm = -65;
%       view(azm,50)
% end
```

4.13.4 Formulations for Free Vibration Problem of Square FSDT Plates

In the free vibration problem of a plate, Equation (4.56) is solved by considering no loading applied to the system. The solutions are assumed to take the form of

$$\begin{aligned} W_{mn}(t) &= W_{mn}e^{i\omega t} \\ X_{mn}(t) &= X_{mn}e^{i\omega t} \\ Y_{mn}(t) &= Y_{mn}e^{i\omega t} \end{aligned} \tag{4.61}$$

where ω_{mn} is the natural frequency of free vibration which is associated with the mth and n^{th} mode shapes of vibration of the plate. The solution of the free vibration problem can be obtained from the following eigenvalue system:

$$\left(\begin{bmatrix} s_{11}-\bar{s}_{11} & s_{12} & s_{13} \\ s_{21} & s_{22} & s_{23} \\ s_{31} & s_{32} & s_{33} \end{bmatrix} - \omega^2 \begin{bmatrix} m_{11} & 0 & 0 \\ 0 & m_{22} & 0 \\ 0 & 0 & m_{33} \end{bmatrix} \right) \begin{Bmatrix} W_{mn} \\ X_{mn} \\ Y_{mn} \end{Bmatrix} = \begin{Bmatrix} 0 \\ 0 \\ 0 \end{Bmatrix} \tag{4.62}$$

By using the eigenvalue analysis, we can compute the natural frequencies ω_{mn} from Equation (4.63).

The smallest value of ω_{mn} is called the *fundamental frequency* of the plate. The non-dimensionalized of the first two fundamental frequencies ($m = n = 2$) of the square isotropic FSDT plate is defined by

$$\bar{\omega}_{mn} = \omega_{mn}\frac{h}{a^2}\sqrt{\frac{E}{\rho}} \tag{4.63}$$

4.13.5 Solution for Free Vibration of a Square FSDT Plate Example

Table 4.2 tabulates the non-dimensionalized natural frequencies $\bar{\omega}_{mn}$ of isotropic ($\nu = 0.3$) plates for $(m,n) = (1,1),(1,2),(2,2)$.

TABLE 4.2
Non-dimensionalized Natural Frequencies of An Isotropic ($\nu = 0.3$) FSDT Plate

b/*h*	$\bar{\omega}_{11}$	$\bar{\omega}_{12} = \bar{\omega}_{21}$	$\bar{\omega}_{22}$
5	5.5920	11.8984	17.3725
10	6.1630	14.3255	22.3678
20	6.3418	15.2664	24.6520
50	6.3953	15.5728	25.4581

4.13.6 FreeVibrationPlateNavier Program List

```
% FreeVibrationPlateNavier.m R1
% Solution of a square FSDT plate example
clear variables; clc;
% Geometrical data
bh = 5;    % 5, 10, 20 or 50
a = 1;
b = 1;
s = b/a;
h = b/bh;
%%
% Series
mmax = 2;
nmax = 2;
%%
% Plate's material
E     = 1.0;
nu    = 0.30;
Io    = h^3/12;
Ixy   = h;
rho   = 10.0;
kappa = 5/6;
Dd    = E*h^3/12/(1-nu^2);    % D Isotropic material
Dxxyy = nu*E*h^3/6/(1-nu^2); % Dxxyy
Dxy   = E*h^3/24/(1+nu);      % Dxy = Dyx
Dxz   = kappa*E*h/2/(1+nu);   % Dxz = Dyz
%
smat=zeros(3,3);
mmat=zeros(3,3);
omega = zeros(mmax,nmax);   % Natural Frequency
% Free Vibration
for n = 1:1:nmax     % y series of solutions
  for m = 1:1:mmax  % x series of solutions
        % stiffness matrix equivalent
        alpham = m*pi/a;
        betan  = n*pi/b;
        smat(1,1) = (alpham^2+betan^2)*Dxz;
        smat(1,2)  = -betan*Dxz;
```

```
        smat(2,1) = smat(1,2);
        smat(2,2) = Dxz+betan^2*Dd+alpham^2*Dxy;
        smat(1,3) = -alpham*Dxz;
        smat(3,1) = smat(1,3);
        smat(2,3) = alpham*betan*(Dxxyy+Dxy);
        smat(3,2) = smat(2,3);
        smat(3,3) = Dxz+alpham^2*Dd+betan^2*Dxy;
        % Mass matrix
        mmat(1,1) = rho*Ixy;
        mmat(2,2) = rho*Io;
        mmat(3,3) = rho*Io;
        % Eigenvalue analysis
        [vec, lam2] = eig(smat,mmat);
        lambda=sort(diag(sqrt(lam2)));
        omega(m,n)=lambda(1);
  end
end
% Normalization
omegaN = h/a^2*sqrt(E/rho);
for n = 1:1:nmax    % y series of solutions
  for m = 1:1:mmax  % x series of solutions
    text=['Normalized the first natural frequency =
',num2str(omega(m,n)/omegaN)];
    disp(text);
  end
end
```

4.13.7 Formulations for Buckling Problem of Square FSDT Plates

In the buckling problem of a square plate subjected to in-plane biaxial compressive loadings at all edges. The applied axial in-plane compressive forces are defined as follows:

$$
\begin{aligned}
N_x &= N \\
N_y &= \lambda N \\
\lambda &= \frac{N_y}{N_x}
\end{aligned}
\tag{4.64}
$$

where λ is the ratio of compressive forces in two directions. The equilibrium equations that consider the effect of buckling of plates in Equation (4.47) become

$$
\left(\begin{bmatrix} s_{11} & s_{12} & s_{13} \\ s_{21} & s_{22} & s_{23} \\ s_{31} & s_{32} & s_{33} \end{bmatrix} - N\begin{bmatrix} \left(\alpha_m^2 + \lambda\beta_n^2\right) & 0 & 0 \\ 0 & 0 & 0 \\ 0 & 0 & 0 \end{bmatrix}\right)\begin{Bmatrix} W_{mn} \\ X_{mn} \\ Y_{mn} \end{Bmatrix} = \begin{Bmatrix} 0 \\ 0 \\ 0 \end{Bmatrix}
\tag{4.65}
$$

TABLE 4.3
Non-dimensionalized Buckling Load of a Simply Supported Square FSDT Plate

$\lambda = 0$		$\lambda = 1$	
b/h	$\bar{N}$	b/h	$\bar{N}$
5	1.6039	5	1.1341
10	0.6167	10	0.4361
20	0.2232	20	0.1578
50	0.0568	50	0.0402

For each combination of m and n, there is a corresponding unique value of buckling load N_{mn}.

By using the eigenvalue analysis, we can compute the values of N_{mn} from Equation (4.65).

The smallest value of N is called the *critical buckling load* (N_{cr}) of a plate. The non-dimensionalized of the first two critical buckling loads $(m = n = 2)$ of the square isotropic FSDT plate is defined by

$$\bar{N}_{mn} = N_{mn}\frac{a^2}{\pi^2 E} \tag{4.66}$$

4.13.8 Solution for Buckling Problem of Square FSDT Plates

Table 4.3 tabulates the effect of the ratio of compressive forces (λ) on the non-dimensionalized critical buckling load of a square FSDT simply supported plate. In all of the cases, the critical buckling mode is considered for $(m,n) = (1,1)$.

Table 4.3 shows the critical buckling loads as a function of the ratio of compressive forces (λ) and the ratio of width to thickness (b/h). The effect of the transverse shear deformation is significant for thick plates having lower aspect width-to-thickness ratios. For thin plates having higher aspect width-to-thickness ratios, the computed critical buckling loads are very close to those of the CPT results.

4.13.9 BucklingPlateNavier Program List

```
% BucklingPlateNavier.m R1
% Solution of a square FSDT plate example
clear variables; clc;
% Geometrical data
bh     = 5;     % 5, 10, 20 or 50
lambda = 1;
a = 1;
```

```
b = 1;
s = b/a;
h = b/bh;
%%
% Series
mmax = 1;
nmax = 1;
%%
% Plate's material
E     = 1.0;
nu    = 0.30;
Io    = h^3/12;
Ixy   = h;
kappa = 5/6;
Dd    = E*h^3/12/(1-nu^2);    % D Isotropic material
Dxxyy = nu*E*h^3/6/(1-nu^2); % Dxxyy
Dxy   = E*h^3/24/(1+nu);     % Dxy = Dyx
Dxz   = kappa*E*h/2/(1+nu);  % Dxz = Dyz
%
smat=zeros(3,3);
fmat=zeros(3,3);
force = zeros(mmax,nmax);  % Natural Frequency
% Free Vibration
for n = 1:1:nmax    % y series of solutions
  for m = 1:1:mmax  % x series of solutions
        % stiffness matrix equivalent
        alpham = m*pi/a;
        betan  = n*pi/b;
        smat(1,1) = (alpham^2+betan^2)*Dxz;
        smat(1,2) = -betan*Dxz;
        smat(2,1) = smat(1,2);
        smat(2,2) = Dxz+betan^2*Dd+alpham^2*Dxy;
        smat(1,3) = -alpham*Dxz;
        smat(3,1) = smat(1,3);
        smat(2,3) = alpham*betan*(Dxxyy+Dxy);
        smat(3,2) = smat(2,3);
        smat(3,3) = Dxz+alpham^2*Dd+betan^2*Dxy;
        % force matrix
        fmat(1,1) = alpham^2+lambda*betan^2;
        % Eigenvalue analysis
        [vec, f2] = eig(smat,fmat);
        f=sort(diag(sqrt(f2)));
        fmat(m,n)=f(1);
  end
end
% Normalization
fN = a^2/pi^2*E;
for n = 1:1:nmax    % y series of solutions
  for m = 1:1:mmax  % x series of solutions
```

```
    text=['Normalized the first buckling force =
',num2str(fmat(m,n)/fN)];
    disp(text);
  end
end
```

References

Carrera E (1995) A class of two-dimensional theories for anisotropic multilayered plates analysis. *Atti Della Accademia Delle Scienze di Torino. Classe di Scienze Fisiche Matematiche e Naturali* 19:1–39.

Carrera E (1997) Requirements-models for the two dimensional analysis of multilayered structures. *Compos Struct* 37(3):373–383.

Carrera E (1998a) Evaluation of layerwise mixed theories for laminated plates analysis. *AIAA J* 36(5):830–839.

Carrera E (1998b) Mixed layer-wise models for multilayered plates analysis. *Compos Struct* 43(1):57–70.

Carrera E (1998c) Layer-wise mixed models for accurate vibrations analysis of multilayered plates. *J Appl Mech* 65(4):820–828.

Carrera E (1999a) Multilayered shell theories accounting for layerwise mixed description, part 2: numerical evaluations. *AIAA J* 37(9):1117–1124.

Carrera E (1999b) Multilayered shell theories accounting for layerwise mixed description, part I. Governing equations. *AIAA J* 37(9):1107–1116.

Carrera E (1999c) A Reissner's mixed variational theorem applied to vibration analysis of multilayered shell. *J Appl Mech* 66(1):69–78.

Carrera E (1999d) A study of transverse normal stress effect on vibration of multilayered plates and shells. *J Sound Vib* 225(5):803–829.

Carrera E (1999e) Transverse normal stress effects in multilayered plates. *J Appl Mech* 66(4):1004–1012.

Carrera E (2000a) An assessment of mixed and classical theories on global and local response of multilayered orthotropic plates. *Compos Struct* 50(2):183–198.

Carrera E (2000b) Single- vs multilayer plate modelings on the basis of reissner's mixed theorem. *AIAA J* 38(2):342–352

Carrera E (2000c) A priori vs. a posteriori evaluation of transverse stresses in multilayered orthotropic plates. *Compos Struct* 48(4):245–260.

Carrera E (2000d) An assessment of mixed and classical theories for the thermal stress analysis of orthotropic multilayered plates. *J Therm Stresses* 23(9):797–831.

Carrera E (2001) Developments, ideas, and evaluations based upon Reissner's mixed variational theorem in the modeling of multilayered plates and shells. *Appl Mech Rev* 54(4):301–329.

Carrera E (2003) Theories and finite elements for multilayered plates and shells: a unified compact formulation with numerical assessment and benchmarking. *Arch Comput Methods Eng* 10(3):215–296.

Carrera E, Giunta G, Petrolo M (2011a) *Beam Structures: Classical and Advanced Theories.* John Wiley & Sons, Chichester.

Carrera E, Brischetto S, Nali P (2011b) *Plates and Shells for Smart Structures: Classical and Advanced Theories for Modeling and Analysis*. John Wiley & Sons, Chichester.
Carrera E, Cinefra M, Petrolo M, Zappino E (2014) *Finite Element Analysis of Structures through Unified Formulation*. John Wiley & Sons, Chichester.
Cheng ZQ, Batra RC (2000) Three-dimensional thermoelastic deformations of a functionally graded elliptic plate. *Compos B Eng* 31(2):97–106.
Demasi L (2009a) Mixed plate theories based on the generalized unified formulation. Part I: governing equations. *Compos Struct* 87(1):1–11.
Demasi L (2009b) Mixed plate theories based on the generalized unified formulation. Part II: layerwise theories. *Compos Struct* 87(1):12–22.
Demasi L (2009c) Mixed plate theories based on the generalized unified formulation. Part III: advanced mixed high order shear deformation theories. *Compos Struct* 87(3):183–194.
Demasi L (2009d) Mixed plate theories based on the generalized unified formulation. Part IV: Zig-zag theories. *Compos Struct* 87(3):195–205.
Demasi L (2009e) Mixed plate theories based on the generalized unified formulation. Part V: results. *Compos Struct* 88(1):1–16.
Ferreira A, Roque C, Jorge R (2005) Analysis of composite plates by trigonometric shear deformation theory and multiquadrics. *Comput Struct* 83(27):2225–2237.
Huffington NJ (1963) Response of elastic columns to axial pulse loading. *AIAA J* 1(9):2099–2104.
Jin G, Su Z, Shi S, Ye T, Gao S (2014) Three-dimensional exact solution for the free vibration of arbitrarily thick functionally graded rectangular plates with general boundary conditions. *Compos Struct* 108:565–577.
Kim SE, Thai HT, Lee J (2009a) Buckling analysis of plates using the two variable refined plate theory. *Thin-Walled Struct* 47(4):455–462.
Kim SE, Thai HT, Lee J (2009b) A two variable refined plate theory for laminated composite plates. *Compos Struct* 89(2):197–205.
Kirchhoff G (1850) Uber das Gleichgewicht und die Bewegung einer elastischen Scheibe. *J Reine Angewandte Mathematik* 40:51–88.
Krishna Murty AV (1987) Flexure of composite plates. *Compos Struct* 7(3):161–177.
Levy M (1877) Memoire sur la theorie des plaques elastiques planes. *J Math Pures Appl* 3:219–306.
Levinson M (1980) An accurate, simple theory of the statics and dynamics of elastic plates. *Mech Res Commun* 7(6):343–350.
Lo KH, Christensen RM, Wu EM (1977a) A high-order theory of plate deformation part 1: Homogeneous plates. *J Appl Mech* 44(4):663–668.
Lo KH, Christensen RM, Wu EM (1977b) A high-order theory of plate deformation part 2: Laminated plates. *J Appl Mech* 44(4):669–676.
Mantari JL, Guedes Soares C (2014) Four-unknown quasi-3D shear deformation theory for advanced composite plates. *Compos Struct* 109:231–239.
Mantari JL, Soares CG (2015) A quasi-3D tangential shear deformation theory with four unknowns for functionally graded plates. *Acta Mech* 226(3):625–642.
Mindlin RD (1951) Influence of rotatory inertia and shear on flexural motions of isotropic, elastic plates. *J Appl Mech (ASME)* 18:31–38.
Murakami H (1986) Laminated composite plate theory with improved in-plane responses. *J Appl Mech* 53(3):661–666.
Reddy JN (1984) A simple higher-order theory for laminated composite plates. *J Appl Mech* 51:745–752.

Reddy JN (2000) Analysis of functionally graded plates. *Int J Numer Methods Eng* 47(1–3):663–684.

Reddy JN (2002) *Energy Principles and Variational Methods in Applied Mechanics*. John Wiley & Sons, Inc, New Jersey.

Reddy JN (2004) *Mechanics of Laminated Composite Plates and Shells: Theory and Analysis*. CRC Press, Boca Raton.

Reddy JN, Cheng ZQ (2001) Three-dimensional thermomechanical deformations of functionally graded rectangular plates. *Eur J Mech-A/Solids* 20(5):841–855.

Reissner E (1945) The effect of transverse shear deformation on the bending of elastic plates. *J Appl Mech (ASME)* 12(2):69–77.

Reissner E (1947) On bending of elastic plates. *Q Appl Math* 5(1):55–68.

Reissner E (1984) On a certain mixed variational theorem and a proposed application. *Int J Numer Methods Eng* 20(7):1366–1368.

Reissner E (1986) On a mixed variational theorem and on shear deformable plate theory. *Int J Numer Methods Eng* 23(2):193–198.

Senthilnathan NR, Chow ST, Lee KH, Lim SP (1987) Buckling of shear-deformable plates. *AIAA J* 25(9):1268–1271.

Shimpi RP (2002) Refined plate theory and its variants. *AIAA J* 40(1):137–146.

Shimpi RP, Patel HG (2006) Free vibrations of plate using two variable refined plate theory. *J Sound Vib* 296(4):979–999.

Stein M (1986) Nonlinear theory for plates and shells including the effects of transverse shearing. *AIAA J* 24(9):1537–1544.

Thai HT, Kim SE (2010) Free vibration of laminated composite plates using two variable refined plate theory. *Int J Mech Sci* 52(4):626–633.

Thai HT, Kim SE (2011) Levy-type solution for buckling analysis of orthotropic plates based on two variable refined plate theory. *Compos Struct* 93(7):1738–1746.

Thai HT, Kim SE (2012a) Analytical solution of a two variable refined plate theory for bending analysis of orthotropic Levy-type plates. *Int J Mech Sci* 54(1):269–276.

Thai HT, Kim SE (2012b) Levy-type solution for free vibration analysis of orthotropic plates based on two variable refined plate theory. *Appl Math Model* 36(8):3870–3882.

Thai HT, Choi DH (2013) Analytical solutions of refined plate theory for bending, buckling and vibration analyses of thick plates. *Appl Math Model* 37(18–19):8310–8323.

Thai HT, Park M, Choi DH (2013) A simple refined theory for bending, buckling, and vibration of thick plates resting on elastic foundation. *Int J Mech Sci* 73:40–52.

Thai HT, Choi DH (2014) Finite element formulation of a refined plate theory for laminated composite plates. *J Compos Mater* 48(28):3521–3538.

Thai HT, Vo TP, Nguyen TK, Lee J (2014) A nonlocal sinusoidal plate model for micro/nanoscale plates. *Proc Inst Mech Eng, Part C: J Mech Eng Sci* 228(14):2652–2660.

Touratier M (1991) An efficient standard plate theory. *Int J Eng Sci* 29(8):901–916.

Tran LV, Thai CH, Le HT, Gan BS, Lee J, Nguyen-Xuan H (2014) Isogeometric analysis of laminated composite plates based on a four-variable refined plate theory. *Eng Anal Boundary Elem* 47:68–81.

Vel SS, Batra RC (2002) Exact solution for thermoelastic deformations of functionally graded thick rectangular plates. *AIAA J* 40(7):1421–1433.

Wang YM, Chen SM, Wu CP (2010) A meshless collocation method based on the differential reproducing kernel interpolation. *Comput Mech* 45(6):585–606.

Wu CP, Chiu KH, Wang YM (2011) RMVT-based meshless collocation and element free Galerkin methods for the quasi-3D analysis of multilayered composite and FGM plates. *Compos Struct* 93(2):923–943.
Wu CP, Chiu KH (2011) RMVT-based meshless collocation and element-free Galerkin methods for the quasi-3D free vibration analysis of multilayered composite and FGM plates. *Compos Struct* 93(5):1433–1448.
Zenkour AM (2004a) Thermal effects on the bending response of fiber-reinforced viscoelastic composite plates using a sinusoidal shear deformation theory. *Acta Mech* 171(3–4):171–187.
Zenkour AM (2004b) Analytical solution for bending of cross-ply laminated plates under thermo-mechanical loading. *Compos Struct* 65(3):367–379.
Zenkour AM (2004c) Buckling of fiber-reinforced viscoelastic composite plates using various plate theories. *J Eng Math* 50(1):75–93.
Zenkour AM (2005a) A comprehensive analysis of functionally graded sandwich plates: Part 1-deflection and stresses. *Int J Solids Struct* 42(18–19):5224–5242.
Zenkour AM (2005b) A comprehensive analysis of functionally graded sandwich plates: part 2-buckling and free vibration. *Int J Solids Struct* 42(18–19):5243–5258.
Zenkour AM (2005c) On vibration of functionally graded plates according to a refined trigonometric plate theory. *Int J Struct Stab Dyn* 5(02):279–297.
Zenkour AM (2006) Generalized shear deformation theory for bending analysis of functionally graded plates. *Appl Math Model* 30(1):67–84.
Zenkour AM (2009) The refined sinusoidal theory for FGM plates on elastic foundations. *Int J Mech Sci* 51(11–12):869–880.
Zenkour AM, Alghamdi NA (2010a) Bending analysis of functionally graded sandwich plates under the effect of mechanical and thermal loads. *Mech Adv Mater Struct* 17(6):419–432.
Zenkour AM, Sobhy M (2010b) Thermal buckling of various types of FGM sandwich plates. *Compos Struct* 93(1):93–102.
Zenkour AM, Sobhy M (2011) Thermal buckling of functionally graded plates resting on elastic foundations using the trigonometric theory. *J Therm Stresses* 34(11):1119–1138.

5

Finite Element Formulation for Plate and Shell

5.1 Finite Element for Solving the Governing Equations

As discussed in previous chapters, the governing equations that were derived analytically require a proper and explicit representation of the functional coordinates. The differential governing equations obtained from the variational methods are ineffective in solving plate and shell problems which are geometrically complex, have discontinuous loadings, or involve nonhomogeneous material or nonlinear geometrical properties. Accordingly, every time the essential boundary conditions or loadings types are changed, the solutions of the governing equations result in different coefficients. The analytical solutions are not available for solving the wide range of complex problems encountered in real situations. Hence, the methods for analytically solving the governing equations are not readily adaptable for computer programming.

Most of the structural components in human-made engineering structures and products are composed of plates and shells. An example to introduce the plate in an engineering structure is a building, as shown in Figure 5.1. In a

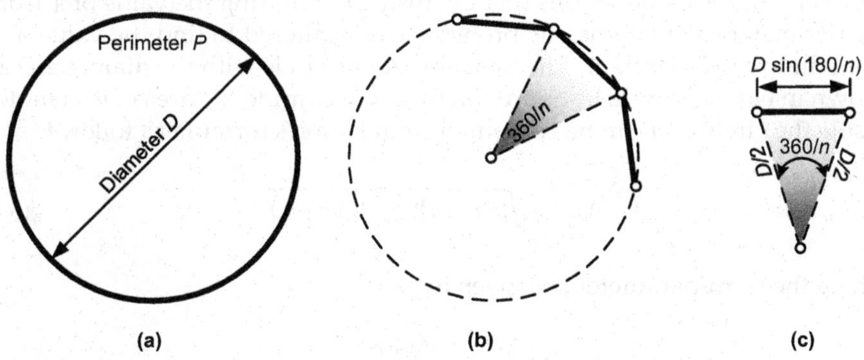

FIGURE 5.1
Illustration of the plate elements in a building.

building structure, a plate is also designed based on the classical plate theory (CPT) with an exception that it can undergo a buckling phenomenon if it is subjected to an excessive compressive axial loading at its four sides. A plate can resist a combination of loading actions. When a plate is supported at least on both of its sides where it is permitted to rotate, we call the supporting type of the plate as simply supported one. If the plate is fixed on one side and freed on the other sides, we call it a cantilever plate.

In a building, the slabs, floor, and wall can be considered as a plate. A plate is a flat structural part of the building with the thickness being much smaller than the plane dimensions. The plate as the structural member has a primary function to support transverse vertical loading and distribute it to the supports. The plate can resist the transverse loading mainly through the bending mechanism. The wall is physically similar to the plate with an exception that it can resist compressive or tensile axial loading mechanism. The compressive zone on one side of the plate depends on the bending mechanism, and the tensile zone on the other side depends on the loading conditions.

5.2 Finite Element: An Explanatory Example

The finite element is a numerical method for solving the governing equations of continuum mechanics problem within acceptable accuracy of solutions.

To illustrate the concept of finite element method (FEM), let us go back to the era of 250 BC when the famous ancient Greek astronomer, mathematician, physicist, engineer, and inventor called Archimedes studied the determination of a circle. The problem was "to find a relationship between the perimeter and the diameter D of a circle." The answer to this problem is to find a numerical value of π by increasing the number of divisions of the perimeter of the circle. In this section, instead of finding the value of π from the perimeter calculations, the problem is now altered to find the value of π from the area of a circle A. The total area A of a circle with the diameter D as shown in Figure 5.2 will be approximated by summing the area of n triangles inside the circle that can be computed from Heron's formula as follows:

$$A_{n=1} = \sqrt{p(p-a)(p-b)(p-c)} \tag{5.1}$$

where the semi-parameter p is given by

$$p = \frac{a+b+c}{2}$$

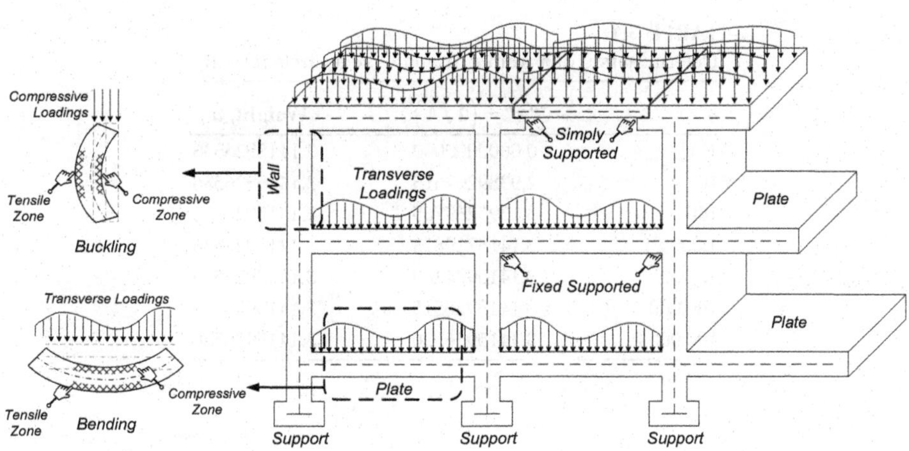

FIGURE 5.2
An explanatory example of the FEM.

where $a, b,$ and c are the three sides of the triangle.

The total area of the circle can be obtained by

$$A = \sum_{1}^{n} A_{n=1} = \sum_{1}^{n} \sqrt{p(p-a)(p-b)(p-c)} \tag{5.2}$$

This total area can be related to the diameter D by using the following formula:

$$A = "\pi"\left(\frac{D}{2}\right)^2 = "\pi"\frac{D^2}{4} \tag{5.3}$$

Substituting $D = 2$ in the above equation, we obtain

$$"\pi" = \frac{4A}{D^2} = A \tag{5.4}$$

where the computed "π" changes according to the number of n triangles inside the circle.

The values of π that are computed from Equation (5.4) are tabulated in Table 5.1. Because the total area A of the circle can be approximated by summing the area of n triangles, we need to increase the value of n to one million divisions to reach the exact value of π up to ten decimal places. The radius of the circle can be assumed to be 1.

TABLE 5.1
Determination of π from n Triangles of a Circle ($D = 2$)

n	$"\pi" = \left(2\sqrt{A}\right)^2$	Weight, w_i
1	0.0000000000	3.1415926536
10	2.9389262615	3.1415926536
100	3.1395259765	3.1415926536
1,000	3.1415719828	3.1415926536
10,000	3.1415924469	3.1415926536
100,000	3.1415926515	3.1415926536
100,000,0	3.1415926536	3.1415926536

5.3 Plate and Shell in the Finite Element Context

The FEM adopts a set of approximation functions to represent the mechanism of a system or structure to be analyzed. The approximation functions are often expressed by algebraic polynomials, which are developed by using the idea of interpolation schemes. The interpolation function can be obtained from the assumed displacement (translation, rotation, or gradient) fields. The approximation function is called the *shape function*. The shape functions are developed for a typical element with geometrically simple shapes that allow the construction of a whole system from subdomains. Finite element models of plates and shells are considerably more complicated because the geometry should be described in two-dimensional domain by using complex partial differential equations.

In the finite element context, a plate or shell can be modeled by a polygon which represents the surface of the plate and shell. There exist a large number of literature on triangular and rectangular plate-bending finite elements of isotropic or orthotropic plates which are based on the CPT (Bazeley et al. 1965; Bogner et al 1965; Clough and Tocher 1965; Fraeijis de Veubeke 1968; Herrmann 1967; Hrabok and Hrudey 1984; Irons 1969; Melosh 1963; Zienkiewicz 1964).

5.4 Shape Function for a Kirchhoff Thin Plate Element

Before we derive the formulations for the shape functions of a plate, we need to understand the concept of *generalized displacements*. The generalized displacements are the quantities of *degrees of freedom* (DOFs) at a node. In a plate considered in this section, the quantities of generalized displacements at a single node are the transverse displacement w, the rotation about the

x-axis θ_x, and the rotation about the y-axis θ_y. Figure 5.3 shows a rectangular plate with four nodes at the corners, whereas there are three DOFs at a node. Thus, the plate is said to have 12-DOF element type. The displacements or rotations at any arbitrary point inside the plate will be interpolated by using the shape functions that we are going to formulate herewith.

In the FEM, through this DOF at each node, the plate can be joined to the adjacent plates sharing the same node. The process of joining the adjacent plates is called *assembling*. The boundary conditions such as pinned or fixed can be constrained to this DOF since the deformation of plate element is controlled by the displacements or rotations at the DOF. This DOF can be interpreted as the *handle* to deform the plate or to receive the effects by the adjacent elements.

The governing equation of a plate given in Equation (3.46) is shown below for convenience purpose in formulating the shape functions of the plate.

$$D\left(\frac{d^4w}{dx^4}\right)+D\left(\frac{d^4w}{dy^4}\right)+2D\left(\frac{d^4w}{dx^2dy^2}\right)+\rho I\left(\frac{d^4w}{dx^2dt^2}\right)+\rho I\left(\frac{d^4w}{dy^2dt^2}\right)+\rho I_{xy}\left(\frac{d^2w}{dt^2}\right)$$
$$=q_z-N_x\left(\frac{d^2w}{dx^2}\right)-N_y\left(\frac{d^2w}{dy^2}\right)-N_{xy}\left(\frac{d^2w}{dxdy}\right)-N_{yx}\left(\frac{d^2w}{dydx}\right) \tag{5.5}$$

where

$$I=\int_{-h/2}^{+h/2}z^2dz=\frac{h^3}{12}$$

$$I_{xy}=\int_{-h/2}^{+h/2}1dz=h$$

The shape functions of the plate can be obtained by solving the homogeneous differential equations of the static equilibrium in the above equation. By omitting the dynamic and loading terms, we can get the homogeneous equations as

$$\left(\frac{d^4w}{dx^4}\right)+\left(\frac{d^4w}{dy^4}\right)+2\left(\frac{d^4w}{dx^2dy^2}\right)=0 \tag{5.6}$$

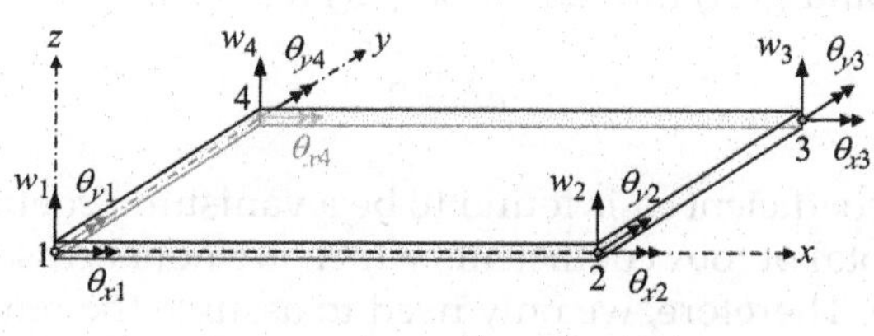

FIGURE 5.3
The generalized displacements in a 12-DOF plate element.

The solution for the above homogeneous equation can be approximated by using the third-order polynomial displacement functions, which resemble the Euler–Bernoulli beam element (Gan 2018). Under the similar principle of bending adopted, the Kirchhoff plate element is an extension of the beam element into the bidirectional form. The vertical displacement of any arbitrary point in the plate can be assumed as

$$w=\left(a_1+a_2x+a_3x^2+a_4x^3\right)\times\left(a_5+a_6y+a_7y^2+a_8y^3\right) \tag{5.7}$$

Expanding the above equation, we get

$$\begin{aligned} w &= c_1+c_2y+c_3x+c_4y^2+c_5yx+c_6x^2+c_7y^3+c_8y^2x+c_9yx^2+c_{10}x^3 \\ &\quad +c_{11}y^3x+c_{12}yx^3+c_{13}y^2x^2+c_{14}y^3x^2+c_{15}y^2x^3+c_{16}y^3x^3 \end{aligned} \tag{5.8}$$

According to Equation (5.6), the fourth derivatives with respect to x and y, the first and second terms should be vanished. The remaining part of the equation is only the third term; after substituting the derivatives of Equation (5.8), we get

$$\left(\frac{d^4w}{dx^2dy^2}\right)=4c_{13}+12c_{14}y+12c_{15}x+36c_{16}xy=0 \tag{5.9}$$

Substituting the coordinate of node 1 $\left(x=0;y=0\right)$ into the above equation results in

$$c_{13}=0 \tag{5.10}$$

Next, substituting the coordinate of node 2 $\left(x=a;y=0\right)$ together with Equation (5.10) into Equation (5.9) results in

$$c_{15}=0 \tag{5.11}$$

Further substitution of the coordinate of node 4 $\left(x=0;y=b\right)$ together with Equations (5.10) and (5.11) into Equation (5.9) results in

$$c_{14}=0 \tag{5.12}$$

Finally, the last coefficient c_{16} is found to be a vanishing coefficient; thus, it is counted as the total of four coefficients which are not necessarily considered in Equation (5.8). Therefore, we only need to assume the remaining 12 coefficients as the unknowns in Equation (5.8):

$$w = c_1 + c_2 y + c_3 x + c_4 y^2 + c_5 yx + c_6 x^2 + c_7 y^3 + c_8 y^2 x + c_9 yx^2$$
$$+ c_{10} x^3 + c_{11} y^3 x + c_{12} yx^3 \tag{5.13}$$

As shown in Figure 5.3, there are three generalized displacements $w_i, \theta_{xi}, \theta_{yi}$ $(i = 1,2,3,4)$ at each node. Hence, we have 12 DOFs in total that can be used to solve the 12 unknown coefficients in Equation (5.13).

For the convenience of numerical integration using the Gaussian quadrature scheme discussed in Chapter 2, Equation (5.13) and its derivatives can be written in local coordinate (ξ, η) as

$$w = c_1 + c_2\eta + c_3\xi + c_4\eta^2 + c_5\eta\xi + c_6\xi^2 + c_7\eta^3 + c_8\eta^2\xi + c_9\eta\xi^2$$
$$+ c_{10}\xi^3 + c_{11}\eta^3\xi + c_{12}\eta\xi^3$$
$$\theta_\xi = \frac{dw}{d\eta} = c_2 + 2c_4\eta + c_5\xi + 3c_7\eta^2 + 2c_8\eta\xi + c_9\xi^2 + 3c_{11}\eta^2\xi + c_{12}\xi^3 \tag{5.14}$$
$$\theta_\eta = \frac{dw}{d\xi} = c_3 + c_5\eta + 2c_6\xi + c_8\eta^2 + 2c_9\eta\xi + 3c_{10}\xi^2 + c_{11}\eta^3 + 3c_{12}\eta\xi^2$$

within the plate domains of $-1 \le \xi \le +1$ and $-1 \le \eta \le +1$.

The generalized displacements can be arranged in a single vector form as

$$\mathbf{d} = \begin{bmatrix} w_1 & \theta_{x1} & \theta_{y1} & w_2 & \theta_{x2} & \theta_{y2} & w_3 & \theta_{x3} & \theta_{y3} & w_4 & \theta_{x4} & \theta_{y4} \end{bmatrix}^T \tag{5.15}$$

The 12 unknown coefficients in Equation (5.14) are arranged into a vector form as

$$\mathbf{c} = \begin{bmatrix} c_1 & c_2 & c_3 & c_4 & c_5 & c_6 & c_7 & c_8 & c_9 & c_{10} & c_{11} & c_{12} \end{bmatrix}^T \tag{5.16}$$

By substituting the local coordinates (ξ, η) of the four nodes of the plate into Equation (5.14) and its derivatives either with respect to ξ or η, we can obtain the relationship between the generalized displacement (DOF) and the unknown coefficients in a matrix form as

$$\{\mathbf{d}\}_{12\times1} = [\Delta]_{12\times12}\{\mathbf{c}\}_{12\times1} \tag{5.17}$$

After substituting the local coordinates of the nodes $(\xi_1 = -1, \eta_1 = -1)$, $(\xi_2 = +1, \eta_2 = -1)$, $(\xi_3 = +1, \eta_3 = +1)$, and $(\xi_4 = -1, \eta_4 = +1)$ into the associated DOFs in Equation (5.14), we can obtain the $[\Delta]$ matrix as follows:

$$[\Delta]=\begin{bmatrix} 1 & -1 & -1 & 1 & 1 & 1 & -1 & -1 & -1 & -1 & 1 & 1 \\ 0 & 1 & 0 & -2 & -1 & 0 & 3 & 2 & 1 & 0 & -3 & -1 \\ 0 & 0 & 1 & 0 & -1 & -2 & 0 & 1 & 2 & 3 & -1 & -3 \\ 1 & -1 & 1 & 1 & -1 & 1 & -1 & 1 & -1 & 1 & -1 & -1 \\ 0 & 1 & 0 & -2 & 1 & 0 & 3 & -2 & 1 & 0 & 3 & 1 \\ 0 & 0 & 1 & 0 & -1 & 2 & 0 & 1 & -2 & 3 & -1 & -3 \\ 1 & 1 & 1 & 1 & 1 & 1 & 1 & 1 & 1 & 1 & 1 & 1 \\ 0 & 1 & 0 & 2 & 1 & 0 & 3 & 2 & 1 & 0 & 3 & 1 \\ 0 & 0 & 1 & 0 & 1 & 2 & 0 & 1 & 2 & 3 & 1 & 3 \\ 1 & 1 & -1 & 1 & -1 & 1 & 1 & -1 & 1 & -1 & -1 & -1 \\ 0 & 1 & 0 & 2 & -1 & 0 & 3 & -2 & 1 & 0 & -3 & -1 \\ 0 & 0 & 1 & 0 & 1 & -2 & 0 & 1 & -2 & 3 & 1 & 3 \end{bmatrix} \tag{5.18}$$

The 12-unknown coefficient vector $\{\mathbf{c}\}$ can be obtained by inverting the $[\Delta]$ matrix and multiplying with the generalized displacement (DOF) vector $\{\mathbf{d}\}$ as follows:

$$\{\mathbf{c}\}_{12\times1}=[\Delta]^{-1}_{12\times12}\{\mathbf{d}\}_{12\times1} \tag{5.19}$$

The above equation will give the 12 unknown coefficients in terms of generalized displacement (DOF).

The transverse displacement $w(\xi,\eta)$ in Equation (5.14) can be written in matrix form as follows:

$$w=[\Phi_w]_{1\times12}\{\mathbf{c}\}_{12\times1} \tag{5.20}$$

where the vector $[\Phi_w]$ contains the local coordinates of an arbitrary point (ξ,η) on the plate.

By substituting the coefficient vector $\{\mathbf{c}\}$ from Equation (5.19) into Equation (5.20), we can get the following equation:

$$w=[\Phi_w]_{1\times12}[\Delta]^{-1}_{12\times12}\{\mathbf{d}\}_{12\times1}=[\mathbf{N}_w]_{1\times12}\{\mathbf{d}\}_{12\times1} \tag{5.21}$$

Because the shape functions given in the above equations are purely from the non-dimensional polynomial functions, they are not consistent with the one-dimensional Euler–Bernoulli beam element which has the coefficient of length in the rotational DOF (Gan 2018). The modifications to the equations of shape functions are necessary in order to have same equations of shape functions of a one-dimensional Euler-Bernoulli beam element.

The elements of the vector $[\mathbf{N}_w]$ are the shape functions associated with the generalized transverse displacements, which are given by

$$[\mathbf{N}_w]=\begin{bmatrix} N_{ww1} & N_{w\theta_x 1} & N_{w\theta_y 1} & N_{ww2} & N_{w\theta_x 2} & N_{w\theta_y 2} & \cdots \\ \cdots & N_{ww3} & N_{w\theta_x 3} & N_{w\theta_y 3} & N_{ww4} & N_{w\theta_x 4} & N_{w\theta_y 4} \end{bmatrix} \tag{5.22}$$

where

$$N_{ww1}=-\frac{1}{8}(-1+\eta)(-1+\xi)(-2+\eta+\eta^2+\xi+\xi^2)$$

$$N_{ww2}=\frac{1}{8}(-1+\eta)(1+\xi)(-2+\eta+\eta^2-\xi+\xi^2)$$

$$N_{ww3}=-\frac{1}{8}(1+\eta)(1+\xi)(-2-\eta+\eta^2-\xi+\xi^2)$$

$$N_{ww4}=\frac{1}{8}(1+\eta)(-1+\xi)(-2-\eta+\eta^2+\xi+\xi^2)$$

$$N_{w\theta_x 1}=-\frac{1}{8}(-1+\eta)^2(1+\eta)(-1+\xi)\times\frac{b}{2}$$

$$N_{w\theta_x 2}=\frac{1}{8}(-1+\eta)^2(1+\eta)(1+\xi)\times\frac{b}{2}$$

$$N_{w\theta_x 3}=\frac{1}{8}(-1+\eta)(1+\eta)^2(1+\xi)\times\frac{b}{2}$$

$$N_{w\theta_x 4}=-\frac{1}{8}(-1+\eta)(1+\eta)^2(-1+\xi)\times\frac{b}{2}$$

$$N_{w\theta_y 1}=-\frac{1}{8}(-1+\eta)(-1+\xi)^2(1+\xi)\times\frac{a}{2}$$

$$N_{w\theta_y 2}=-\frac{1}{8}(-1+\eta)(-1+\xi)(1+\xi)^2\times\frac{a}{2}$$

$$N_{w\theta_y 3}=\frac{1}{8}(1+\eta)(-1+\xi)(1+\xi)^2\times\frac{a}{2}$$

$$N_{w\theta_\eta 1}=\frac{1}{8}(1+\eta)(-1+\xi)^2(1+\xi)\times\frac{a}{2}$$

The shape functions of the transverse displacement N_{ww1}, N_{ww2}, N_{ww3}, and N_{ww4} are shown in Figure 5.4.

The shape functions of the transverse displacement $N_{w\theta_x 1}$, $N_{w\theta_x 2}$, $N_{w\theta_x 3}$, and $N_{w\theta_x 4}$ are shown in Figure 5.5.

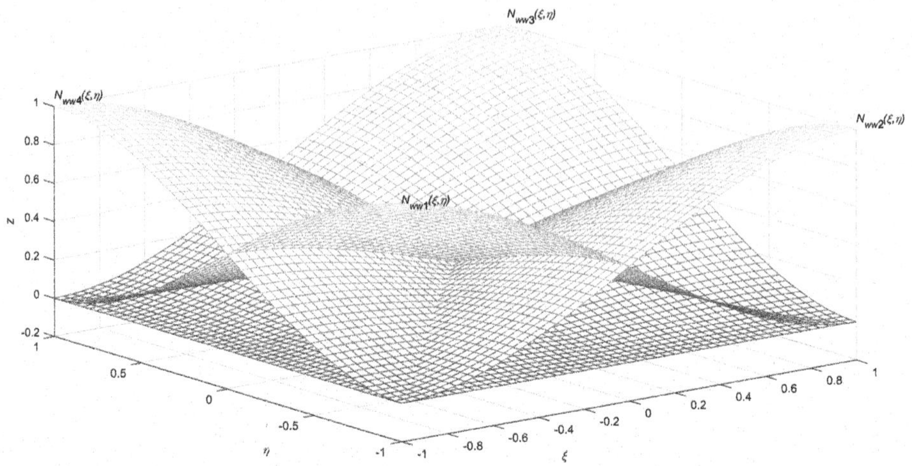

FIGURE 5.4
Basis functions of the transverse displacement N_{ww1}, N_{ww2}, N_{ww3}, and N_{ww4}.

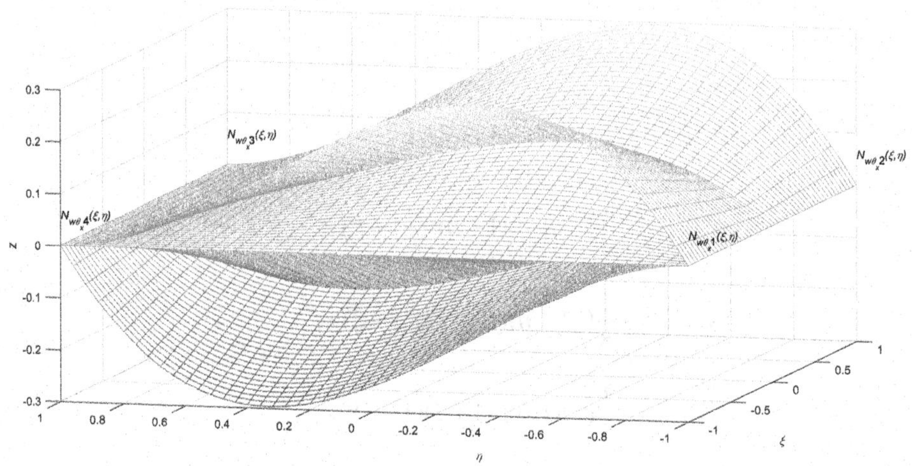

FIGURE 5.5
Basis functions of the transverse displacement $N_{w\theta_x 1}$, $N_{w\theta_x 2}$, $N_{w\theta_x 3}$, and $N_{w\theta_x 4}$.

The shape functions of the transverse displacement $N_{w\theta_y 1}$, $N_{w\theta_y 2}$, $N_{w\theta_y 3}$, and $N_{w\theta_y 4}$ are shown in Figure 5.6.

Next, the rotation about the x-axis $\theta_x(\xi,\eta)$ in Equation (5.14) can be written in matrix form as follows:

$$\theta_x = \left[\Phi_{\theta_x}\right]_{1\times 12}\{\mathbf{c}\}_{12\times 1} \tag{5.23}$$

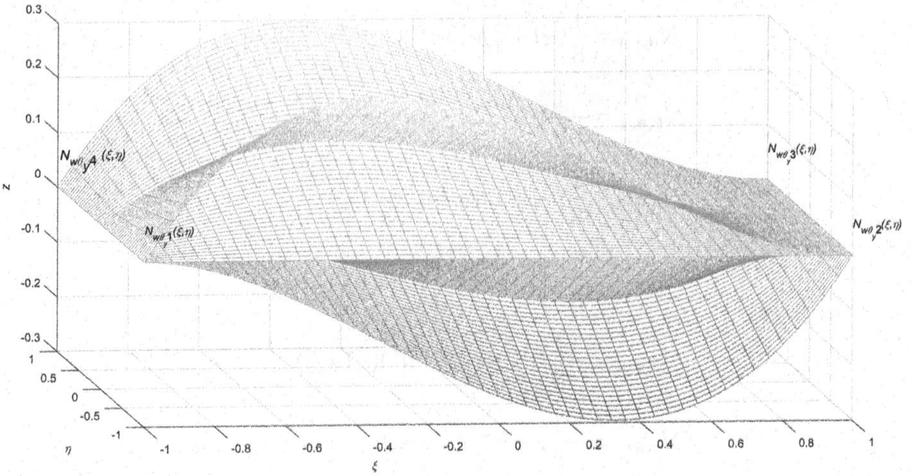

FIGURE 5.6
Basis functions of the transverse displacement $N_{w\theta_y 1}$, $N_{w\theta_y 2}$, $N_{w\theta_y 3}$, and $N_{w\theta_y 4}$.

where the vector $\left[\Phi_{\theta_x}\right]$ contains the local coordinates of an arbitrary point (ξ,η) on the plate.

Substituting the coefficient vector $\{\mathbf{c}\}$ from Equation (5.19) into Equation (5.23), we can get the following equation:

$$\theta_x = \left[\Phi_{\theta_x}\right]_{1\times12}\left[\Delta\right]_{12\times12}^{-1}\{\mathbf{d}\}_{12\times1} = \left[\mathbf{N}_{\theta_x}\right]_{1\times12}\{\mathbf{d}\}_{12\times1} \tag{5.24}$$

The elements of the vector $\left[\mathbf{N}_{\theta_x}\right]$ are the shape functions associated with the generalized transverse displacements, which are given by

$$\left[N_{\theta_x}\right] = \begin{bmatrix} N_{\theta_x w1} & N_{\theta_x\theta_x 1} & N_{\theta_x\theta_y 1} & N_{\theta_x w2} & N_{\theta_x\theta_x 2} & N_{\theta_x\theta_y 2} & \cdots \\ \cdots & N_{\theta_x w3} & N_{\theta_x\theta_x 3} & N_{\theta_x\theta_y 3} & N_{\theta_x w4} & N_{\theta_x\theta_x 4} & N_{\theta_x\theta_y 4} \end{bmatrix} \tag{5.25}$$

where

$$N_{\theta_x w1} = -\frac{1}{8}(-1+\xi)(-3+3\eta^2+\xi+\xi^2)\times\frac{2}{b}$$
$$N_{\theta_x w2} = \frac{1}{8}(1+\xi)(-3+3\eta^2-\xi+\xi^2)\times\frac{2}{b}$$
$$N_{\theta_x w3} = -\frac{1}{8}(1+\xi)(-3+3\eta^2-\xi+\xi^2)\times\frac{2}{b}$$

$$N_{\theta_x w4} = \frac{1}{8}(-1+\xi)(-3+3\eta^2+\xi+\xi^2)\times\frac{2}{b}$$

$$N_{\theta_x\theta_x 1} = -\frac{1}{8}(-1-2\eta+3\eta^2)(-1+\xi)$$

$$N_{\theta_x\theta_x 2} = \frac{1}{8}(-1-2\eta+3\eta^2)(1+\xi)$$

$$N_{\theta_x\theta_x 3} = \frac{1}{8}(-1+2\eta+3\eta^2)(1+\xi)$$

$$N_{\theta_x\theta_x 4} = -\frac{1}{8}(-1+2\eta+3\eta^2)(-1+\xi)$$

$$N_{\theta_x\theta_y 1} = -\frac{1}{8}(-1+\xi)^2(1+\xi)$$

$$N_{\theta_x\theta_y 2} = -\frac{1}{8}(-1+\xi)(1+\xi)^2$$

$$N_{\theta_x\theta_y 3} = \frac{1}{8}(-1+\xi)(1+\xi)^2$$

$$N_{\theta_x\theta_y 4} = \frac{1}{8}(-1+\xi)^2(1+\xi)$$

The shape functions of the transverse displacement $N_{\theta_x w1}$, $N_{\theta_x w2}$, $N_{\theta_x w3}$, and $N_{\theta_x w4}$ are shown in Figure 5.7.

The shape functions of the transverse displacement $N_{\theta_x\theta_x 1}$, $N_{\theta_x\theta_x 2}$, $N_{\theta_x\theta_x 3}$, and $N_{\theta_x\theta_x 4}$ are shown in Figure 5.8.

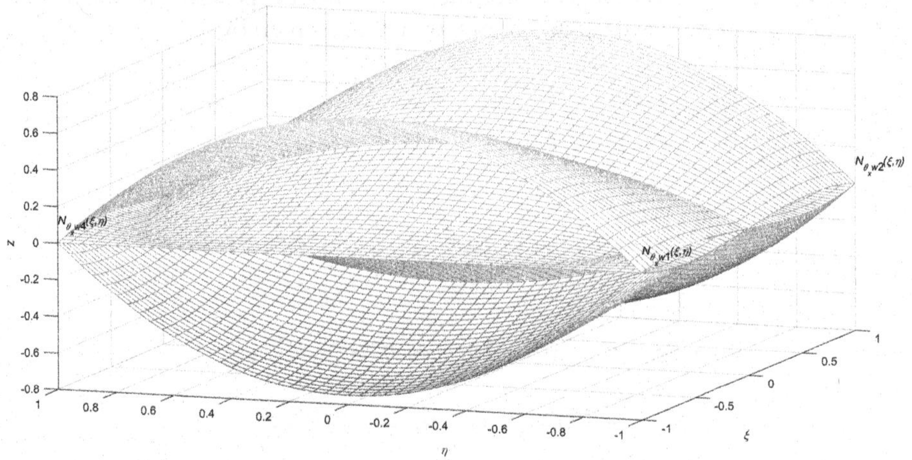

FIGURE 5.7
Basis functions of the transverse displacement $N_{\theta_x w1}$, $N_{\theta_x w2}$, $N_{\theta_x w3}$, and $N_{\theta_x w4}$.

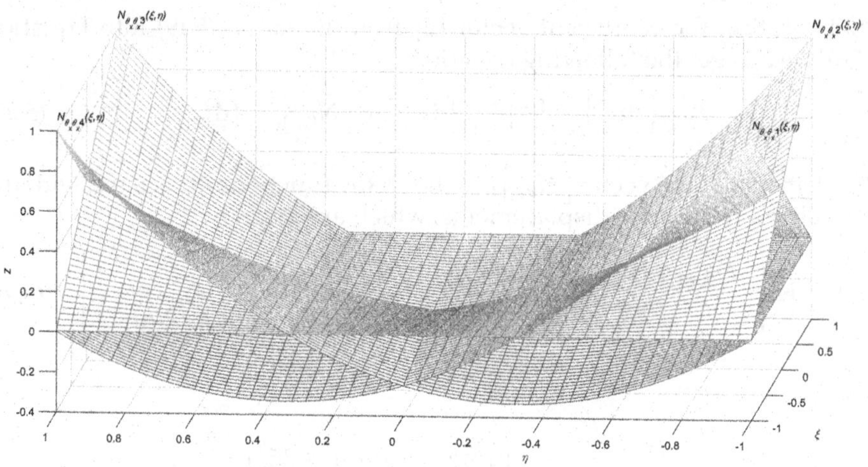

FIGURE 5.8
Basis functions of the transverse displacement $N_{\theta_x\theta_x 1}$, $N_{\theta_x\theta_x 2}$, $N_{\theta_x\theta_x 3}$, and $N_{\theta_x\theta_x 4}$.

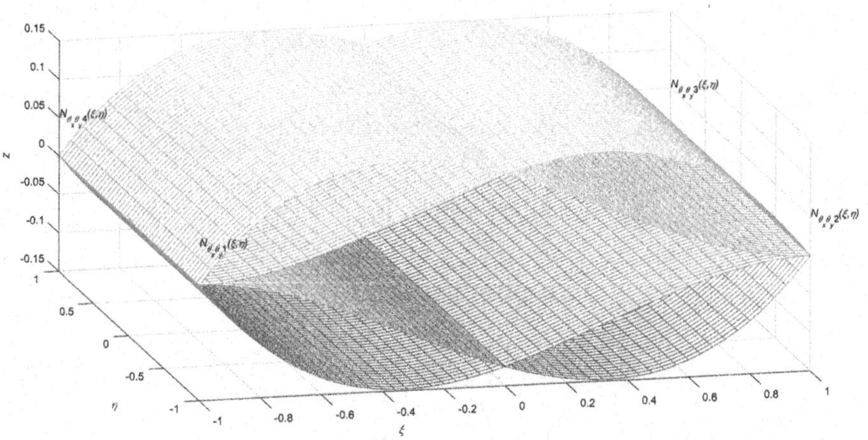

FIGURE 5.9
Basis functions of the transverse displacement $N_{\theta_x\theta_y 1}$, $N_{\theta_x\theta_y 2}$, $N_{\theta_x\theta_y 3}$, and $N_{\theta_x\theta_y 4}$.

The shape functions of the transverse displacement $N_{\theta_x\theta_y 1}$, $N_{\theta_x\theta_y 2}$, $N_{\theta_x\theta_y 3}$, and $N_{\theta_x\theta_y 4}$ are shown in Figure 5.9.

Finally, the rotation about the y-axis $\theta_y(\xi,\eta)$ in Equation (5.14) can be written in matrix form as follows:

$$\theta_y = \left[\Phi_{\theta_y}\right]_{1\times 12}\{\mathbf{c}\}_{12\times 1} \tag{5.26}$$

where the vector $\left[\Phi_{\theta_y}\right]$ contains the local coordinates of an arbitrary point (ξ,η) on the plate.

Substituting the coefficient vector $\{\mathbf{c}\}$ from Equation (5.19) into Equation (5.26), we can get the following equation:

$$\theta_y = \left[\Phi_{\theta_y}\right]_{1\times12}[\Delta]^{-1}_{12\times12}\{\mathbf{d}\}_{12\times1} = \left[\mathbf{N}_{\theta_y}\right]_{1\times12}\{\mathbf{d}\}_{12\times1} \tag{5.27}$$

The elements of the vector $\left[\mathbf{N}_{\theta_y}\right]$ are the shape functions associated with the generalized transverse displacements, which are given by

$$\left[N_{\theta_y}\right] = \begin{bmatrix} N_{\theta_y w1} & N_{\theta_y\theta_x 1} & N_{\theta_y\theta_y 1} & N_{\theta_y w2} & N_{\theta_y\theta_x 2} & N_{\theta_y\theta_x 2} & \cdots \\ \cdots & N_{\theta_y w3} & N_{\theta_y\theta_x 3} & N_{\theta_y\theta_y 3} & N_{\theta_y w4} & N_{\theta_y\theta_x 4} & N_{\theta_y\theta_y 4} \end{bmatrix} \tag{5.28}$$

where

$$N_{\theta_y w1} = -\frac{1}{8}(-1+\eta)(-3+\eta+\eta^2+3\xi^2)\times\frac{2}{a}$$

$$N_{\theta_y w2} = \frac{1}{8}(-1+\eta)(-3+\eta+\eta^2+3\xi^2)\times\frac{2}{a}$$

$$N_{\theta_y w3} = -\frac{1}{8}(1+\eta)(-3-\eta+\eta^2+3\xi^2)\times\frac{2}{a}$$

$$N_{\theta_y w4} = \frac{1}{8}(1+\eta)(-3-\eta+\eta^2+3\xi^2)\times\frac{2}{a}$$

$$N_{\theta_y\theta_x 1} = -\frac{1}{8}(-1+\eta)^2(1+\eta)$$

$$N_{\theta_y\theta_x 2} = \frac{1}{8}(-1+\eta)^2(1+\eta)$$

$$N_{\theta_y\theta_x 3} = \frac{1}{8}(-1+\eta)(1+\eta)^2$$

$$N_{\theta_y\theta_x 4} = -\frac{1}{8}(-1+\eta)(1+\eta)^2$$

$$N_{\theta_y\theta_y 1} = -\frac{1}{8}(-1+\eta)(-1-2\xi+3\xi^2)$$

$$N_{\theta_y\theta_y 2} = -\frac{1}{8}(-1+\eta)(-1+2\xi+3\xi^2)$$

$$N_{\theta_y\theta_y 3} = \frac{1}{8}(1+\eta)(-1+2\xi+3\xi^2)$$

$$N_{\theta_y\theta_y 4} = \frac{1}{8}(1+\eta)(-1-2\xi+3\xi^2)$$

The shape functions of the transverse displacement $N_{\theta_y w1}$, $N_{\theta_y w2}$, $N_{\theta_y w3}$, and $N_{\theta_y w4}$ are shown in Figure 5.10.

The shape functions of the transverse displacement $N_{\theta_y \theta_x 1}$, $N_{\theta_y \theta_x 2}$, $N_{\theta_y \theta_x 3}$, and $N_{\theta_y \theta_x 4}$ are shown in Figure 5.11.

The shape functions of the transverse displacement $N_{\theta_y \theta_y 1}$, $N_{\theta_y \theta_y 2}$, $N_{\theta_y \theta_y 3}$, and $N_{\theta_y \theta_y 4}$ are shown in Figure 5.12.

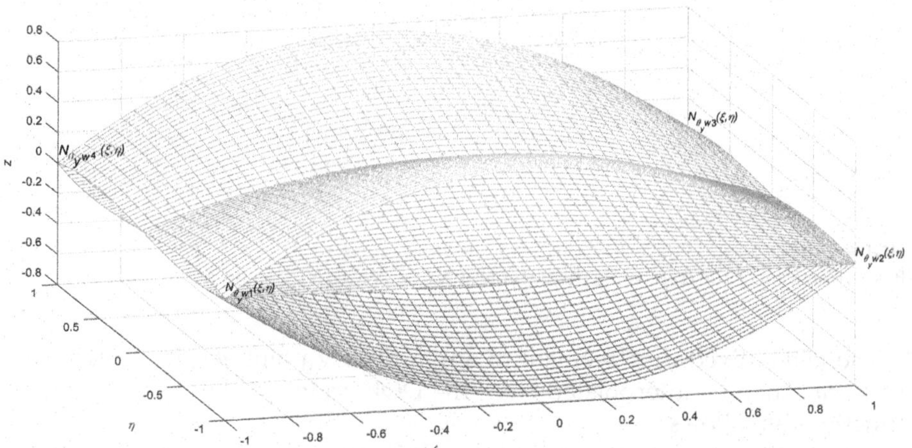

FIGURE 5.10
Basis functions of the transverse displacement $N_{\theta_y w1}$, $N_{\theta_y w2}$, $N_{\theta_y w3}$, and $N_{\theta_y w4}$.

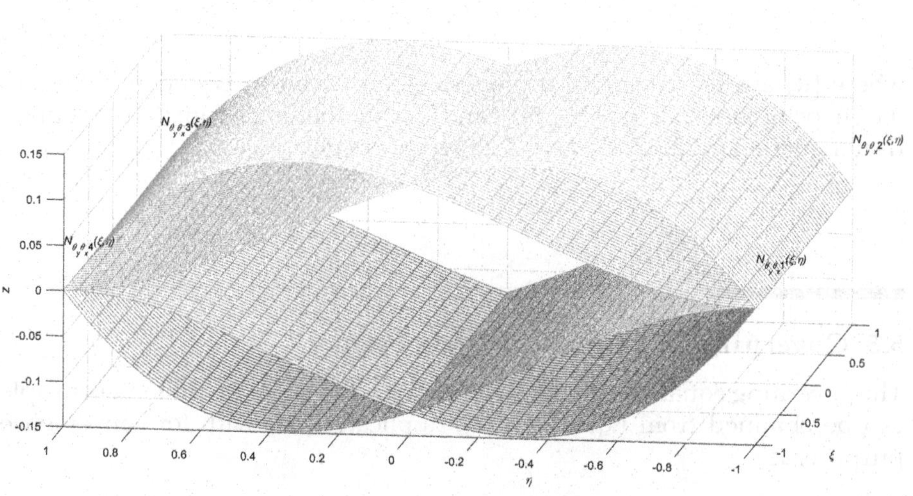

FIGURE 5.11
Basis functions of the transverse displacement $N_{\theta_y \theta_x 1}$, $N_{\theta_y \theta_x 2}$, $N_{\theta_y \theta_x 3}$, and $N_{\theta_y \theta_x 4}$.

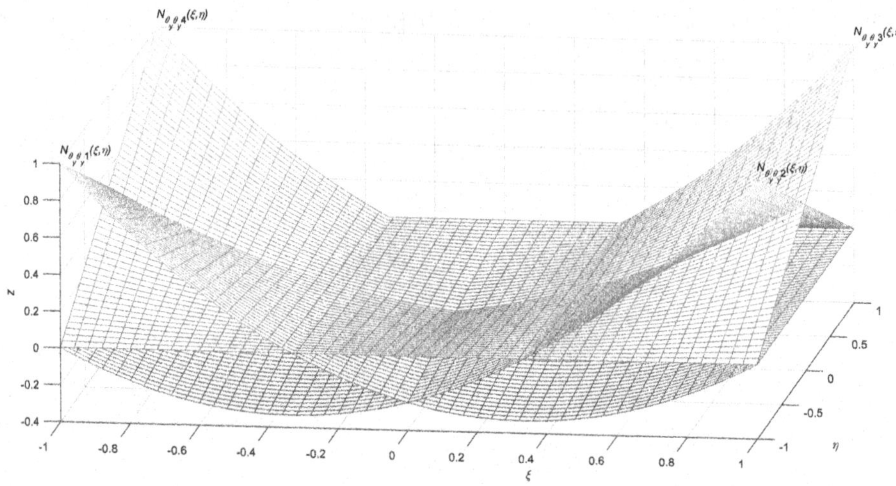

FIGURE 5.12
Basis functions of the transverse displacement $N_{\theta_y\theta_y1}$, $N_{\theta_y\theta_y2}$, $N_{\theta_y\theta_y3}$, and $N_{\theta_y\theta_y4}$.

The generalized displacement of any arbitrary point at any location in a plate can then be interpolated from the DOF vector $\{\mathbf{d}\}$ by using the shape functions as follows:

$$\begin{Bmatrix} w \\ \theta_x \\ \theta_y \end{Bmatrix} = \begin{bmatrix} \mathbf{N}_w \\ \mathbf{N}_{\theta_x} \\ \mathbf{N}_{\theta_y} \end{bmatrix} \{\mathbf{d}\} \tag{5.29}$$

where $\{\mathbf{d}\}$ are the generalized displacements given in Equation (5.15) and the shape function vectors $\mathbf{N}_w$, $\mathbf{N}_{\theta_x}$, and $\mathbf{N}_{\theta_y}$ are the shape functions obtained from Equation (5.22), (5.25), and (5.28), respectively.

5.5 Governing Equation in Matrix Form

The governing equation expressed in matrix form using Hamilton's principle can be obtained from Equation (3.6), duplicated herewith for convenience purpose, as

$$\delta H = \int_{t_1}^{t_2} \left(\delta S_E - \delta K_E - \delta W_E\right) dt = 0 \tag{5.30}$$

For static analysis where the instantaneous time $t_1 \approx t_2$, the above equation becomes

$$\delta S_E - \delta K_E - \delta W_E = 0 \tag{5.31}$$

The strain energy formula of a plate with constant thickness h which is used for developing the matrix form can be obtained from Equation (3.10), by considering $D_{xxyy} + D_{xy} = 2D$ (see Equation 3.46) and $\delta\left(\frac{d^2w}{dx^2}\right)\left(\frac{d^2w}{dy^2}\right) = \delta\left(\frac{d^2w}{dxdy}\right)\left(\frac{d^2w}{dxdy}\right) + BC$ (see Equation 3.19):

$$\delta S_E = D\int_0^b\int_0^a \delta\left(\frac{d^2w}{dx^2}\right)\left(\frac{d^2w}{dx^2}\right)dx\,dy + D\int_0^b\int_0^a \delta\left(\frac{d^2w}{dy^2}\right)\left(\frac{d^2w}{dy^2}\right)dx\,dy$$
$$+ D_{xxyy}\int_0^b\int_0^a \delta\left(\frac{d^2w}{dx^2}\right)\left(\frac{d^2w}{dy^2}\right)dx\,dy + D_{xy}\int_0^b\int_0^a \delta\left(\frac{d^2w}{dxdy}\right)\left(\frac{d^2w}{dxdy}\right)dx\,dy \tag{5.32}$$

where $D = \frac{Eh^3}{12(1-\nu^2)}$, $D_{xxyy} = \frac{\nu Eh^3}{6(1-\nu^2)}$, and $D_{xy} = \frac{Eh^3}{6(1+\nu)}$.

In Equation (5.32), there are three types of the second derivative with respect to x or y terms which can be written in matrix form by using Equation (5.29) as,

$$\begin{aligned}
\frac{d^2w}{dx^2} &= \frac{d\theta_y}{dx} = \left[\mathbf{N}_{\theta_{y,x}}\right] = \left[\mathbf{N}_{w,xx}\right] \\
\frac{d^2w}{dy^2} &= \frac{d\theta_x}{dy} = \left[\mathbf{N}_{\theta_{x,y}}\right] = \left[\mathbf{N}_{w,yy}\right] \\
\frac{d^2w}{dxdy} &= \frac{d\theta_x}{dx} = \frac{d\theta_y}{dy} = \left[\mathbf{N}_{\theta_{x,x}}\right] = \left[\mathbf{N}_{\theta_{y,y}}\right] = \left[\mathbf{N}_{w,xy}\right] = \left[\mathbf{N}_{w,yx}\right]
\end{aligned} \tag{5.33}$$

where the notation "," inside the index subscript stands for "derivative with respect to."

By substituting Equations (5.29) and (5.33), the energy variation of the strain energy in Equation (5.32) can be expressed in matrix form as follows:

$$\delta S_E = \delta\{\mathbf{d}\}^T \left(\begin{aligned} &\int_0^b\int_0^a \left(\mathbf{N}_{w,xx}^T D \mathbf{N}_{w,xx}\right)dx\,dy + \int_0^b\int_0^a \left(\mathbf{N}_{w,yy}^T D \mathbf{N}_{w,yy}\right)dx\,dy \\ &+\int_0^b\int_0^a \left(\mathbf{N}_{w,xy}^T 2D \mathbf{N}_{w,xy}\right)dx\,dy \end{aligned}\right)\{\mathbf{d}\} \tag{5.34}$$

Next, the kinetic energy formula of a plate with constant thickness h which is used for developing the matrix form can be obtained from Equation (3.25), which can be written as

$$\delta K_E = \rho I \int_0^b \int_0^a \delta\left(\frac{d\theta_x}{dt}\right)\left(\frac{d\theta_x}{dt}\right) dx\,dy + \rho I \int_0^b \int_0^a \delta\left(\frac{d\theta_y}{dt}\right)\left(\frac{d\theta_y}{dt}\right) dx\,dy$$
$$+ \rho I_{xy} \int_0^b \int_0^a \delta\left(\frac{dw}{dt}\right)\left(\frac{dw}{dt}\right) dx\,dy \quad (5.35)$$

Because the shape functions are not the function of time, only the generalized displacement vector $\{\mathbf{d}\}$ will be derived with respect to time. The energy variation of kinetic energy in Equation (5.35) can be written in matrix form by using Equation (5.29) as follows:

$$\delta K_E = \delta\{\ddot{\mathbf{d}}\}^T \begin{pmatrix} \int_0^b \int_0^a \left(\mathbf{N}_{\theta_x}^T \rho I \mathbf{N}_{\theta_x}\right) dx\,dy + \int_0^b \int_0^a \left(\mathbf{N}_{\theta_y}^T \rho I \mathbf{N}_{\theta_y}\right) dx\,dy \\ + \int_0^b \int_0^a \left(\mathbf{N}_w^T \rho I_{xy} \mathbf{N}_w\right) dx\,dy \end{pmatrix} \{\ddot{\mathbf{d}}\} \quad (5.36)$$

The energy done by the external work formula of a plate with constant thickness h can be obtained from Equation (3.36) as

$$\delta W_E = \int_0^b \int_0^a \delta\left(\frac{dw}{dx}\right) N_x \left(\frac{dw}{dx}\right) dx\,dy + \int_0^b \int_0^a \delta\left(\frac{dw}{dy}\right) N_y \left(\frac{dw}{dy}\right) dx\,dy$$
$$+ \int_0^b \int_0^a \delta\left(\frac{dw}{dx}\right) N_{xy} \left(\frac{dw}{dy}\right) dx\,dy + \int_0^b \int_0^a \delta\left(\frac{dw}{dy}\right) N_{yx} \left(\frac{dw}{dx}\right) dx\,dy$$
$$+ \int_0^b \int_0^a q_z\, \delta w\, dx\,dy \quad (5.37)$$

In Equation (5.37), there are two types of the first derivative with respect to x and y terms, which can be written in matrix form by using Equation (5.29) as

$$\frac{dw}{dx} = \theta_y = \left[\mathbf{N}_{\theta_y}\right] = \left[\mathbf{N}_{w,x}\right]$$
$$\frac{dw}{dy} = \theta_x = \left[\mathbf{N}_{\theta_x}\right] = \left[\mathbf{N}_{w,y}\right] \quad (5.38)$$

By substituting Equations (5.29) and (5.38), the energy variation of the external work in Equation (5.37) can be expressed in matrix form as follows:

$$\delta W_E = \delta\{\mathbf{d}\}^T \begin{pmatrix} \int_0^b\int_0^a \left(\mathbf{N}_{w,x}^T\left(N_x + N_{xy}\right)\mathbf{N}_{w,x}\right)dx\,dy + \int_0^b\int_0^a \\ \left(\mathbf{N}_{w,y}^T\left(N_x + N_{xy}\right)\mathbf{N}_{w,y}\right)dxdy + \int_0^b\int_0^a \left(\mathbf{N}_w^T q_z\right)dx\,dy \end{pmatrix}\{\mathbf{d}\} \tag{5.39}$$

Finally, the governing equation of a plate element in the FEM can be developed conveniently from the theory of minimum potential energy principle as follows:

$$\begin{aligned} \Pi &= \frac{1}{2}\{\mathbf{d}\}^T[\mathbf{K}]\{\mathbf{d}\} + \frac{1}{2}\{\mathbf{d}\}^T[\mathbf{F}]\{\mathbf{d}\} + \frac{1}{2}\{\ddot{\mathbf{d}}\}^T[\mathbf{M}]\{\ddot{\mathbf{d}}\} - \{\mathbf{d}\}^T\{\mathbf{f}\} \\ \frac{d\Pi}{d\{\mathbf{d}\}^T} &= [\mathbf{K}]\{\mathbf{d}\} + [\mathbf{F}]\{\mathbf{d}\} + [\mathbf{M}]\{\ddot{\mathbf{d}}\} - \{\mathbf{f}\} = \{\mathbf{0}\} \end{aligned} \tag{5.40}$$

The stiffness matrix $[\mathbf{K}]$ can be obtained from Equation (5.34) as follows:

$$\begin{aligned} [\mathbf{K}] = &\int_0^b\int_0^a \left(\mathbf{N}_{w,xx}^T D\mathbf{N}_{w,xx}\right)dx\,dy + \int_0^b\int_0^a \left(\mathbf{N}_{w,yy}^T D\mathbf{N}_{w,yy}\right)dx\,dy \\ &+ \int_0^b\int_0^a \left(\mathbf{N}_{w,xy}^T 2D\mathbf{N}_{w,xy}\right)dx\,dy \end{aligned} \tag{5.41}$$

where the elements of the symmetric stiffness matrix $K_{ij} = K_{ji}$ are given by

$$K_{0101} = K_{0202} = K_{0303} = K_{0404} = \frac{128(10a^4 + 7a^2b^2 + 10b^4)D}{5a^3b^3}$$

$$K_{0102} = K_{0304} = \frac{128(5a^4 - 7a^2b^2 - 10b^4)D}{5a^3b^3}$$

$$K_{0103} = K_{0204} = -\frac{128(5a^4 - 7a^2b^2 + 5b^4)D}{5a^3b^3}$$

$$K_{0104} = K_{0203} = -\frac{128(10a^4 + 7a^2b^2 - 5b^4)D}{5a^3b^3}$$

$$K_{0105} = K_{0108} = K_{0206} = K_{0207} = -K_{0306} = -K_{0307} = -K_{0405} = -K_{0408}$$

$$= \frac{32(20a^3 + ab^2 + b^3)D}{5ab^3}$$

$$K_{0106} = K_{0107} = K_{0205} = K_{0208} = -K_{0305} = -K_{0308} = -K_{0406} = -K_{0407}$$

$$= -\frac{32(-10a^2 + b^2)D}{5b^3}$$

$$K_{0109} = K_{0110} = -K_{0209} = -K_{0210} = -K_{0311} = -K_{0312} = K_{0411} = K_{0412}$$
$$= \frac{32(a^2 + ab + 20b^2)D}{5a^2 b}$$

$$K_{0111} = K_{0112} = -K_{0211} = -K_{0212} = -K_{0309} = -K_{0310} = K_{0409} = K_{0410}$$
$$= -\frac{32(a - 10b)D}{5a^2}$$

$$K_{0505} = K_{0606} = K_{0707} = K_{0808} = \frac{256(5a^3 + b^3)D}{15b^3}$$

$$K_{0909} = K_{1010} = K_{1111} = K_{1212} = \frac{128(a + 10b)D}{15a}$$

$$K_{0506} = K_{0708} = \frac{128\left(5a^3 - 2b^3\right)D}{15b^3}; \quad K_{0507} = K_{0608} = \frac{256(5a^2 + b^2)D}{15ab^3}$$

$$K_{0508} = K_{0607} = \frac{64(10a^3 - b^3)D}{15b^3}; \quad K_{0910} = K_{1112} = -\frac{32(a - 20b)D}{15a}$$

$$K_{0911} = K_{1012} = \frac{32(a + 10b)D}{15a}; \quad K_{0912} = K_{1011} = -\frac{128(a - 5b)D}{15a}$$

$$K_{0509} = K_{0510} = K_{0511} = K_{0512} = K_{0609} = K_{0610} = K_{0611} = K_{0612} = K_{0709}$$
$$= K_{0710} = K_{0711} = K_{0712} = K_{0809} = K_{0810} = K_{0811} = K_{0812} = 0$$

The mass matrix $[\mathbf{M}]$ can be obtained from Equation (5.36) as follows:

$$[\mathbf{M}] = \int_0^b \int_0^a \left(\mathbf{N}_{\theta_x}^T \rho I \mathbf{N}_{\theta_x}\right) dx\,dy + \int_0^b \int_0^a \left(\mathbf{N}_{\theta_y}^T \rho I \mathbf{N}_{\theta_y}\right) dx\,dy$$
$$+ \int_0^b \int_0^a \left(\mathbf{N}_w^T \rho I_{xy} \mathbf{N}_{\theta_w}\right) dx\,dy \tag{5.42}$$

where the elements of the symmetric mass matrix $M_{ij} = M_{ji}$ are given by

$$M_{0101} = M_{0202} = M_{0303} = M_{0404} = \left(\frac{11776a}{105b^3} + \frac{11776b}{105a^3}\right)\rho I + \frac{13816ab}{1575}\rho I_{xy}$$

$$M_{0102} = M_{0304} = \left(\frac{4352a}{105b^3} - \frac{11776b}{105a^3}\right)\rho I + \frac{4904ab}{1575}\rho I_{xy}$$

$$M_{0103} = M_{0204} = \left(-\frac{4352a}{105b^3} - \frac{4352b}{105a^3}\right)\rho I + \frac{1576ab}{1575}\rho I_{xy}$$

$$M_{0104} = M_{0203} = \left(-\frac{11776a}{105b^3} + \frac{4352b}{105a^3}\right)\rho I + \frac{4904ab}{1575}\rho I_{xy}$$

$$M_{0105} = M_{0206} = -M_{0307} = -M_{0408} = \left(\frac{128a}{15b^2} + \frac{1408b}{105a^2}\right)\rho I + \frac{1844ab^2}{1575}\rho I_{xy}$$

$$M_{0106} = M_{0205} = -M_{0308} = -M_{0407} = \left(\frac{64a}{15b^2} - \frac{1408b}{105a^2}\right)\rho I + \frac{796ab^2}{1575}\rho I_{xy}$$

$$M_{0107} = M_{0208} = -M_{0305} = -M_{0406} = \left(\frac{64a}{15b^2} + \frac{832b}{105a^2}\right)\rho I - \frac{464ab^2}{1575}\rho I_{xy}$$

$$M_{0108} = M_{0207} = -M_{0306} = -M_{0405} = \left(\frac{128a}{15b^2} - \frac{832b}{105a^2}\right)\rho I - \frac{1096ab^2}{1575}\rho I_{xy}$$

$$M_{0109} = -M_{0210} = -M_{0311} = M_{0412} = \left(\frac{1408a}{105b^2} + \frac{128b}{15a^2}\right)\rho I + \frac{1844a^2b}{1575}\rho I_{xy}$$

$$M_{0110} = -M_{0209} = -M_{0312} = M_{0411} = \left(-\frac{832a}{105b^2} + \frac{128b}{15a^2}\right)\rho I - \frac{1096a^2b}{1575}\rho I_{xy}$$

$$M_{0111} = -M_{0212} = -M_{0309} = M_{0410} = \left(\frac{832a}{105b^2} + \frac{64b}{15a^2}\right)\rho I - \frac{464a^2b}{1575}\rho I_{xy}$$

$$M_{0112} = -M_{0211} = -M_{0310} = M_{0409} = \left(-\frac{1408a}{105b^2} + \frac{64b}{15a^2}\right)\rho I + \frac{796a^2b}{1575}\rho I_{xy}$$

$$M_{0505} = M_{0606} = M_{0707} = M_{0808} = \left(\frac{512a}{45b} + \frac{256b}{105a}\right)\rho I + \frac{64}{315}ab^3\rho I_{xy}$$

$$M_{0506} = M_{0708} = \left(\frac{256a}{45b} - \frac{256b}{105a}\right)\rho I + \frac{32}{315}ab^3\rho I_{xy}$$

$$M_{0507} = M_{0608} = \left(-\frac{64a}{45b} + \frac{64b}{35a}\right)\rho I - \frac{8}{105}ab^3\rho I_{xy}$$

$$M_{0508} = M_{0607} = \left(-\frac{128a}{45b} - \frac{64b}{35a}\right)\rho I - \frac{16}{105}ab^3\rho I_{xy}$$

$$M_{0509} = -M_{0610} = M_{0711} = -M_{0812} = \frac{4}{25}a^2b^2\rho I_{xy}$$

$$M_{0510} = -M_{0512} = -M_{0609} = M_{0611} = -M_{0710} = M_{0712} = M_{0809} = -M_{0811}$$

$$= -\frac{8}{75}a^2b^2\rho I_{xy}$$

$$M_{0511} = -M_{0612} = M_{0709} = -M_{0810} = -\frac{16}{225}a^2b^2\rho I_{xy}$$

$$M_{0909} = M_{1010} = M_{1111} = M_{1212} = \left(\frac{256a}{105b} + \frac{512b}{45a}\right)\rho I + \frac{64}{315}a^3 b\rho I_{xy}$$

$$M_{0910} = M_{1112} = \left(-\frac{64a}{35b} - \frac{128b}{45a}\right)\rho I - \frac{16}{105}a^3 b\rho I_{xy}$$

$$M_{0911} = M_{1012} = \left(\frac{64a}{35b} - \frac{64b}{45a}\right)\rho I - \frac{8}{105}a^3 b\rho I_{xy}$$

$$M_{0912} = M_{1011} = \left(-\frac{256a}{105b} + \frac{256b}{45a}\right)\rho I + \frac{32}{315}a^3 b\rho I_{xy}$$

The axial loading matrix $[\mathbf{F}]$ and external loading vector $\{\mathbf{f}\}$ can be obtained from Equation (5.39) as follows:

$$[\mathbf{F}] = \int_0^b \int_0^a \left(\mathbf{N}_{w,x}^T\left(N_x + N_{xy}\right)\mathbf{N}_{w,x}\right)dx\,dy + \int_0^b \int_0^a \left(\mathbf{N}_{w,y}^T\left(N_x + N_{xy}\right)\mathbf{N}_{w,y}\right)dx\,dy$$

$$\{\mathbf{f}\} = \int_0^b \int_0^a \left(\mathbf{N}_w^T q_z\right)dx\,dy \tag{5.43}$$

It is worth noting that the axial loading matrix $[\mathbf{F}]$ can be used to determine the buckling problem of the plate.

5.6 Gaussian Quadrature for Integrating the Matrices in FEM

Equation (5.41) is numerically integrated by using the Gaussian quadrature as follows:

$$[\mathbf{K}] = \sum_{i=0}^{m}\sum_{j=0}^{n}\left\{\left(\mathbf{N}_{w,xx}^T D\mathbf{N}_{w,xx}\right) + \left(\mathbf{N}_{w,yy}^T D\mathbf{N}_{w,yy}\right) + \left(\mathbf{N}_{w,xy}^T 2D\mathbf{N}_{w,xy}\right)\right\}Jw_j w_i \tag{5.44}$$

where

$$\mathbf{N}_{w,xx} = \frac{dN_w^2}{dx^2} = \frac{dN_w^2}{d\xi^2}\frac{d\xi^2}{dx^2} = \frac{dN_w^2}{d\xi^2}\frac{1}{J_x^2}$$

$$\mathbf{N}_{w_{yy}} = \frac{dN_w^2}{dy^2} = \frac{dN_w^2}{d\eta^2}\frac{d\eta^2}{dy^2} = \frac{dN_w^2}{d\eta^2}\frac{1}{J_y^2}$$

$$\mathbf{N}_{w_{xy}} = \frac{dN_w^2}{dxdy} = \frac{dN_w^2}{d\xi d\eta}\frac{d\xi d\eta}{dxdy} = \frac{dN_w^2}{d\xi d\eta}\frac{1}{J_x J_y}$$

where $J_x = dx(\xi,\eta)/d\xi$ and $J_y = dx(\xi,\eta)/d\eta$ are the Jacobian operators used for transforming from global to natural coordinates.

Referring to Equation (5.42), the mass matrix of a CPT plate is given by

$$[M] = \sum_{i=0}^{m}\sum_{j=0}^{n}\left\{\left(N_{\theta_x}^T \rho I N_{\theta_x}\right) + \left(N_{\theta_y}^T \rho I N_{\theta_y}\right) + \left(N_w^T \rho I_{xy} N_w\right)\right\} J w_j w_i \tag{5.45}$$

Referring to Equation (5.43), the loading vector of a CPT plate is given by

$$[\mathbf{F}] = \sum_{i=0}^{m}\sum_{j=0}^{n}\left\{\left(N_{w,x}^T\left(N_x + N_{xy}\right)N_{w,x}\right) + \left(N_{w,y}^T\left(N_x + N_{xy}\right)N_{w,y}\right)\right\} J w_j w_i$$

$$\{\mathbf{f}\} = \sum_{i=0}^{m}\sum_{j=0}^{n}\left(N_w^T q_z\right) J w_j w_i \tag{5.45a}$$

where J is the determinant value of Jacobian operator defined in Equation (2.7).

The generalized displacement vectors $\{\mathbf{d}\}^T$ are given as

$$\mathbf{w} = \begin{bmatrix} w_1 & w_2 & w_3 & w_4 \end{bmatrix}^T; \quad \ddot{\mathbf{w}} = \begin{bmatrix} \ddot{w}_1 & \ddot{w}_2 & \ddot{w}_3 & \ddot{w}_4 \end{bmatrix}^T$$

$$\boldsymbol{\theta}_x = \begin{bmatrix} \theta_{x1} & \theta_{x2} & \theta_{x3} & \theta_{x4} \end{bmatrix}^T; \quad \ddot{\boldsymbol{\theta}}_x = \begin{bmatrix} \ddot{\theta}_{x1} & \ddot{\theta}_{x2} & \ddot{\theta}_{x3} & \ddot{\theta}_{x4} \end{bmatrix}^T$$

$$\boldsymbol{\theta}_y = \begin{bmatrix} \theta_{y1} & \theta_{y2} & \theta_{y3} & \theta_{y4} \end{bmatrix}^T; \quad \ddot{\boldsymbol{\theta}}_y = \begin{bmatrix} \ddot{\theta}_{y1} & \ddot{\theta}_{y2} & \ddot{\theta}_{y3} & \ddot{\theta}_{y4} \end{bmatrix}^T \tag{5.46}$$

where the notation $(\ddot{\ })$ means the second derivative with respect to time. The subscripts p and q are the orders of the plate surface. The subscripts m and n are the number of integration points of the Gaussian quadrature in the x- and y-directions, respectively.

5.7 Making the Stiffness and Mass Matrices of the Kirchhoff Plate Element

MATLAB® code KMmatrixKirchhoffFEM.m uses the functions NShapeKirchhoffFEM.m and Legendre.m to construct the stiffness and mass matrices of the Kirchhoff plate theory (CPT) of an isotropic plate shown in Figure 5.13.

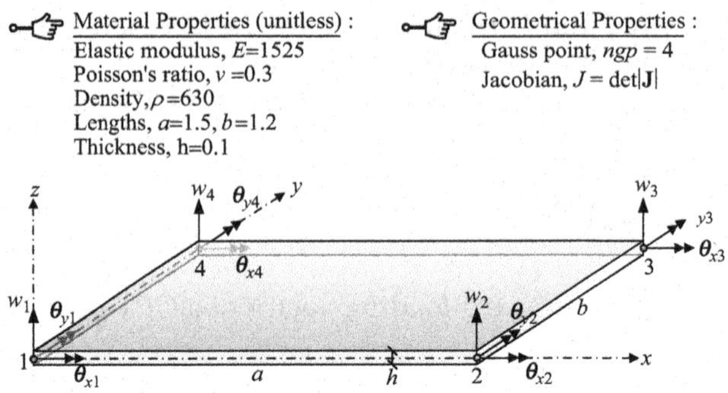

FIGURE 5.13
A rectangular isotropic flat Kirchhoff plate example.

5.7.1 KMmatrixKirchhoffFEM Program List

```
% KMmatrixKirchhoffFEM.m (R7)
% Construct matrices K and M of a four-node, 12 DOF plate
element
% Based on the Kirchhoff Plate Theory (CPT) using Polynomial
functions
% Consistent with one D Euler-Bernoulli beam theory
clear all; clc;
a   = 1.5;    b   = 1.2;    h   = 0.1;
E   = 1525;   nu  = 0.3;    rho = 630;
I   = h^3/12; Ixy = h;
ngpx = 4;    ngpy = 4;    ndof=12;
D    = E*h^3/12/(1-nu^2);    % isotropic material
Dxxyy = nu*E*h^3/6/(1-nu^2);
Dxy   = E*h^3/6/(1+nu);
Gpwx = Legendre(ngpx);    Gpwy = Legendre(ngpy);
Jac = a/2*b/2;
Jx  = a/2;
Jy  = b/2;
Kmtx = zeros(ndof,ndof);
Mmtx = zeros(ndof,ndof);
Nw   = zeros(ngpx,ngpy,ndof);
Ntx = zeros(ngpx,ngpy,ndof);
Nty = zeros(ngpx,ngpy,ndof);
Nwx = zeros(ngpx,ngpy,ndof);
Nwy = zeros(ngpx,ngpy,ndof);
Nwxx = zeros(ngpx,ngpy,ndof);
Nwyy = zeros(ngpx,ngpy,ndof);
Nwxy = zeros(ngpx,ngpy,ndof);
Nwyx = zeros(ngpx,ngpy,ndof);
for m=1:ngpx
  s = Gpwx(m,1);
  for n=1:ngpy
   t = Gpwy(n,1);
```

```
   NKirchhoff = NShapeKirchhoff(0,s,t,Jx,Jy,a,b); % 0th
derivative
   Nw(m,n,:) = NKirchhoff(:,1);
   Ntx(m,n,:) = NKirchhoff(:,2);
   Nty(m,n,:) = NKirchhoff(:,3);
   NKirchhoff = NShapeKirchhoff(2,s,t,Jx,Jy,a,b); % 2nd
derivative
   Nwxx(m,n,:) = NKirchhoff(:,1);
   Nwyy(m,n,:) = NKirchhoff(:,2);
   Nwxy(m,n,:) = NKirchhoff(:,3);
   Nwyx(m,n,:) = NKirchhoff(:,4);
  end
end
% Stiffness matrix K
for m=1:ngpx
 for n=1:ngpy
  for j=1:ndof
   for k=1:ndof
    Kmtx(j,k) = Kmtx(j,k) ...
      + Nwxx(m,n,k)*D*Nwxx(m,n,j)*Jac*Gpwx(m,2)*Gpwy(n,2)...
      + Nwyy(m,n,k)*D*Nwyy(m,n,j)*Jac*Gpwx(m,2)*Gpwy(n,2)...
      + Nwxx(m,n,k)*Dxxyy*Nwyy(m,n,j)*Jac*Gpwx(m,2)*Gpw
y(n,2)...
      + Nwxy(m,n,k)*Dxy*Nwyx(m,n,j)*Jac*Gpwx(m,2)*Gpwy(n,2);
   end
  end
 end
end
% disp(Kmtx); % before swapping
% Swapping the matrices to the general displacement vector
order
Kmtx=Kmtx([1,4,7,10,2,5,8,11,3,6,9,12],:); % row
Kmtx=Kmtx(:,[1,4,7,10,2,5,8,11,3,6,9,12]); % column
disp('Stiffness Matrix of Kirchhoff Plate');
disp(Kmtx);
% Mass matrix M
for m=1:ngpx
 for n=1:ngpy
  for j=1:ndof
   for k=1:ndof
    Mmtx(j,k) = Mmtx(j,k) ...
      + Ntx(m,n,k)*rho*I*Ntx(m,n,j)*Jac*Gpwx(m,2)*Gpwy(n,2)...
      + Nty(m,n,k)*rho*I*Nty(m,n,j)*Jac*Gpwx(m,2)*Gpwy(n,2)...
      + Nw(m,n,k)*rho*Ixy*Nw(m,n,j)*Jac*Gpwx(m,2)*Gpwy(n,2);
   end
  end
 end
end
% Swapping the matrices to the general displacement vector
Mmtx=Mmtx([1,4,7,10,2,5,8,11,3,6,9,12],:); % row
```

```
Mmtx=Mmtx(:,[1,4,7,10,2,5,8,11,3,6,9,12]); % column
disp('Mass Matrix of Kirchhoff Plate');
disp(Mmtx);
```

5.7.2 NShapeKirchhoffFEM Function List

```
function NKirchhoff = NShapeKirchhoff(i,s,t,Jx,Jy,a,b)  % for R7
% i  = 0~p derivative Shape Function, 0 not derived
% xi = Gauss coordinate
% Jx  = Jacobian operator x
% Jy  = Jacobian operator y
%----------------------------------------------------
NKirchhoff=zeros(12,3); % (Nw,Ntx,Nty=12DOF, type=1,2,3)
switch i
  case 0  % 0-derivative
    % Nw shape functions
    NKirchhoff(1,1)  = -(-1+t)*(-1+s)*(-2+t+t^2+s+s^2)/8;
    NKirchhoff(2,1)  = (-1+t)*(1+s)*(-2+t+t^2-s+s^2)/8;
    NKirchhoff(3,1)  = -(1+t)*(1+s)*(-2-t+t^2-s+s^2)/8;
    NKirchhoff(4,1)  = (1+t)*(-1+s)*(-2-t+t^2+s+s^2)/8;
    NKirchhoff(5,1)  = -(-1+t)^2*(1+t)*(-1+s)/8*b/2;
    NKirchhoff(6,1)  = (-1+t)^2*(1+t)*(1+s)/8*b/2;
    NKirchhoff(7,1)  = (-1+t)*(1+t)^2*(1+s)/8*b/2;
    NKirchhoff(8,1)  = -(-1+t)*(1+t)^2*(-1+s)/8*b/2;
    NKirchhoff(9,1)  = -(-1+t)*(-1+s)^2*(1+s)/8*a/2;
    NKirchhoff(10,1)  = -(-1+t)*(-1+s)*(1+s)^2/8*a/2;
    NKirchhoff(11,1)  = (1+t)*(-1+s)*(1+s)^2/8*a/2;
    NKirchhoff(12,1)  = (1+t)*(-1+s)^2*(1+s)/8*a/2;
    % Ntx = dNw/dy shape functions
    NKirchhoff(1,2)  = -(-1+s)*(-3+3*t^2+s+s^2)/8/Jy*2/b;
    NKirchhoff(2,2)  = (1+s)*(-3+3*t^2-s+s^2)/8/Jy*2/b;
    NKirchhoff(3,2)  = -(1+s)*(-3+3*t^2-s+s^2)/8/Jy*2/b;
    NKirchhoff(4,2)  = (-1+s)*(-3+3*t^2+s+s^2)/8/Jy*2/b;
    NKirchhoff(5,2)  = -(-1-2*t+3*t^2)*(-1+s)/8/Jy;
    NKirchhoff(6,2)  = (-1-2*t+3*t^2)*(1+s)/8/Jy;
    NKirchhoff(7,2)  = (-1+2*t+3*t^2)*(1+s)/8/Jy;
    NKirchhoff(8,2)  = -(-1+2*t+3*t^2)*(-1+s)/8/Jy;
    NKirchhoff(9,2)  = -(-1+s)^2*(1+s)/8/Jy;
    NKirchhoff(10,2)  = -(-1+s)*(1+s)^2/8/Jy;
    NKirchhoff(11,2)  = (-1+s)*(1+s)^2/8/Jy;
    NKirchhoff(12,2)  = (-1+s)^2*(1+s)/8/Jy;
    % Nty = dNw/dx shape functions
    NKirchhoff(1,3)  = -(-1+t)*(-3+t+t^2+3*s^2)/8/Jx*2/a;
    NKirchhoff(2,3)  = (-1+t)*(-3+t+t^2+3*s^2)/8/Jx*2/a;
    NKirchhoff(3,3)  = -(1+t)*(-3-t+t^2+3*s^2)/8/Jx*2/a;
    NKirchhoff(4,3)  = -(1+t)*(-3-t+t^2+3*s^2)/8/Jx*2/a;
    NKirchhoff(5,3)  = -(-1+t)^2*(1+t)/8/Jx;
    NKirchhoff(6,3)  = (-1+t)^2*(1+t)/8/Jx;
    NKirchhoff(7,3)  = (-1+t)*(1+t)^2/8/Jx;
    NKirchhoff(8,3)  = -(-1+t)*(1+t)^2/8/Jx;
```

```
    NKirchhoff(9,3) = -(-1+t)*(-1-2*s+3*s^2)/8/Jx;
    NKirchhoff(10,3) = -(-1+t)*(-1+2*s+3*s^2)/8/Jx;
    NKirchhoff(11,3) = (1+t)*(-1+2*s+3*s^2)/8/Jx;
    NKirchhoff(12,3) = (1+t)*(-1-2*s+3*s^2)/8/Jx;
  case 1
    % Nw,x = dNw/dx shape functions
    NKirchhoff(1,1) = -(-1+t)*(-3+t+t^2+3*s^2)/8/Jx;
    NKirchhoff(2,1) = (-1+t)*(-3+t+t^2+3*s^2)/8/Jx;
    NKirchhoff(3,1) = -(1+t)*(-3-t+t^2+3*s^2)/8/Jx;
    NKirchhoff(4,1) = -(1+t)*(-3-t+t^2+3*s^2)/8/Jx;
    NKirchhoff(5,1) = -(-1+t)^2*(1+t)/8/Jx*b/2;
    NKirchhoff(6,1) = (-1+t)^2*(1+t)/8/Jx*b/2;
    NKirchhoff(7,1) = (-1+t)*(1+t)^2/8/Jx*b/2;
    NKirchhoff(8,1) = -(-1+t)*(1+t)^2/8/Jx*b/2;
    NKirchhoff(9,1) = -(-1+t)*(-1-2*s+3*s^2)/8/Jx*a/2;
    NKirchhoff(10,1) = -(-1+t)*(-1+2*s+3*s^2)/8/Jx*a/2;
    NKirchhoff(11,1) = (1+t)*(-1+2*s+3*s^2)/8/Jx*a/2;
    NKirchhoff(12,1) = (1+t)*(-1-2*s+3*s^2)/8/Jx*a/2;
    % Nw,y = dNw/dy shape functions
    NKirchhoff(1,2) = -(-1+s)*(-3+3*t^2+s+s^2)/8/Jy;
    NKirchhoff(2,2) = (1+s)*(-3+3*t^2-s+s^2)/8/Jy;
    NKirchhoff(3,2) = -(1+s)*(-3+3*t^2-s+s^2)/8/Jy;
    NKirchhoff(4,2) = (-1+s)*(-3+3*t^2+s+s^2)/8/Jy;
    NKirchhoff(5,2) = -(-1-2*t+3*t^2)*(-1+s)/8/Jy*a/2;
    NKirchhoff(6,2) = (-1-2*t+3*t^2)*(1+s)/8/Jy*a/2;
    NKirchhoff(7,2) = (-1+2*t+3*t^2)*(1+s)/8/Jy*a/2;
    NKirchhoff(8,2) = -(-1+2*t+3*t^2)*(-1+s)/8/Jy*a/2;
    NKirchhoff(9,2) = -(-1+s)^2*(1+s)/8/Jy*b/2;
    NKirchhoff(10,2) = -(-1+s)*(1+s)^2/8/Jy*b/2;
    NKirchhoff(11,2) = (-1+s)*(1+s)^2/8/Jy*b/2;
    NKirchhoff(12,2) = (-1+s)^2*(1+s)/8/Jy*b/2;
  case 2
    % Nw,xx shape functions
    NKirchhoff(1,1) = -6*(-1+t)*s/Jx/Jx;
    NKirchhoff(2,1) = 6*(-1+t)*s/Jx/Jx;
    NKirchhoff(3,1) = -6*(1+t)*s/Jx/Jx;
    NKirchhoff(4,1) = 6*(1+t)*s/Jx/Jx;
    NKirchhoff(5,1) = 0;
    NKirchhoff(6,1) = 0;
    NKirchhoff(7,1) = 0;
    NKirchhoff(8,1) = 0;
    NKirchhoff(9,1) = -2*(-1+t)*(-1+3*s)/Jx/Jx*a/2;
    NKirchhoff(10,1) = -2*(-1+t)*(1+3*s)/Jx/Jx*a/2;
    NKirchhoff(11,1) = 2*(1+t)*(1+3*s)/Jx/Jx*a/2;
    NKirchhoff(12,1) = 2*(1+t)*(-1+3*s)/Jx/Jx*a/2;
    % Nw,yy shape functions
    NKirchhoff(1,2) = -6*t*(-1+s)/Jy/Jy;
    NKirchhoff(2,2) = 6*t*(1+s)/Jy/Jy;
    NKirchhoff(3,2) = -6*t*(1+s)/Jy/Jy;
    NKirchhoff(4,2) = 6*t*(-1+s)/Jy/Jy;
```

```
    NKirchhoff(5,2)  = -2*(-1+3*t)*(-1+s)/Jy/Jy*a/2;
    NKirchhoff(6,2)  = 2*(-1+3*t)*(1+s)/Jy/Jy*a/2;
    NKirchhoff(7,2)  = 2*(1+3*t)*(1+s)/Jy/Jy*a/2;
    NKirchhoff(8,2)  = -2*(1+3*t)*(-1+s)/Jy/Jy*a/2;
    NKirchhoff(9,2)  = 0;
    NKirchhoff(10,2)  = 0;
    NKirchhoff(11,2)  = 0;
    NKirchhoff(12,2)  = 0;
    % Nw,xy shape functions
    NKirchhoff(1,3)  = (4-3*t^2-3*s^2)/Jx/Jy;
    NKirchhoff(2,3)  = (-4+3*t^2+3*s^2)/Jx/Jy;
    NKirchhoff(3,3)  = (4-3*t^2-3*s^2)/Jx/Jy;
    NKirchhoff(4,3)  = (-4+3*t^2+3*s^2)/Jx/Jy;
    NKirchhoff(5,3)  = (1+2*t-3*t^2)/Jx/Jy*b/2;
    NKirchhoff(6,3)  = (-1-2*t+3*t^2)/Jx/Jy*b/2;
    NKirchhoff(7,3)  = (-1+2*t+3*t^2)/Jx/Jy*b/2;
    NKirchhoff(8,3)  = (1-2*t-3*t^2)/Jx/Jy*b/2;
    NKirchhoff(9,3)  = (1+2*s-3*s^2)/Jx/Jy*a/2;
    NKirchhoff(10,3)  = (1-2*s-3*s^2)/Jx/Jy*a/2;
    NKirchhoff(11,3)  = (-1+2*s+3*s^2)/Jx/Jy*a/2;
    NKirchhoff(12,3)  = (-1-2*s+3*s^2)/Jx/Jy*a/2;
    % Nw,yx shape functions
    NKirchhoff(1,4)  = (4-3*s^2-3*t^2)/Jx/Jy;
    NKirchhoff(2,4)  = (-4+3*s^2+3*t^2)/Jx/Jy;
    NKirchhoff(3,4)  = (4-3*s^2-3*t^2)/Jx/Jy;
    NKirchhoff(4,4)  = (-4+3*s^2+3*t^2)/Jx/Jy;
    NKirchhoff(5,4)  = (1+2*s-3*s^2)/Jx/Jy*a/2;
    NKirchhoff(6,4)  = (-1-2*s+3*s^2)/Jx/Jy*a/2;
    NKirchhoff(7,4)  = (-1+2*s+3*s^2)/Jx/Jy*a/2;
    NKirchhoff(8,4)  = (1-2*s-3*s^2)/Jx/Jy*a/2;
    NKirchhoff(9,4)  = (1+2*t-3*t^2)/Jx/Jy*b/2;
    NKirchhoff(10,4)  = (1-2*t-3*t^2)/Jx/Jy*b/2;
    NKirchhoff(11,4)  = (-1+2*t+3*t^2)/Jx/Jy*b/2;
    NKirchhoff(12,4)  = (-1-2*t+3*t^2)/Jx/Jy*b/2;
end
return
```

5.8 Shape Function for a Mindlin Thick Plate Element

The Mindlin (1951) thick plate theory (first-order shear deformation theory [FSDT]) is a particular case of the theory developed by Reissner (1945, 1947) where it employs a linear variation of shear deformation effect in the governing equation. The thick plate theory derived in Chapter 4 discussed the derivations of the FSDT governing equations of a plate. The shape functions of the Mindlin plate element can be obtained from the static homogeneous partial differential equations of the governing equations in Equation (4.48), which can be written as

$$\delta w: -D_{xz}\left(\frac{d^2w}{dx^2}-\frac{d\theta_y}{dx}\right)-D_{yz}\left(\frac{d^2w}{dy^2}-\frac{d\theta_x}{dy}\right)=0 \tag{5.47}$$

$$\delta\theta_x: -D\left(\frac{d^2\theta_x}{dy^2}\right)-D_{xxyy}\left(\frac{d^2\theta_y}{dxdy}\right)-D_{xy}\left(\frac{d^2\theta_x}{dx^2}+\frac{d^2\theta_x}{dxdy}\right)-D_{yz}\left(\frac{dw}{dy}-\theta_x\right)=0 \tag{5.48}$$

$$\delta\theta_y: -D\left(\frac{d^2\theta_y}{dx^2}\right)-D_{xxyy}\left(\frac{d^2\theta_x}{dxdy}\right)-D_{xy}\left(\frac{d^2\theta_y}{dy^2}+\frac{d^2\theta_y}{dxdy}\right)-D_{xz}\left(\frac{dw}{dx}-\theta_y\right)=0 \tag{5.49}$$

where

$$D=\frac{Eh^3}{12(1-\nu^2)}, D_{xxyy}=\frac{\nu Eh^3}{6(1-\nu^2)}, D_{xy}=\frac{Eh^3}{24(1+\nu)}, D_{xz}=D_{yz}$$

$$=\kappa Gh, G=\frac{E}{2(1+\nu)}$$

As shown in Equations (5.47)–(5.49), we have three governing equations corresponding to the variational DOF at a node in the plate element. The solutions of the static governing equations can be approximated by using the second order for rotational displacements θ_x, θ_y and the third order for vertical displacement w, which resembles the Timoshenko beam element (Gan 2018). The shape functions of the Mindlin thick plate can be solved from the above three differential equations by considering two uncoupled variables (ξ and η).

The first step to obtain all the shape functions is by assuming a two-dimensional second-order multivariate complete polynomial for the shape function of the rotational displacements θ_x, θ_y and a two-dimensional third-order multivariate complete polynomial for the shape function of the vertical displacements w in the following form:

$$\begin{aligned}
\theta_x &= a_1+a_2\xi+a_3\eta+a_4\eta\xi+a_5\xi^2+a_6\eta^2+a_7\xi^2\eta+a_8\eta^2\xi+a_9\eta^2\xi^2\\
\theta_y &= b_1+b_2\xi+b_3\eta+b_4\eta\xi+b_5\xi^2+b_6\eta^2+b_7\xi^2\eta+b_8\eta^2\xi+b_9\eta^2\xi^2\\
w &= c_1+c_2\xi+c_3\eta+c_4\eta\xi+c_5\xi^2+c_6\eta^2+c_7\xi^2\eta+c_8\eta^2\xi+c_9\xi^3\\
&\quad +c_{10}\eta^3+c_{11}\eta\xi^3+c_{12}\xi\eta^3+c_{13}\eta^2\xi^2+c_{14}\eta^2\xi^3+c_{15}\eta^3\xi^2+c_{16}\eta^3\xi^3
\end{aligned} \tag{5.50}$$

The solutions can be obtained by representing the coefficients a's and b's in θ_x^0, θ_y^0 in terms of c's in w and further equating both Equations (5.48) and (5.49) after substituted by Equation (5.50) which results in,

$$c_{13} = c_{14} = c_{15} = 0 \tag{5.51a}$$

It can be found that Equation (5.47) leads to a relationship where c_{11} and c_{12} must not vanish in finding the shape function.

By equating all the coefficients of coordinate variables in Equations (5.48) and (5.49), the following relationships can be obtained:

$$\begin{aligned}
a_1 &= \frac{D_{xz}c_3 + 2D_{xy}c_7 + 2(D_{xxyy} + D_{xy})c_8 + 6Dc_{10}}{D_{xz}} \\
a_2 &= c_4 \\
a_3 &= 2c_6 \\
a_4 &= 2c_8 \\
a_5 &= c_7 \\
a_6 &= 3c_{10} \\
a_7 &= a_8 = a_9 = 0
\end{aligned} \tag{5.51b}$$

and

$$\begin{aligned}
b_1 &= \frac{D_{xz}c_2 + 2D_{xy}c_7 + 2(D_{xxyy} + D_{xy})c_8 + 6Dc_9}{D_{xz}} \\
b_2 &= 2c_5 \\
b_3 &= c_4 \\
b_4 &= 2c_7 \\
b_5 &= 3c_9 \\
b_6 &= c_8 \\
b_8 &= b_7 = b_9 = 0
\end{aligned} \tag{5.51c}$$

Further substituting all the coefficients c's into Equation (5.50), the rotational displacements θ_x and θ_y and the vertical displacement w can be obtained in terms of coefficients as c's as follows:

$$w = c_1 + c_2\xi + c_3\eta + c_4\eta\xi + c_5\xi^2 + c_6\eta^2 + c_7\xi^2\eta + c_8\eta^2\xi + c_9\xi^3 + c_{10}\eta^3 + c_{11}\eta\xi^3 + c_{12}\xi\eta^3$$

$$\theta_x = c_3 + c_4\xi + 2c_6\eta + c_7\left(\frac{2(D_{xxyy} + D_{xy})}{D_{xz}} + \xi^2\right) + c_8\left(\frac{2D_{xy}}{D_{xz}} + 2\xi\eta\right) + c_{10}\left(\frac{6D}{D_{xz}} + 3\eta^2\right) \tag{5.52}$$

$$\theta_y = c_2 + c_4\eta + 2c_5\xi + c_7\left(\frac{2D_{xy}}{D_{xz}} + 2\xi\eta\right) + c_8\left(\frac{2(D_{xxyy} + D_{xy})}{D_{xz}} + \eta^2\right) + c_9\left(\frac{6D}{D_{xz}} + 3\xi^2\right)$$

within the plate domains of $-1 \le \xi \le +1$ and $-1 \le \eta \le +1$.

The generalized displacements can be arranged in a single vector form as

$$\mathbf{d} = \begin{bmatrix} w_1 & \theta_{x1} & \theta_{y1} & w_2 & \theta_{x2} & \theta_{y2} & w_3 & \theta_{x3} & \theta_{y3} & w_4 & \theta_{x4} & \theta_{y4} \end{bmatrix}^T \tag{5.53}$$

The 12 unknown coefficients in Equation (5.52) are arranged into a vector form as

$$\mathbf{c} = \begin{bmatrix} c_1 & c_2 & c_3 & c_4 & c_5 & c_6 & c_7 & c_8 & c_9 & c_{10} & c_{11} & c_{12} \end{bmatrix}^T \tag{5.54}$$

By substituting the local coordinates (ξ, η) of the four nodes of the plate into Equation (5.52), we can obtain the relationship between the generalized displacement (DOF) and the unknown coefficients in a matrix form as

$$\{\mathbf{d}\}_{12\times1} = [\Delta]_{12\times12}\{\mathbf{c}\}_{12\times1} \tag{5.55}$$

After substituting the local coordinates of the nodes $(\xi_1 = -1, \eta_1 = -1)$; $(\xi_2 = +1, \eta_2 = -1)$, $(\xi_3 = +1, \eta_3 = +1)$, and $(\xi_4 = -1, \eta_4 = +1)$ into the associated DOFs in Equation (5.53), we can obtain the $[\Delta]$ matrix as follows:

$$[\Delta]=\begin{bmatrix}1&-1&-1&1&1&1&-1&-1&-1&-1&1&1\\0&0&1&-1&0&-2&(1+\chi+\mu)&2+\mu&0&3+3\lambda&0&0\\0&1&0&-1&-2&0&2+\mu&(1+\chi+\mu)&3+3\lambda&0&0&0\\1&1&-1&-1&1&1&-1&1&1&-1&-1&-1\\0&0&1&1&0&-2&(1+\chi+\mu)&-2+\mu&0&3+3\lambda&0&0\\0&1&0&-1&2&0&-2+\mu&(1+\chi+\mu)&3+3\lambda&0&0&0\\1&1&1&1&1&1&1&1&1&1&1&1\\0&0&1&1&0&2&(1+\chi+\mu)&2+\mu&0&3+3\lambda&0&0\\0&1&0&1&2&0&2+\mu&(1+\chi+\mu)&3+3\lambda&0&0&0\\1&-1&1&-1&1&1&1&-1&-1&1&-1&-1\\0&0&1&-1&0&2&(1+\chi+\mu)&-2+\mu&0&3+3\lambda&0&0\\0&1&0&1&-2&0&-2+\mu&(1+\chi+\mu)&3+3\lambda&0&0&0\end{bmatrix} \tag{5.56}$$

where

$$\lambda=\frac{2D}{D_{xz}},\chi=\frac{2D_{xxyy}}{D_{xz}},\mu=\frac{2D_{xy}}{D_{xz}}$$

The 12-unknown coefficient vector $\{\mathbf{c}\}$ can be obtained by inverting the $[\Delta]$ matrix and multiplying with the generalized displacement (DOF) vector $\{\mathbf{d}\}$. However, it is revealed that the $[\Delta]$ matrix is not invertible, because it has one rank deficiency; hence, there is a singularity problem inside the matrix. Since the relationship between the general displacement vector and the coefficient vector is not mathematically based on the one-by-one correlation principle, we can still solve the matrix by using the pseudo-inverse technique developed by Moore and Penrose (Ben-Israel and Greville, 2003) to obtain the inverse matrix which can be used to compute the coefficient vector:

$$\{\mathbf{c}\}_{12\times1}=[\Delta]^{-1}_{12\times12}\{\mathbf{d}\}_{12\times1} \tag{5.57}$$

The vertical $w(\xi,\eta)$ and rotational displacements θ_x and θ_y in Equation (5.52) can be written in matrix form as follows:

$$\begin{Bmatrix}w\\\theta_x\\\theta_y\end{Bmatrix}=\begin{bmatrix}\Phi_w\\\Phi_{\theta_x}\\\Phi_{\theta_y}\end{bmatrix}_{3\times12}\{\mathbf{c}\}_{12\times1} \tag{5.58}$$

where the vector $\begin{bmatrix}\Phi_w & \Phi_{\theta_x} & \Phi_{\theta_y}\end{bmatrix}^T$ contains the local coordinates of an arbitrary point (ξ,η) on the plate.

Substituting the coefficient vector $\{\mathbf{c}\}$ from Equation (5.57) into Equation (5.58), we can get the following equation:

$$\left\{\begin{matrix} w \\ \theta_x \\ \theta_y \end{matrix}\right\} = \begin{bmatrix} \mathbf{\Phi}_w \\ \mathbf{\Phi}_{\theta_x} \\ \mathbf{\Phi}_{\theta_y} \end{bmatrix}_{3\times12} [\Delta]^{-1}_{12\times12}\{\mathbf{d}\}_{12\times1} = \left\{\begin{matrix} \mathbf{N}_w \\ \mathbf{N}_{\theta_x} \\ \mathbf{N}_{\theta_y} \end{matrix}\right\}_{3\times12} \{\mathbf{d}\}_{12\times1} \tag{5.59}$$

The vertical displacement w of any arbitrary point on the plate can be analyzed as follows:

$$w = [\mathbf{\Phi}_w]_{1\times12}[\Delta]^{-1}_{12\times12}\{\mathbf{d}\}_{12\times1} = [\mathbf{N}_w]_{1\times12}\{\mathbf{d}\}_{12\times1} \tag{5.60}$$

The elements of the vector $[\mathbf{N}_w]$ are the shape functions associated with the generalized transverse displacements, which are given by

$$[\mathbf{N}_w] = \begin{bmatrix} N_{ww1} & N_{w\theta_x1} & N_{w\theta_y1} & N_{ww2} & N_{w\theta_x2} & N_{w\theta_y2} & \cdots \\ \cdots & N_{ww3} & N_{w\theta_x3} & N_{w\theta_y3} & N_{ww4} & N_{w\theta_x4} & N_{w\theta_y4} \end{bmatrix} \tag{5.61}$$

where

$$N_{ww1} = \frac{2(-1+\eta)(-2+\eta+\eta^2-3\lambda)+\xi^3(2+\eta(2+3\lambda))+\xi(-6(1+\lambda)+\eta^3(2+3\lambda))}{8(2+3\lambda)}$$

$$N_{ww2} = \frac{2(-1+\eta)(-2+\eta+\eta^2-3\lambda)-\xi^3(2+\eta(2+3\lambda))+\xi(6(1+\lambda)-\eta^3(2+3\lambda))}{8(2+3\lambda)}$$

$$N_{ww3} = \frac{4-2\eta^3+6\lambda+6\eta(1+\lambda)+\xi^3(-2+\eta(2+3\lambda))+\xi(6(1+\lambda)+\eta^3(2+3\lambda))}{8(2+3\lambda)}$$

$$N_{ww4} = \frac{4-2\eta^3+6\lambda+6\eta(1+\lambda)+\xi^3(2-\eta(2+3\lambda))-\xi(6(1+\lambda)+\eta^3(2+3\lambda))}{8(2+3\lambda)}$$

$$N_{w\theta_x1} = \frac{1}{32+48\lambda}\left\{\begin{bmatrix} (-2(-1+\eta^2)(2+3\lambda+\eta(-2+\mu))+\xi^3(\eta(2+3\lambda)-2(\mu+\chi)) \\ +\xi(-2\eta(2+3\lambda)+\eta^3(2+3\lambda)+\eta^2(4+6\lambda) \\ +2(-2-3\lambda+\mu+\chi))) \end{bmatrix}\right\}$$

$$N_{w\theta_x2} = \frac{1}{32+48\lambda}\left\{\begin{bmatrix} (2(-1+\eta^2)(-2-3\lambda+\eta(2+\mu))+\xi^3(-\eta(2+3\lambda)+2(\mu+\chi)) \\ -\xi(-2\eta(2+3\lambda)+\eta^3(2+3\lambda)+\eta^2(4+6\lambda) \\ +2(-2-3\lambda+\mu+\chi))) \end{bmatrix}\right\}$$

$$N_{w\theta_x 3} = \frac{1}{32+48\lambda} \times \begin{Bmatrix} (-2(-1+\eta^2)(-2-3\lambda+\eta(-2+\mu)) - \xi^3(\eta(2+3\lambda)+2(\mu+\chi)) \\ +\xi(-\eta^3(2+3\lambda)+\eta(4+6\lambda)+\eta^2(4+6\lambda)+2(-2-3\lambda+\mu+\chi))) \end{Bmatrix}$$

$$N_{w\theta_x 4} = \frac{1}{32+48\lambda} \times \begin{Bmatrix} (2(-1+\eta^2)(2+3\lambda+\eta(2+\mu)) + \xi(6\lambda-2\eta(2+3\lambda)-2\eta^2(2+3\lambda) \\ +\eta^3(2+3\lambda)-2(-2+\mu+\chi)) + \xi^3(\eta(2+3\lambda)+2(\mu+\chi))) \end{Bmatrix}$$

$$N_{w\theta_y 1} = \frac{1}{32+48\lambda} \times \begin{Bmatrix} (2\xi^2(-1+\eta)(2+3\lambda) + \xi(-2\eta(2+3\lambda)+\eta^3(2+3\lambda)+2(-2+\mu)) \\ +\xi^3(4+\eta(2+3\lambda)-2\mu) - 2(-1+\eta)(2+3\lambda+\eta(\mu+\chi)+\eta^2(\mu+\chi))) \end{Bmatrix}$$

$$N_{w\theta_x 2} = \frac{1}{32+48\lambda} \times \begin{Bmatrix} (2(-1+\eta^2)(-2-3\lambda+\eta(2+\mu)) + \xi^3(-\eta(2+3\lambda)+2(\mu+\chi)) \\ -\xi(-2\eta(2+3\lambda)+\eta^3(2+3\lambda)+\eta^2(4+6\lambda)+2(-2-3\lambda+\mu+\chi))) \end{Bmatrix}$$

$$N_{w\theta_x 3} = \frac{1}{32+48\lambda} \times \begin{Bmatrix} (-2(-1+\eta^2)(-2-3\lambda+\eta(-2+\mu)) - \xi^3(\eta(2+3\lambda)+2(\mu+\chi)) \\ +\xi(-\eta^3(2+3\lambda)+\eta(4+6\lambda)+\eta^2(4+6\lambda)+2(-2-3\lambda+\mu+\chi))) \end{Bmatrix}$$

$$N_{w\theta_x 3} = \frac{1}{32+48\lambda} \times \begin{Bmatrix} (-2(-1+\eta^2)(-2-3\lambda+\eta(-2+\mu)) - \xi^3(\eta(2+3\lambda)+2(\mu+\chi)) \\ +\xi(-\eta^3(2+3\lambda)+\eta(4+6\lambda)+\eta^2(4+6\lambda)+2(-2-3\lambda+\mu+\chi))) \end{Bmatrix}$$

The shape functions of the transverse displacement N_{ww1}, N_{ww2}, N_{ww3}, and N_{ww4} of the Mindlin thick plate are shown in Figure 5.14.

The shape functions of the transverse displacement $N_{w\theta_x 1}$, $N_{w\theta_x 2}$, $N_{w\theta_x 3}$, and $N_{w\theta_x 4}$ of the Mindlin thick plate are shown in Figure 5.15.

The shape functions of the transverse displacement $N_{w\theta_y 1}$, $N_{w\theta_y 2}$, $N_{w\theta_y 3}$, and $N_{w\theta_y 4}$ of the Mindlin thick plate are shown in Figure 5.16.

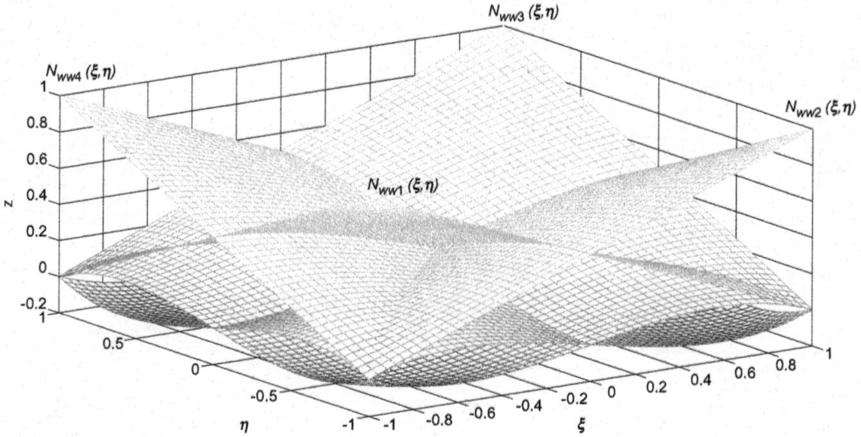

FIGURE 5.14
Basis functions of the transverse displacement N_{ww1}, N_{ww2}, N_{ww3}, and N_{ww4}.

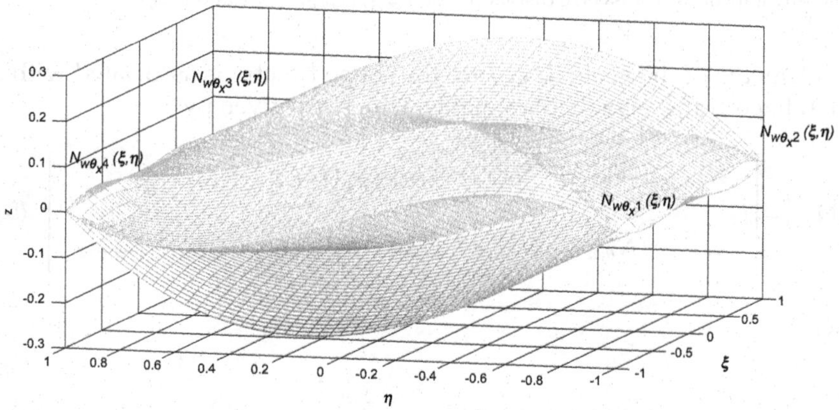

FIGURE 5.15
Basis functions of the transverse displacement $N_{w\theta_x 1}$, $N_{w\theta_x 2}$, $N_{w\theta_x 3}$, and $N_{w\theta_y 4}$.

Next, the rotation about the x-axis $\theta_x(\xi,\eta)$ in Equation (5.52) can be written in matrix form as follows:

$$\theta_x = \left[\Phi_{\theta_x}\right]_{1\times12}\{\mathbf{c}\}_{12\times1} \tag{5.62}$$

where the vector $\left[\Phi_{\theta_x}\right]$ contains the local coordinates of an arbitrary point (ξ,η) on the plate.

Substituting the coefficient vector $\{\mathbf{c}\}$ from Equation (5.59) into Equation (5.62), we can get the following equation:

$$\theta_x = \left[\Phi_{\theta_x}\right]_{1\times12}\left[\Delta\right]^{-1}_{12\times12}\{\mathbf{d}\}_{12\times1} = \left[\mathbf{N}_{\theta_x}\right]_{1\times12}\{\mathbf{d}\}_{12\times1} \tag{5.63}$$

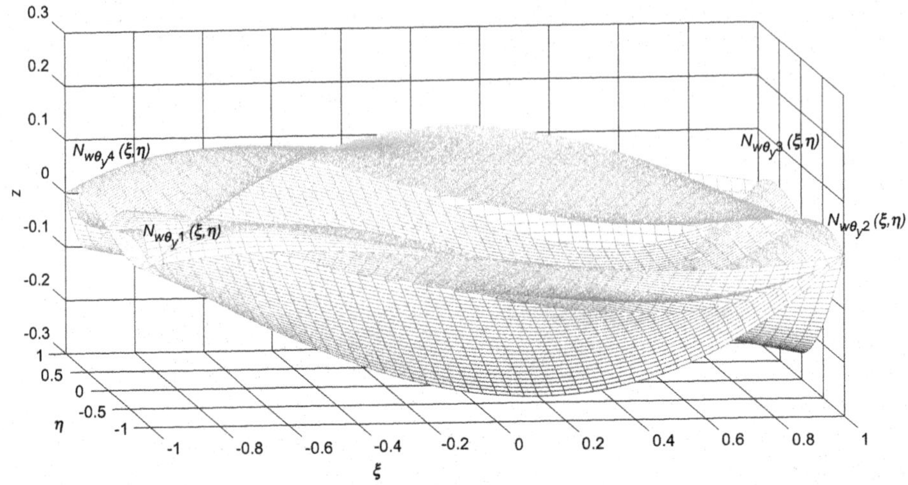

FIGURE 5.16
Basis functions of the transverse displacement $N_{w\theta_y1}$, $N_{w\theta_y2}$, $N_{w\theta_y3}$, and $N_{w\theta_y4}$.

The elements of the vector $\left[\mathbf{N}_{\theta_x}\right]$ are the shape functions associated with the generalized transverse displacements, which are given by

$$\left[\mathbf{N}_{\theta_x}\right]=\left[\begin{array}{cccccc c} N_{\theta_x w1} & N_{\theta_x\theta_x 1} & N_{\theta_x\theta_y 1} & N_{\theta_x w2} & N_{\theta_x\theta_x 2} & N_{\theta_x\theta_y 2} & \cdots \\ \cdots & N_{\theta_x w3} & N_{\theta_x\theta_x 3} & N_{\theta_x\theta_y 3} & N_{\theta_x w4} & N_{\theta_x\theta_x 4} & N_{\theta_x\theta_y 4} \end{array}\right] \tag{5.64}$$

where

$$N_{\theta_x w1}=\frac{3(-1+\eta^2)}{8+12\lambda}$$

$$N_{\theta_x w2}=\frac{3(-1+\eta^2)}{8+12\lambda}$$

$$N_{\theta_x w3}=\frac{3-3\eta^2}{8+12\lambda}$$

$$N_{\theta_x w4}=\frac{3-3\eta^2}{8+12\lambda}$$

$$N_{\theta_x\theta_x 1}=\frac{\xi(-1+2\eta)(2+3\lambda)-(-1+\eta)(-2+6\lambda+3\eta(-2+\mu)+3\mu)}{8(2+3\lambda)}$$

$$N_{\theta_x\theta_x 2}=\frac{-\xi(-1+2\eta)(2+3\lambda)+(-1+\eta)(2-6\lambda+3\mu+3\eta(2+\mu))}{8(2+3\lambda)}$$

$$N_{\theta_x\theta_x 3} = \frac{\xi(1+2\eta)(2+3\lambda)-(1+\eta)(2-6\lambda+3\eta(-2+\mu)-3\mu)}{8(2+3\lambda)}$$

$$N_{\theta_x\theta_x 4} = \frac{-\xi(1+2\eta)(2+3\lambda)+(1+\eta)(-2+6\lambda-3\mu+3\eta(2+\mu))}{8(2+3\lambda)}$$

$$N_{\theta_x\theta_y 1} = -\frac{2+3\lambda+\xi(2+3\lambda)-\xi^2(2+3\lambda)-3\mu+3\eta^2\mu-3\chi+3\eta^2\chi}{8(2+3\lambda)}$$

$$N_{\theta_x\theta_y 2} = -\frac{-2-3\lambda+\xi(2+3\lambda)+\xi^2(2+3\lambda)+3\mu-3\eta^2\mu+3\chi-3\eta^2\chi}{8(2+3\lambda)}$$

$$N_{\theta_x\theta_y 3} = \frac{-2-3\lambda+\xi(2+3\lambda)+\xi^2(2+3\lambda)+3\mu-3\eta^2\mu+3\chi-3\eta^2\chi}{8(2+3\lambda)}$$

$$N_{\theta_x\theta_y 4} = \frac{2+3\lambda+\xi(2+3\lambda)-\xi^2(2+3\lambda)-3\mu+3\eta^2\mu-3\chi+3\eta^2\chi}{8(2+3\lambda)}$$

The shape functions of the transverse displacement $N_{\theta_x w1}$, $N_{\theta_x w2}$, $N_{\theta_x w3}$, and $N_{\theta_x w4}$ of the Mindlin plate are shown in Figure 5.17.

The shape functions of the transverse displacement $N_{\theta_x\theta_x 1}$, $N_{\theta_x\theta_x 2}$, $N_{\theta_x\theta_x 3}$, and $N_{\theta_x\theta_x 4}$ of the Mindlin plate are shown in Figure 5.18.

The shape functions of the transverse displacement $N_{\theta_x\theta_y 1}$, $N_{\theta_x\theta_y 2}$, $N_{\theta_x\theta_y 3}$, and $N_{\theta_x\theta_y 4}$ of the Mindlin plate are shown in Figure 5.19.

Finally, the rotation about the y-axis $\theta_y(\xi,\eta)$ in Equation (5.52) can be written in matrix form as follows:

$$\theta_y = \left[\Phi_{\theta_y}\right]_{1\times 12}\{\mathbf{c}\}_{12\times 1} \tag{5.65}$$

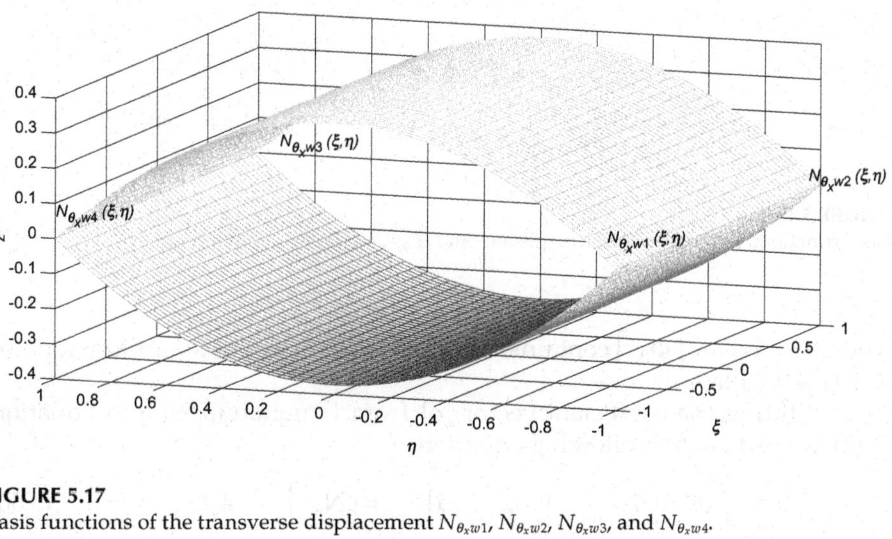

FIGURE 5.17
Basis functions of the transverse displacement $N_{\theta_x w1}$, $N_{\theta_x w2}$, $N_{\theta_x w3}$, and $N_{\theta_x w4}$.

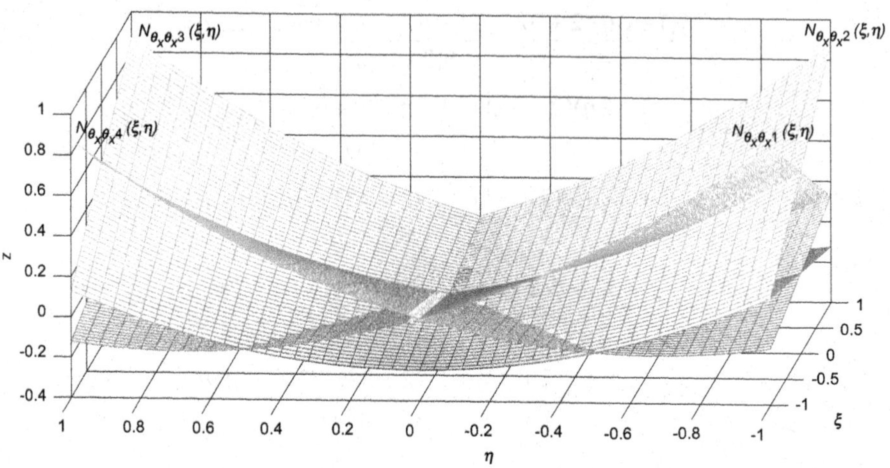

FIGURE 5.18
Basis functions of the transverse displacement $N_{\theta_x\theta_x1}$, $N_{\theta_x\theta_x2}$, $N_{\theta_x\theta_x3}$, and $N_{\theta_x\theta_x4}$.

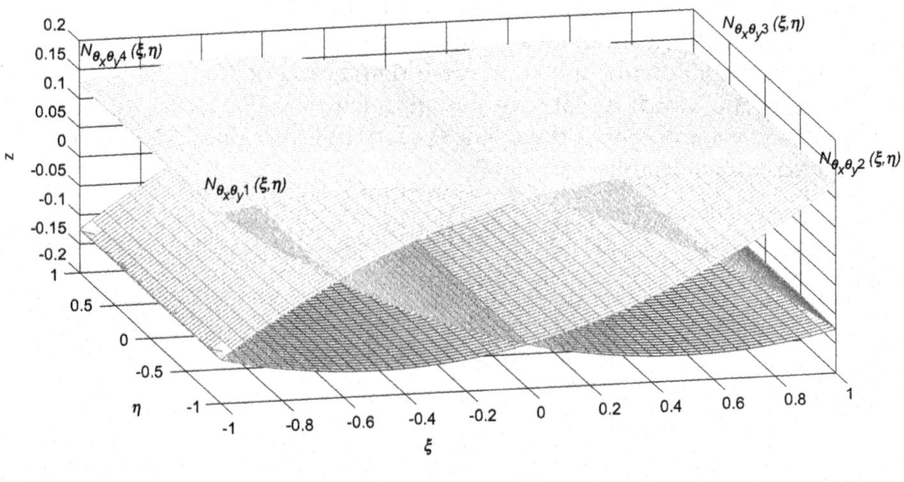

FIGURE 5.19
Basis functions of the transverse displacement $N_{\theta_x\theta_y1}$, $N_{\theta_x\theta_y2}$, $N_{\theta_x\theta_y3}$, and $N_{\theta_x\theta_y4}$.

where the vector $\left[\Phi_{\theta_y}\right]$ contains the local coordinates of an arbitrary point (ξ,η) on the plate.

Substituting the coefficient vector $\{\mathbf{c}\}$ from Equation (5.59) into Equation (5.62), we can get the following equation:

$$\theta_y = \left[\Phi_{\theta_y}\right]_{1\times12}[\Delta]^{-1}_{12\times12}\{\mathbf{d}\}_{12\times1} = \left[\mathbf{N}_{\theta_y}\right]_{1\times12}\{\mathbf{d}\}_{12\times1} \tag{5.66}$$

The elements of the vector $\left[\mathbf{N}_{\theta_y}\right]$ are the shape functions associated with the generalized transverse displacements, which are given by

$$\left[\mathbf{N}_{\theta_y}\right]=\left[\begin{array}{cccccccc} N_{\theta_y w1} & N_{\theta_y \theta_x 1} & N_{\theta_y \theta_y 1} & N_{\theta_y w2} & N_{\theta_y \theta_x 2} & N_{\theta_y \theta_y 2} & \cdots & \\ \cdots & N_{\theta_y w3} & N_{\theta_y \theta_x 3} & N_{\theta_y \theta_y 3} & N_{\theta_y w4} & N_{\theta_y \theta_x 4} & N_{\theta_y \theta_y 4} & \end{array}\right] \tag{5.67}$$

where

$$N_{\theta_y w1} = \frac{3(-1+\xi^2)}{8+12\lambda}$$

$$N_{\theta_y w2} = \frac{3-3\xi^2}{8+12\lambda}$$

$$N_{\theta_y w3} = \frac{3-3\xi^2}{8+12\lambda}$$

$$N_{\theta_y w4} = \frac{3(-1+\xi^2)}{8+12\lambda}$$

$$N_{\theta_y \theta_x 1} = -\frac{2+3\lambda+\eta(2+3\lambda)-\eta^2(2+3\lambda)-3\mu+3\xi^2\mu-3\chi+3\xi^2\chi}{8(2+3\lambda)}$$

$$N_{\theta_y \theta_x 2} = \frac{2+3\lambda+\eta(2+3\lambda)-\eta^2(2+3\lambda)-3\mu+3\xi^2\mu-3\chi+3\xi^2\chi}{8(2+3\lambda)}$$

$$N_{\theta_y \theta_x 3} = \frac{-2-3\lambda+\eta(2+3\lambda)+\eta^2(2+3\lambda)+3\mu-3\xi^2\mu+3\chi-3\xi^2\chi}{8(2+3\lambda)}$$

$$N_{\theta_y \theta_x 4} = -\frac{-2-3\lambda+\eta(2+3\lambda)+\eta^2(2+3\lambda)+3\mu-3\xi^2\mu+3\chi-3\xi^2\chi}{8(2+3\lambda)}$$

$$N_{\theta_y \theta_y 1} = \frac{-2+6\lambda+2\xi(-1+\eta)(2+3\lambda)-\eta(2+3\lambda)-3\xi^2(-2+\mu)+3\mu}{8(2+3\lambda)}$$

$$N_{\theta_y \theta_y 2} = -\frac{2-6\lambda+2\xi(-1+\eta)(2+3\lambda)+\eta(2+3\lambda)+3\mu-3\xi^2(2+\mu)}{8(2+3\lambda)}$$

$$N_{\theta_y \theta_y 3} = \frac{-2+6\lambda+\eta(2+3\lambda)+2\xi(1+\eta)(2+3\lambda)-3\xi^2(-2+\mu)+3\mu}{8(2+3\lambda)}$$

$$N_{\theta_y \theta_y 4} = \frac{-2+6\lambda+\eta(2+3\lambda)-2\xi(1+\eta)(2+3\lambda)-3\mu+3\xi^2(2+\mu)}{8(2+3\lambda)}$$

The shape functions of the transverse displacement $N_{\theta_y w1}$, $N_{\theta_y w2}$, $N_{\theta_y w3}$, and $N_{\theta_y w4}$ of the Mindlin plate are shown in Figure 5.20.

The shape functions of the transverse displacement $N_{\theta_y \theta_x 1}$, $N_{\theta_y \theta_x 2}$, $N_{\theta_y \theta_x 3}$, and $N_{\theta_y \theta_x 4}$ of the Mindlin plate are shown in Figure 5.21.

The shape functions of the transverse displacement $N_{\theta_y \theta_y 1}$, $N_{\theta_y \theta_y 2}$, $N_{\theta_y \theta_y 3}$, and $N_{\theta_y \theta_y 4}$ of the Mindlin plate are shown in Figure 5.22.

The generalized displacement of any arbitrary point at any location in a plate can then be interpolated from the DOF vector $\{\mathbf{d}\}$ by using the shape functions as follows:

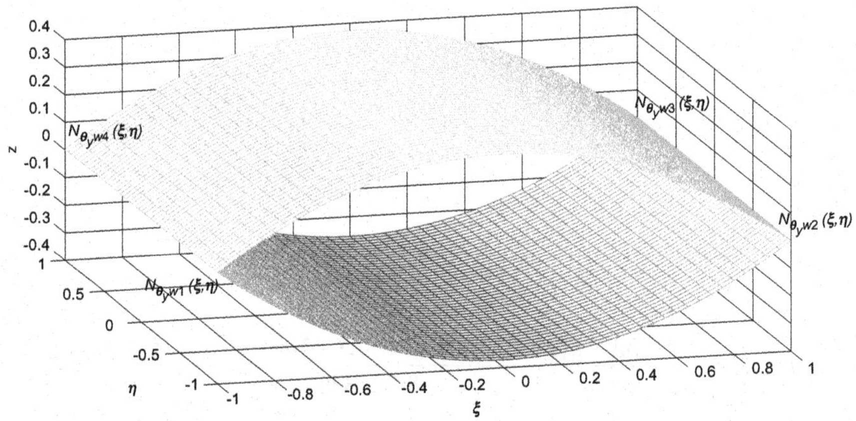

FIGURE 5.20
Basis functions of the transverse displacement $N_{\theta_y w1}$, $N_{\theta_y w2}$, $N_{\theta_y w3}$, and $N_{\theta_y w4}$.

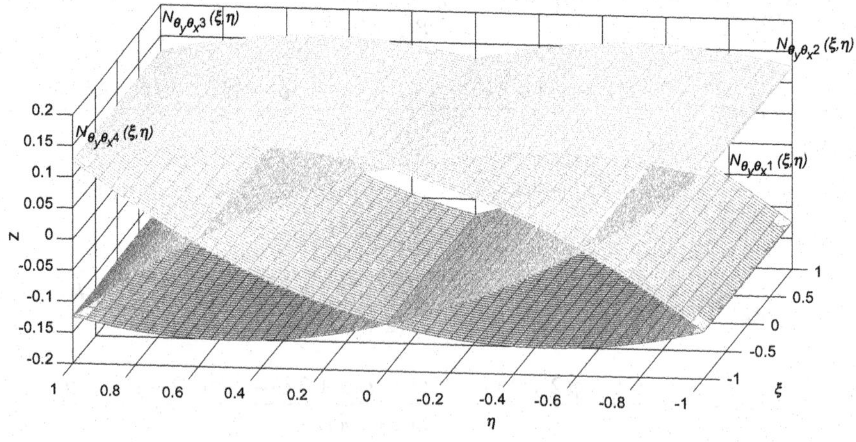

FIGURE 5.21
Basis functions of the transverse displacement $N_{\theta_y \theta_x 1}$, $N_{\theta_y \theta_x 2}$, $N_{\theta_y \theta_x 3}$, and $N_{\theta_y \theta_x 4}$.

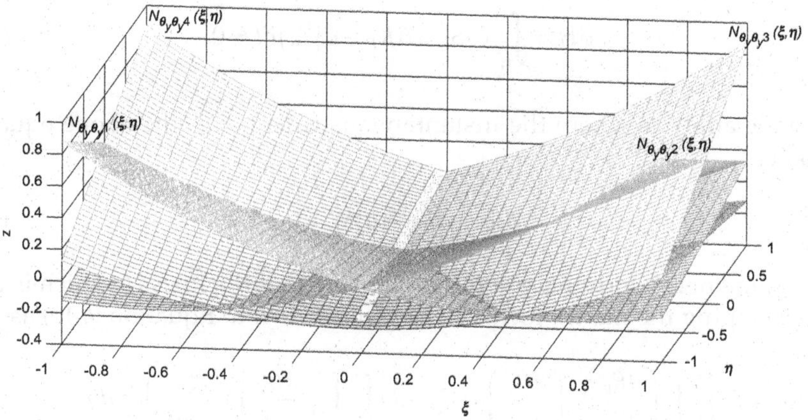

FIGURE 5.22
Basis functions of the transverse displacement $N_{\theta_y\theta_y1}$, $N_{\theta_y\theta_y2}$, $N_{\theta_y\theta_y3}$, and $N_{\theta_y\theta_y4}$.

$$\begin{Bmatrix} w \\ \theta_x \\ \theta_y \end{Bmatrix} = \begin{bmatrix} \mathbf{N}_w \\ \mathbf{N}_{\theta_x} \\ \mathbf{N}_{\theta_y} \end{bmatrix} \{\mathbf{d}\} \tag{5.68}$$

The elements of the vector $[\mathbf{N}_w]$, $[\mathbf{N}_{\theta_x}]$, and $[\mathbf{N}_{\theta_y}]$ are the shape functions for a four-node Mindlin thick plate element, which are derived from the governing equations of Mindlin thick plate as shown in Equation (5.52). Thus, the shape functions are accurate and consistent. Hence, the shape functions developed in this book can be used to formulate a finite element model of the Mindlin thick plate by using only one element, because it already satisfies the static governing equations during the development.

It is worth noting that the shape functions derived above are not yet written in other books on the finite element of Mindlin thick plate theory. Most of the shape functions shown in other books are not consistent with the governing equations, because of no bending and shear rigidities inside the formulation in the development; usually only Lagrange polynomial or isoparametric equations contain only coordinate information on the plate element.

5.9 Governing Equation in Matrix Form

The governing equation expressed in matrix form using Hamilton's principle can be obtained from Equation (4.6), duplicated herewith for convenience purpose, as

$$\delta H = \int_{t_1}^{t_2} (\delta S_E - \delta K_E - \delta W_E) dt = 0 \tag{5.69}$$

For static analysis where the instantaneous time $t_1 \approx t_2$, the above equation becomes

$$\delta S_E - \delta K_E - \delta W_E = 0 \tag{5.70}$$

The strain energy formula of a plate with constant thickness h which is used for developing the matrix form can be obtained from Equation (4.10) as

$$\begin{aligned}
\delta S_E = {} & D\int_0^b\int_0^a \left(\frac{d\theta_y}{dx}\right)\delta\left(\frac{d\theta_y}{dx}\right)dx\,dy + D\int_0^b\int_0^a \left(\frac{d\theta_x}{dy}\right)\delta\left(\frac{d\theta_x}{dy}\right)dx\,dy \\
& + D_{xxyy}\int_0^b\int_0^a \left(\frac{d\theta_y}{dx}\right)\delta\left(\frac{d\theta_x}{dy}\right)dx\,dy + D_{xxyy}\int_0^b\int_0^a \left(\frac{d\theta_x}{dy}\right)\delta\left(\frac{d\theta_y}{dx}\right)dx\,dy \\
& + D_{xy}\int_0^b\int_0^a \left(\frac{d\theta_x}{dx}+\frac{d\theta_y}{dy}\right)\delta\left(\frac{d\theta_x}{dx}+\frac{d\theta_y}{dy}\right)dx\,dy \\
& + D_{xz}\int_0^b\int_0^a \left(\frac{dw}{dx}-\theta_y\right)\delta\left(\frac{dw}{dx}-\theta_y\right)dx\,dy \\
& + D_{yz}\int_0^b\int_0^a \left(\frac{dw}{dy}-\theta_x\right)\delta\left(\frac{dw}{dy}-\theta_x\right)dx\,dy
\end{aligned} \tag{5.71}$$

By substituting Equation (5.68), the energy variation of the strain energy can now be expressed in matrix form as follows:

$$\delta S_E = \delta\{\mathbf{d}\}^T \left(\begin{aligned}
& \int_0^b\int_0^a \left(\mathbf{N}_{\theta_{y,x}}^T D \mathbf{N}_{\theta_{y,x}}\right)dx\,dy + \int_0^b\int_0^a \left(\mathbf{N}_{\theta_{x,y}}^T D \mathbf{N}_{\theta_{x,y}}\right)dx\,dy \\
& + \int_0^b\int_0^a \left(\mathbf{N}_{\theta_{y,x}}^T D_{xxyy} \mathbf{N}_{\theta_{x,y}}\right)dx\,dy + \int_0^b\int_0^a \left(\mathbf{N}_{\theta_{x,y}}^T D_{xxyy} \mathbf{N}_{\theta_{y,x}}\right)dx\,dy \\
& + \int_0^b\int_0^a \left(\left[\mathbf{N}_{\theta_{x,x}} + \mathbf{N}_{\theta_{y,y}}\right]^T D_{xy} \left[\mathbf{N}_{\theta_{x,x}} + \mathbf{N}_{\theta_{y,y}}\right]\right)dx\,dy \\
& + \int_0^b\int_0^a \left(\left[\mathbf{N}_{w,x} - \mathbf{N}_{\theta_y}\right]^T D_{xz} \left[\mathbf{N}_{w,x} - \mathbf{N}_{\theta_y}\right]\right)dx\,dy \\
& + \int_0^b\int_0^a \left(\left[\mathbf{N}_{w,y} - \mathbf{N}_{\theta_x}\right]^T D_{yz} \left[\mathbf{N}_{w,y} - \mathbf{N}_{\theta_x}\right]\right)dx\,dy
\end{aligned}\right)\{\mathbf{d}\} \tag{5.72}$$

Next, the kinetic energy formula of a plate with constant thickness h which is used for developing the matrix form can be obtained from Equation (4.26):

$$\delta K_E = \rho I \int_0^b \int_0^a \delta\left(\frac{d\theta_x}{dt}\right)\left(\frac{d\theta_x}{dt}\right) dx\,dy + \rho I \int_0^b \int_0^a \delta\left(\frac{d\theta_y}{dt}\right)\left(\frac{d\theta_y}{dt}\right) dx\,dy$$

$$+ \rho I_{xy} \int_0^b \int_0^a \delta\left(\frac{dw}{dt}\right)\left(\frac{dw}{dt}\right) dx\,dy \tag{5.73}$$

Because the shape functions are not the function of time, only the generalized displacement vector $\{\mathbf{d}\}$ will be derived with respect to time. The energy variation of kinetic energy can now be written in matrix form as follow,

$$\delta K_E = \delta\{\ddot{\mathbf{d}}\}^T \begin{pmatrix} \int_0^b \int_0^a \left(\mathbf{N}_{\theta x}^T \rho I \mathbf{N}_{\theta x}\right) dx\,dy + \int_0^b \int_0^a \left(\mathbf{N}_{\theta y}^T \rho I \mathbf{N}_{\theta y}\right) dx\,dy \\ + \int_0^b \int_0^a \left(\mathbf{N}_w^T \rho I_{xy} \mathbf{N}_w\right) dx\,dy \end{pmatrix} \{\ddot{\mathbf{d}}\} \tag{5.74}$$

The energy done by the external works formulas of a plate with constant thickness h can be obtained from Equation (4.36) as,

$$\delta W_E = \int_0^b \int_0^a \delta\left(\frac{dw}{dx}\right) N_x \left(\frac{dw}{dx}\right) dx\,dy + \int_0^b \int_0^a \delta\left(\frac{dw}{dy}\right) N_y \left(\frac{dw}{dy}\right) dx\,dy$$

$$+ \int_0^b \int_0^a \delta\left(\frac{dw}{dx}\right) N_{xy} \left(\frac{dw}{dy}\right) dx\,dy + \int_0^b \int_0^a \delta\left(\frac{dw}{dy}\right) N_{yx} \left(\frac{dw}{dx}\right) dx\,dy$$

$$+ \int_0^b \int_0^a q_z\, \delta w\, dx\,dy \tag{5.75}$$

By substituting Equation (5.68), the energy variation of the external work can then be expressed in matrix form as follows:

$$\delta W_E = \delta\{\mathbf{d}\}^T \begin{pmatrix} \int_0^b \int_0^a \left(\mathbf{N}_{w,x}^T \left(N_x + N_{xy}\right) \mathbf{N}_{w,x}\right) dx\,dy \\ + \int_0^b \int_0^a \left(\mathbf{N}_{w,y}^T \left(N_x + N_{xy}\right) \mathbf{N}_{w,y}\right) dx\,dy + \int_0^b \int_0^a \left(\mathbf{N}_w^T q_z\right) dx\,dy \end{pmatrix} \{\mathbf{d}\} \tag{5.76}$$

Finally, the governing equation of a plate element in the FEM can be developed conveniently from the theory of minimum potential energy principle as follows:

$$\Pi = \frac{1}{2}\{\mathbf{d}\}^T [\mathbf{K}]\{\mathbf{d}\} + \frac{1}{2}\{\mathbf{d}\}^T [\mathbf{F}]\{\mathbf{d}\} + \frac{1}{2}\{\ddot{\mathbf{d}}\}^T [\mathbf{M}]\{\ddot{\mathbf{d}}\} - \{\mathbf{d}\}^T \{\mathbf{f}\}$$
$$\frac{d\Pi}{d\{\mathbf{d}\}^T} = [\mathbf{K}]\{\mathbf{d}\} + [\mathbf{F}]\{\mathbf{d}\} + [\mathbf{M}]\{\ddot{\mathbf{d}}\} - \{\mathbf{f}\} = \{\mathbf{0}\} \tag{5.77}$$

The stiffness matrix $[\mathbf{K}]$ can be obtained from Equation (5.72) as follows:

$$[\mathbf{K}] = \int_0^b \int_0^a \left\{ \begin{matrix} \mathbf{N}_{\theta_{y,x}}^T D \mathbf{N}_{\theta_{y,x}} + \mathbf{N}_{\theta_{x,y}}^T D \mathbf{N}_{\theta_{x,y}} + \mathbf{N}_{\theta_{y,x}}^T D_{xxyy} \mathbf{N}_{\theta_{x,y}} + \mathbf{N}_{\theta_{x,y}}^T D_{xxyy} \\ \mathbf{N}_{\theta_{y,x}} + \left(\mathbf{N}_{\theta_{x,x}} + \mathbf{N}_{\theta_{y,y}}\right)^T D_{xy} \left(\mathbf{N}_{\theta_{x,x}} + \mathbf{N}_{\theta_{y,y}}\right) + \left(\mathbf{N}_{w,y} - \mathbf{N}_{\theta_x}\right)^T \\ D_{xz} \left(\mathbf{N}w_{,y} - \mathbf{N}_{\theta_x}\right) + \left(\mathbf{N}_{w,x} - \mathbf{N}_{\theta_y}\right)^T D_{xz} \left(\mathbf{N}_{w,x} - \mathbf{N}_{\theta_y}\right) \end{matrix} \right\} dx\,dy \tag{5.78}$$

where the elements of the symmetric stiffness matrix $K_{ij} = K_{ji}$ are too complex, thus being abbreviated.

The mass matrix $[\mathbf{M}]$ can be obtained from Equation (5.74) as follows:

$$[\mathbf{M}] = \int_0^b \int_0^a \left\{ \mathbf{N}_w^T \rho I_{xy} \mathbf{N}_w + \mathbf{N}_{\theta_x}^T \rho I \mathbf{N}_{\theta_x} + \mathbf{N}_{\theta_y}^T \rho I \mathbf{N}_{\theta_y} \right\} dx\,dy \tag{5.79}$$

The axial loading matrix $[\mathbf{F}]$ and external loading vector $\{\mathbf{f}\}$ can be obtained from Equation (5.76) as follows:

$$[\mathbf{F}] = \int_0^b \int_0^a \left(\mathbf{N}_{w,x}^T \left(N_x + N_{xy}\right) \mathbf{N}_{w,x} \right) dx\,dy + \int_0^b \int_0^a \left(\mathbf{N}_{w,y}^T \left(N_x + N_{xy}\right) \mathbf{N}_{w,y} \right) dx\,dy$$
$$\{\mathbf{f}\} = \int_0^b \int_0^a \left(\mathbf{N}_w^T q_z \right) dx\,dy \tag{5.80}$$

5.10 Gaussian Quadrature for Integrating the Matrices in FEM

The stiffness matrix of a Mindlin thick plate (FSDT) in Equation (5.78) is numerically integrated by using the Gaussian quadrature as follows:

$$[\mathbf{K}]=\sum_{i=0}^{m}\sum_{j=0}^{n}\begin{bmatrix}\mathbf{N}_{\theta_{y,x}}^{T}D\mathbf{N}_{\theta_{y,x}}+\mathbf{N}_{\theta_{x,y}}^{T}D\mathbf{N}_{\theta_{x,y}}+\mathbf{N}_{\theta_{y,x}}^{T}D_{xxyy}\mathbf{N}_{\theta_{x,y}}+\mathbf{N}_{\theta_{x,y}}^{T}D_{xxyy}\\ \mathbf{N}_{\theta_{y,x}}+\left(\mathbf{N}_{\theta_{x,x}}+\mathbf{N}_{\theta_{y,y}}\right)^{T}D_{xy}\left(\mathbf{N}_{\theta_{x,x}}+\mathbf{N}_{\theta_{y,y}}\right)+\left(\mathbf{N}_{w,y}-\mathbf{N}_{\theta_{x}}\right)^{T}\\ D_{xz}\left(\mathbf{N}_{w,y}-\mathbf{N}_{\theta_{x}}\right)+\left(\mathbf{N}_{w,x}-\mathbf{N}_{\theta_{y}}\right)^{T}D_{xz}\left(\mathbf{N}_{w,x}-\mathbf{N}_{\theta_{y}}\right)\end{bmatrix}Jw_{j}w_{i} \tag{5.81}$$

where

$$\mathbf{N}_{w,x}=\frac{dN_w}{dx}=\frac{dN_w}{d\xi}\frac{d\xi}{dx}=\frac{dN_w}{d\xi}\frac{1}{J_x},\quad \mathbf{N}_{w,y}=\frac{dN_w}{dy}=\frac{dN_w}{d\eta}\frac{d\eta}{dy}=\frac{dN_w}{d\eta}\frac{1}{J_y}$$

$$\mathbf{N}_{\theta_{y,x}}=\frac{dN_{\theta_y}}{dx}=\frac{dN_{\theta_y}}{d\xi}\frac{d\xi}{dx}=\frac{dN_{\theta_y}}{d\xi}\frac{1}{J_x},\quad \mathbf{N}_{\theta_{x,y}}=\frac{dN_{\theta_x}}{dy}=\frac{dN_{\theta_x}}{d\eta}\frac{d\eta}{dy}=\frac{dN_{\theta_x}}{d\eta}\frac{1}{J_y}$$

$$\mathbf{N}_{\theta_{x,x}}=\frac{dN_{\theta_x}}{dx}=\frac{dN_{\theta_x}}{d\xi}\frac{d\xi}{dx}=\frac{dN_{\theta_x}}{d\xi}\frac{1}{J_x},\quad \mathbf{N}_{\theta_{y,y}}=\frac{dN_{\theta_y}}{dy}=\frac{dN_{\theta_y}}{d\eta}\frac{d\eta}{dy}=\frac{dN_{\theta_y}}{d\eta}\frac{1}{J_y}$$

Referring to Equation (5.79), the mass matrix of a Mindlin thick plate (FSDT) is given by

$$[\mathbf{M}]=\sum_{i=0}^{m}\sum_{j=0}^{n}\left\{\mathbf{N}_{w}^{T}\rho I_{xy}\mathbf{N}_{w}+\mathbf{N}_{\theta_x}^{T}\rho I\mathbf{N}_{\theta_x}+\mathbf{N}_{\theta_y}^{T}\rho I\mathbf{N}_{\theta_y}\right\}Jw_{j}w_{i} \tag{5.82}$$

Referring to Equation (5.80), the loading vector of a Mindlin thick plate (FSDT) is given by

$$[\mathbf{F}]=\sum_{i=0}^{m}\sum_{j=0}^{n}\left\{\left(\mathbf{N}_{w,x}^{T}\left(N_x+N_{xy}\right)\mathbf{N}_{w,x}\right)+\left(\mathbf{N}_{w,y}^{T}\left(N_x+N_{xy}\right)\mathbf{N}_{w,y}\right)\right\}Jw_{j}w_{i}$$

$$\{\mathbf{f}\}=\sum_{i=0}^{m}\sum_{j=0}^{n}\left(\mathbf{N}_{w}^{T}q_z\right)Jw_{j}w_{i} \tag{5.83}$$

where J is the determinant value of Jacobian operator defined in Equation (2.7).

The generalized displacement vectors $\{\mathbf{d}\}^T$ are given as

$$\mathbf{w} = \begin{bmatrix} w_1 & w_2 & w_3 & w_4 \end{bmatrix}^T; \quad \ddot{\mathbf{w}} = \begin{bmatrix} \ddot{w}_1 & \ddot{w}_2 & \ddot{w}_3 & \ddot{w}_4 \end{bmatrix}^T$$

$$\theta_x = \begin{bmatrix} \theta_{x1} & \theta_{x2} & \theta_{x3} & \theta_{x4} \end{bmatrix}^T; \quad \ddot{\theta}_x = \begin{bmatrix} \ddot{\theta}_{x1} & \ddot{\theta}_{x2} & \ddot{\theta}_{x3} & \ddot{\theta}_{x4} \end{bmatrix}^T$$

$$\theta_y = \begin{bmatrix} \theta_{y1} & \theta_{y2} & \theta_{y3} & \theta_{y4} \end{bmatrix}^T; \quad \ddot{\theta}_y = \begin{bmatrix} \ddot{\theta}_{y1} & \ddot{\theta}_{y2} & \ddot{\theta}_{y3} & \ddot{\theta}_{y4} \end{bmatrix}^T \tag{5.84}$$

where the notation $(\ddot{\ })$ means the second derivative with respect to time. The subscripts p and q are the orders of the plate surface. The subscripts m and n are the number of integration points of the Gaussian quadrature in the x- and y-directions, respectively.

5.11 Making the Stiffness and Mass Matrices of the Mindlin Plate Element

MATLAB code KMmatrixMindlin.m uses the functions NShapeMindlin-Consistent.m and Legendre.m to construct the stiffness and mass matrices of the Mindlin plate theory (FSDT) of an isotropic plate shown in Figure 5.23.

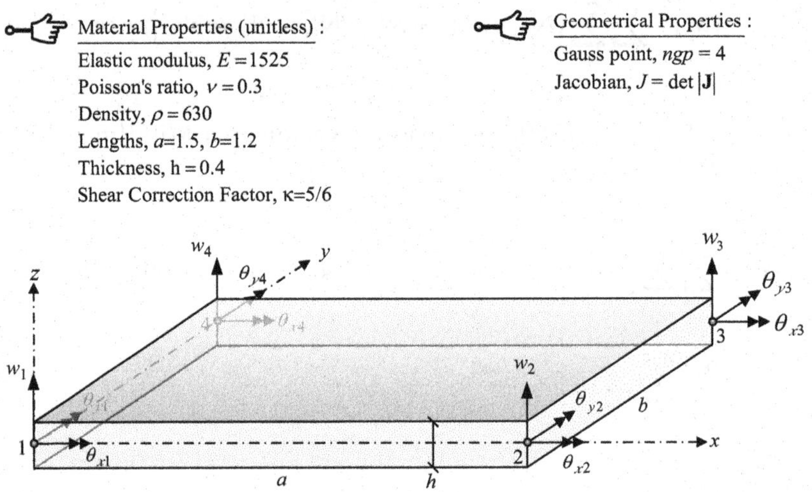

FIGURE 5.23
A rectangular isotropic flat Mindlin plate example.

5.11.1 KMmatrixMindlin Program List

```
% KMmatrixMindlin.m (R3)
% Construct matrices K and M of a four-node, 12 DOF plate
element
% Based on the Mindlin Plate Theory (CPT) using Polynomial
functions
clear all; clc;
a     = 1.5;      b     = 1.2;      h     = 0.1;
E     = 1525;     nu    = 0.3;      rho   = 630;
I     = h^3/12;   Ixy   = h;        kappa = 5/6;
ngpx  = 4;        ngpy  = 4;        ndof=12;
D     = E*h^3/12/(1-nu^2);    % D          Isotropic material
Dxxyy = nu*E*h^3/6/(1-nu^2); % Dxxyy
Dxy   = E*h^3/24/(1+nu);      % Dxy = Dyx
Dxz   = kappa*E*h/2/(1+nu);   % Dxz = Dyz
Gpwx  = Legendre(ngpx);      Gpwy  = Legendre(ngpy);
Jac   = a/2*b/2;
Jx    = a/2;
Jy    = b/2;
Kmat = zeros(ndof,ndof);
Mmat = zeros(ndof,ndof);
Nw   = zeros(ngpx,ngpy,ndof);
Ntx  = zeros(ngpx,ngpy,ndof);
Nty  = zeros(ngpx,ngpy,ndof);
Nwx  = zeros(ngpx,ngpy,ndof);
Nwy  = zeros(ngpx,ngpy,ndof);
Ntxx = zeros(ngpx,ngpy,ndof);
Ntxy = zeros(ngpx,ngpy,ndof);
Ntyx = zeros(ngpx,ngpy,ndof);
Ntyy = zeros(ngpx,ngpy,ndof);
for m=1:ngpx
   s = Gpwx(m,1);
   for n=1:ngpy
      t = Gpwy(n,1);
      % 0th derivative
      NMindlinCon = NShapeMindlinConsistent(0,s,t,Jx,Jy,E,nu,h,
kappa);
      Nw(m,n,:)  = NMindlinCon(:,1);
      Ntx(m,n,:) = NMindlinCon(:,2);
      Nty(m,n,:) = NMindlinCon(:,3);
      % 1st derivative
      NMindlinCon = NShapeMindlinConsistent(1,s,t,Jx,Jy,E,nu,h,
kappa);
      Nwx(m,n,:)  = NMindlinCon(:,1);
      Nwy(m,n,:)  = NMindlinCon(:,2);
      Ntxx(m,n,:) = NMindlinCon(:,3);
      Ntxy(m,n,:) = NMindlinCon(:,4);
      Ntyx(m,n,:) = NMindlinCon(:,5);
      Ntyy(m,n,:) = NMindlinCon(:,6);
```

```
   end
end
% Stiffness matrix K
for m=1:ngpx
  for n=1:ngpy
    for j=1:ndof
      for k=1:ndof
        Kmat(j,k) = Kmat(j,k) ...
           + Ntyx(m,n,k)*D*Ntyx(m,n,j)*Jac*Gpwx(m,2)*Gpw
y(n,2)...
           + Ntxy(m,n,k)*D*Ntxy(m,n,j)*Jac*Gpwx(m,2)*Gpw
y(n,2)...
           + Ntyx(m,n,k)*Dxxyy*Ntxy(m,n,j)*Jac*Gpwx(m,2)*Gpw
y(n,2)...
           + Ntxy(m,n,k)*Dxxyy*Ntyx(m,n,j)*Jac*Gpwx(m,2)*Gpw
y(n,2)...
           + (Ntxx(m,n,k)+Ntyy(m,n,k))*Dxy*(Ntxx(m,n,j)+Ntyy
(m,n,j))...
             *Jac*Gpwx(m,2)*Gpwy(n,2)...
           + (Nwx(m,n,k)-Nty(m,n,k))*Dxz*
(Nwx(m,n,j)-Nty(m,n,j))...
             *Jac*Gpwx(m,2)*Gpwy(n,2)...
           + (Nwy(m,n,k)-Ntx(m,n,k))*Dxz*
(Nwy(m,n,j)-Ntx(m,n,j))...
             *Jac*Gpwx(m,2)*Gpwy(n,2);
      end
    end
  end
end
% disp(Kmtx); % before swapping
% Swapping the matrices to the general displacement vector
order
Kmat=Kmat([1,4,7,10,2,5,8,11,3,6,9,12],:); % row
Kmat=Kmat(:,[1,4,7,10,2,5,8,11,3,6,9,12]); % column
disp('Stiffness Matrix of Mindlin Plate');
disp(Kmat);
% Mass matrix M
for m=1:ngpx
  for n=1:ngpy
    for j=1:ndof
      for k=1:ndof
        Mmat(j,k) = Mmat(j,k) ...
           + Ntx(m,n,k)*rho*I*Ntx(m,n,j)*Jac*Gpwx(m,2)*Gpw
y(n,2)...
           + Nty(m,n,k)*rho*I*Nty(m,n,j)*Jac*Gpwx(m,2)*Gpw
y(n,2)...
           + Nw(m,n,k)*rho*Ixy*Nw(m,n,j)*Jac*Gpwx(m,2)*Gpw
y(n,2);
      end
    end
```

```
  end
end
% Swapping the matrices to the general displacement vector
Mmat=Mmat([1,4,7,10,2,5,8,11,3,6,9,12],:); % row
Mmat=Mmat(:,[1,4,7,10,2,5,8,11,3,6,9,12]); % column
disp('Mass Matrix of Mindlin Plate');
disp(Mmat);
```

5.11.2 NShapeMindlinConsistent Function List

```
function NMindlinCon = NShapeMindlinConsistent(i,s,t,Jx,Jy,E,n
u,h,kappa)
% i  = 0~p derivative Shape Function, 0 not derived
% xi = Gauss coordinate
% Jx  = Jacobian operator x
% Jy  = Jacobian operator y
% E = Isotropic elastic modulus
% nu = Poisson' Ratio
% h = plate thickness
% kappa = shear correction factor
%--------------------------------------------------------
NMindlinCon=zeros(12,6); % (Nw,Ntx,Nty=12DOF, type=1,2,3)
G=E*h^3/12/(1-nu^2); A=nu*E*h^3/6/(1-nu^2)+E*h^3/24/(1+nu);
B=E*h^3/24/(1+nu);    H=kappa*E*h/2/(1+nu);
P=2*A/H; %Chi
M=2*B/H; %Myu
L=2*G/H; %Lambda
switch i
  case 0  % 0-derivative
    % Nw shape functions
    NMindlinCon(1,1) =(2*(-1+t)*(-2+t+t^2-3*L)+s^3*
(2+t*(2+3*L))+...
        s*(-6*(1+L)+t^3*(2+3*L)))/(8*(2+3*L));
    NMindlinCon(2,1) =(2*(-1+t)*(-2+t+t^2-3*L)-s^3*
(2+t*(2+3*L))+...
        s*(6*(1+L)-t^3*(2+3*L)))/(8*(2+3*L));
    NMindlinCon(3,1) =(4-2*t^3+6*L+6*t*(1+L)+s^3*
(-2+t*(2+3*L))+...
        s*(6*(1+L)+t^3*(2+3*L)))/(8*(2+3*L));
    NMindlinCon(4,1) =(4-2*t^3+6*L+6*t*(1+L)+s^3*
(2-t*(2+3*L))-...
        s*(6*(1+L)+t^3*(2+3*L)))/(8*(2+3*L));
    NMindlinCon(5,1) =(1/(32+48*L))*(-2*(-1+t^2)*
(2+3*L+t*(-2+M))+...
        s^3*(t*(2+3*L)-2*(M+P))+s*(-2*t*(2+3*L)+t^3*(2+3*L)+...
        t^2*(4+6*L)+2*(-2-3*L+M+P)));
    NMindlinCon(6,1) =(1/(32+48*L))*(2*(-1+t^2)*
(-2-3*L+t*(2+M))+...
        s^3*(-t*(2+3*L)+2*(M+P))-s*(-2*t*(2+3*L)+t^3*(2+3*L)+...
```

```
        t^2*(4+6*L)+2*(-2-3*L+M+P)));
    NMindlinCon(7,1) =(1/(32+48*L))*(-2*(-1+t^2)*
(-2-3*L+t*(-2+M))-...
        s^3*(t*(2+3*L)+2*(M+P))+s*(-t^3*(2+3*L)+t*(4+6*L)+...
        t^2*(4+6*L)+2*(-2-3*L+M+P)));
    NMindlinCon(8,1) =(1/(32+48*L))*(2*(-1+t^2)*
(2+3*L+t*(2+M))+...
    s*(6*L-2*t*(2+3*L)-2*t^2*(2+3*L)+t^3*(2+3*L)-2*
(-2+M+P))+...
        s^3*(t*(2+3*L)+2*(M+P)));
    NMindlinCon(9,1) =(1/(32+48*L))*(-2*(-1+t^2)*
(-2-3*L+t*(-2+M))-...
        s^3*(t*(2+3*L)+2*(M+P))+s*(-t^3*(2+3*L)+t*(4+6*L)
+t^2*(4+6*L)+...
        2*(-2-3*L+M+P)));
    NMindlinCon(10,1)=(1/(32+48*L))*(2*s^2*(-1+t)*(2+3*L)+...
        s*(-2*t*(2+3*L)+t^3*(2+3*L)+2*(-2+M))+s^3*
(4+t*(2+3*L)-2*M)-...
        2*(-1+t)*(2+3*L+t*(M+P)+t^2*(M+P)));
    NMindlinCon(11,1)=(1/(32+48*L))*(-2*s^2*(-1+t)*(2+3*L)+...
        s^3*(4+2*t+3*t*L+2*M)+s*(-2*t*(2+3*L)+t^3*(2+3*L)-2*
(2+M))+...
        2*(-1+t)*(2+3*L+t*(M+P)+t^2*(M+P)));
    NMindlinCon(12,1)=(1/(32+48*L))*(2*s^2*(1+t)*(2+3*L)+...
        s*(-t^3*(2+3*L)+t*(4+6*L)+2*(-2+M))-s^3*
(-4+2*t+3*t*L+2*M)-...
        2*(1+t)*(2+3*L-t*(M+P)+t^2*(M+P)));
    % Ntx shape functions
    NMindlinCon(1,2) =(6*G-3*H+3*H*t^2-3*H*L)/(8*H+12*H*L);
    NMindlinCon(2,2) =(6*G-3*H+3*H*t^2-3*H*L)/(8*H+12*H*L);
    NMindlinCon(3,2) =(3*(-2*G+H*(1-t^2+L)))/(4*H*(2+3*L));
    NMindlinCon(4,2) =(3*(-2*G+H*(1-t^2+L)))/(4*H*(2+3*L));
    NMindlinCon(5,2) =(B*(4+6*L)-6*G*(-2+M)+H*
(-2-2*t*(2+3*L)+...
        s*(-1+2*t)*(2+3*L)-3*t^2*(-2+M)+M))/(8*H*(2+3*L));
    NMindlinCon(6,2) =-((2*B*(2+3*L)-6*G*(2+M)+H*
(2+s*(-1+2*t)*(2+3*L)+...
        t*(4+6*L)+M-3*t^2*(2+M)))/(8*H*(2+3*L)));
    NMindlinCon(7,2) =(B*(4+6*L)-6*G*(-2+M)+H*
(-2+s*(1+2*t)*(2+3*L)+...
        t*(4+6*L)-3*t^2*(-2+M)+M))/(8*H*(2+3*L));
    NMindlinCon(8,2) =-((2*B*(2+3*L)-6*G*(2+M)+H*
(2-2*t*(2+3*L)+...
        s*(1+2*t)*(2+3*L)+M-3*t^2*(2+M)))/(8*H*(2+3*L)));
    NMindlinCon(9,2) =1/8*(-s+s^2-(3*t^2*(M+P))/(2+3*L)+...
        (2*A+2*B-(6*G*(M+P))/(2+3*L)+(H*(-2-3*L+M+P))/
(2+3*L))/H);
    NMindlinCon(10,2)=1/8*(-s-s^2+(3*t^2*(M+P))/(2+3*L)-...
        (2*A+2*B-(6*G*(M+P))/(2+3*L)+(H*(-2-3*L+M+P))/
(2+3*L))/H);
```

```
    NMindlinCon(11,2)=1/8*(s+s^2-(3*t^2*(M+P))/(2+3*L)+...
         (4*A+4*B-2*H+6*A*L+6*B*L-3*H*L-6*G*M+H*M-6*G*P+H*P)/
(2*H+3*H*L));
    NMindlinCon(12,2)=1/8*(s-s^2+(3*t^2*(M+P))/(2+3*L)-...
         (2*A+2*B-(6*G*(M+P))/(2+3*L)+(H*(-2-3*L+M+P))/
(2+3*L))/H);
    % Nty shape functions
    NMindlinCon(1,3) =(6*G-3*H+3*H*s^2-3*H*L)/(8*H+12*H*L);
    NMindlinCon(2,3) =(3*(-2*G+H*(1-s^2+L)))/(4*H*(2+3*L));
    NMindlinCon(3,3) =(3*(-2*G+H*(1-s^2+L)))/(4*H*(2+3*L));
    NMindlinCon(4,3) =(6*G-3*H+3*H*s^2-3*H*L)/(8*H+12*H*L);
    NMindlinCon(5,3) =1/8*(-t+t^2-(3*s^2*(M+P))/(2+3*L)+...
         (2*A+2*B-(6*G*(M+P))/(2+3*L)+(H*(-2-3*L+M+P))/
(2+3*L))/H);
    NMindlinCon(6,3) =1/8*(t-t^2+(3*s^2*(M+P))/(2+3*L)-...
         (2*A+2*B-(6*G*(M+P))/(2+3*L)+(H*(-2-3*L+M+P))/
(2+3*L))/H);
    NMindlinCon(7,3) =1/8*(t+t^2-(3*s^2*(M+P))/(2+3*L)+...
         (4*A+4*B-2*H+6*A*L+6*B*L-3*H*L-6*G*M+H*M-6*G*P+H*P)/
(2*H+3*H*L));
    NMindlinCon(8,3) =1/8*(-t-t^2+(3*s^2*(M+P))/(2+3*L)-...
         (2*A+2*B-(6*G*(M+P))/(2+3*L)+(H*(-2-3*L+M+P))/
(2+3*L))/H);
    NMindlinCon(9,3) =(B*(4+6*L)-6*G*(-2+M)+...
        H*(-2-2*t-3*t*L+2*s*(-1+t)*(2+3*L)-3*s^2*(-2+M)+M))/
(8*H*(2+3*L));
    NMindlinCon(10,3)=(-2*B*(2+3*L)+6*G*(2+M)+...
        H*(-2-2*t-3*t*L-2*s*(-1+t)*(2+3*L)-M+3*s^2*(2+M)))/
(8*H*(2+3*L));
    NMindlinCon(11,3)=(B*(4+6*L)-6*G*(-2+M)+...
        H*(-2+2*t+3*t*L+2*s*(1+t)*(2+3*L)-3*s^2*(-2+M)+M))/
(8*H*(2+3*L));
    NMindlinCon(12,3)=(-2*B*(2+3*L)+6*G*(2+M)+...
        H*(-2+2*t+3*t*L-2*s*(1+t)*(2+3*L)-M+3*s^2*(2+M)))/
(8*H*(2+3*L));
  case 1
    % dNw/dx shape functions
    NMindlinCon(1,1) = (-6*(1+L)+t^3*(2+3*L)+...
        s^2*(6+t*(6+9*L)))/(8*(2+3*L));
    NMindlinCon(2,1) = -((-6*(1+L)+t^3*(2+3*L)+...
        s^2*(6+t*(6+9*L)))/(8*(2+3*L)));
    NMindlinCon(3,1) = (6*(1+L)+t^3*(2+3*L)+...
        s^2*(-6+t*(6+9*L)))/(8*(2+3*L));
    NMindlinCon(4,1) = (-6*(1+L)-t^3*(2+3*L)+...
        s^2*(6-3*t*(2+3*L)))/(8*(2+3*L));
    NMindlinCon(5,1) = 1/(32+48*L)*(-2*t*(2+3*L)+t^3*
(2+3*L)+...
        t^2*(4+6*L)+2*(-2-3*L+M+P)+3*s^2*(t*(2+3*L)-2*(M+P)));
    NMindlinCon(6,1) = 1/(32+48*L)*(2*t*(2+3*L)-2*t^2*
(2+3*L)-...
```

```
        t^3*(2+3*L)-2*(-2-3*L+M+P)+3*s^2*(-t*(2+3*L)+2*(M+P)));
    NMindlinCon(7,1) = 1/(32+48*L)*(-t^3*(2+3*L)+t*(4+6*L)+...
        t^2*(4+6*L)+2*(-2-3*L+M+P)-3*s^2*(t*(2+3*L)+2*(M+P)));
    NMindlinCon(8,1) = 1/(32+48*L)*(6*L-2*t*(2+3*L)-2*t^2*
(2+3*L)+...
        t^3*(2+3*L)-2*(-2+M+P)+3*s^2*(t*(2+3*L)+2*(M+P)));
    NMindlinCon(9,1) = 1/(32+48*L)*(4*s*(-1+t)*(2+3*L)-2*t*
(2+3*L)+...
        t^3*(2+3*L)+3*s^2*(4+t*(2+3*L)-2*M)+2*(-2+M));
    NMindlinCon(10,1) =1/(32+48*L)*(-4*s*(-1+t)*(2+3*L)-2*t*
(2+3*L)+...
        t^3*(2+3*L)-2*(2+M)+3*s^2*(4+2*t+3*t*L+2*M));
    NMindlinCon(11,1) =(-t^3*(2+3*L)+4*s*(1+t)*(2+3*L)+t*(4+
6*L)+...
        2*(-2+M)-3*s^2*(-4+2*t+3*t*L+2*M))/(32+48*L);
    NMindlinCon(12,1) =1/(32+48*L)*(-t^3*(2+3*L)-4*s*(1+t)*
(2+3*L)+...
        t*(4+6*L)-2*(2+M)+s^2*(-3*t*(2+3*L)+6*(2+M)));
    % dNw/dy shape functions
    NMindlinCon(1,2) = (6*(-1+t^2-L)+s^3*(2+3*L)+...
        3*s*t^2*(2+3*L))/(8*(2+3*L));
    NMindlinCon(2,2) = -((6-6*t^2+6*L+s^3*(2+3*L)+...
        3*s*t^2*(2+3*L))/(8*(2+3*L)));
    NMindlinCon(3,2) = (6-6*t^2+6*L+s^3*(2+3*L)+...
        3*s*t^2*(2+3*L))/(8*(2+3*L));
    NMindlinCon(4,2) = -((-6+6*t^2-6*L+s^3*(2+3*L)+...
        3*s*t^2*(2+3*L))/(8*(2+3*L)));
    NMindlinCon(5,2) = (s^3*(2+3*L)+s*(-2+4*t+3*t^2)*
(2+3*L)-...
        2*(2+t*(4+6*L)+3*t^2*(-2+M)-M))/(32+48*L);
    NMindlinCon(6,2) = -((s^3*(2+3*L)+s*(-2+4*t+3*t^2)*
(2+3*L)+...
        2*(2+t*(4+6*L)+M-3*t^2*(2+M)))/(32+48*L));
    NMindlinCon(7,2) = -((s^3*(2+3*L)+s*(-2-4*t+3*t^2)*
(2+3*L)-...
        2*(-2+t*(4+6*L)-3*t^2*(-2+M)+M))/(32+48*L));
    NMindlinCon(8,2) = (s^3*(2+3*L)+s*(-2-4*t+3*t^2)*
(2+3*L)+...
        2*(-2+t*(4+6*L)-M+3*t^2*(2+M)))/(32+48*L);
    NMindlinCon(9,2) = (s^3*(2+3*L)+s*(-2+3*t^2)*(2+3*L)+...
        s^2*(4+6*L)+2*(-2-3*L+M-3*t^2*M+P-3*t^2*P))/(32+48*L);
    NMindlinCon(10,2) =(4+6*L-2*s^2*(2+3*L)+s^3*(2+3*L)+...
        s*(-2+3*t^2)*(2+3*L)-2*M+6*t^2*M-2*P+6*t^2*P)/
(32+48*L);
    NMindlinCon(11,2) =(-s^3*(2+3*L)-s*(-2+3*t^2)*(2+3*L)+...
        s^2*(4+6*L)+2*(-2-3*L+M-3*t^2*M+P-3*t^2*P))/(32+48*L);
    NMindlinCon(12,2) =-((s^3*(2+3*L)+s*(-2+3*t^2)*(2+3*L)+...
        s^2*(4+6*L)+2*(-2-3*L+M-3*t^2*M+P-3*t^2*P))/
(32+48*L));
    % dNtx/dx shape functions
```

```
    NMindlinCon(1,3) = 0;
    NMindlinCon(2,3) = 0;
    NMindlinCon(3,3) = 0;
    NMindlinCon(4,3) = 0;
    NMindlinCon(5,3) = 1/8*(-1+2*t);
    NMindlinCon(6,3) = 1/8*(1-2*t);
    NMindlinCon(7,3) = 1/8*(1+2*t);
    NMindlinCon(8,3) = 1/8*(-1-2*t);
    NMindlinCon(9,3) = 1/8*(-1+2*s);
    NMindlinCon(10,3) =1/8*(-1-2*s);
    NMindlinCon(11,3) =1/8*(1+2*s);
    NMindlinCon(12,3) =1/8*(1-2*s);
    % dNtx/dy shape functions
    NMindlinCon(1,4) = (3*t)/(4+6*L);
    NMindlinCon(2,4) = (3*t)/(4+6*L);
    NMindlinCon(3,4) = -((3*t)/(4+6*L));
    NMindlinCon(4,4) = -((3*t)/(4+6*L));
    NMindlinCon(5,4) = (-2-3*L+s*(2+3*L)-3*t*(-2+M))/(8+12*L);
    NMindlinCon(6,4) = -((2+3*L+s*(2+3*L)-3*t*(2+M))/
(8+12*L));
    NMindlinCon(7,4) = (2+2*s+6*t+3*L+3*s*L-3*t*M)/(8+12*L);
    NMindlinCon(8,4) = (2+3*L-s*(2+3*L)+3*t*(2+M))/(8+12*L);
    NMindlinCon(9,4) = -((3*t*(M+P))/(8+12*L));
    NMindlinCon(10,4) = (3*t*(M+P))/(8+12*L);
    NMindlinCon(11,4) = -((3*t*(M+P))/(8+12*L));
    NMindlinCon(12,4) = (3*t*(M+P))/(8+12*L);
    % dNty/dx shape functions
    NMindlinCon(1,5) = (3*s)/(4+6*L);
    NMindlinCon(2,5) = -((3*s)/(4+6*L));
    NMindlinCon(3,5) = -((3*s)/(4+6*L));
    NMindlinCon(4,5) = (3*s)/(4+6*L);
    NMindlinCon(5,5) = -((3*s*(M+P))/(8+12*L));
    NMindlinCon(6,5) = (3*s*(M+P))/(8+12*L);
    NMindlinCon(7,5) = -((3*s*(M+P))/(8+12*L));
    NMindlinCon(8,5) = (3*s*(M+P))/(8+12*L);
    NMindlinCon(9,5) = ((-1+t)*(2+3*L)-3*s*(-2+M))/(8+12*L);
    NMindlinCon(10,5) =(-(-1+t)*(2+3*L)+3*s*(2+M))/(8+12*L);
    NMindlinCon(11,5) =((1+t)*(2+3*L)-3*s*(-2+M))/(8+12*L);
    NMindlinCon(12,5) =(-(1+t)*(2+3*L)+3*s*(2+M))/(8+12*L);
    % dNty/dy shape functions
    NMindlinCon(1,6) = 0;
    NMindlinCon(2,6) = 0;
    NMindlinCon(3,6) = 0;
    NMindlinCon(4,6) = 0;
    NMindlinCon(5,6) = 1/8*(-1+2*t);
    NMindlinCon(6,6) = 1/8*(1-2*t);
    NMindlinCon(7,6) = 1/8*(1+2*t);
    NMindlinCon(8,6) = 1/8*(-1-2*t);
    NMindlinCon(9,6) = 1/8*(-1+2*s);
    NMindlinCon(10,6) =1/8*(-1-2*s);
```

```
    NMindlinCon(11,6) =1/8*(1+2*s);
    NMindlinCon(12,6) =1/8*(1-2*s);
    % Jacobian operator multipled
    for n=1:12
        NMindlinCon(n,1) = NMindlinCon(n,1)/Jx;
        NMindlinCon(n,2) = NMindlinCon(n,2)/Jy;
        NMindlinCon(n,3) = NMindlinCon(n,3)/Jx;
        NMindlinCon(n,4) = NMindlinCon(n,4)/Jy;
        NMindlinCon(n,5) = NMindlinCon(n,5)/Jx;
        NMindlinCon(n,6) = NMindlinCon(n,6)/Jy;
    end
end
return
```

References

Bazeley GP, Cheung YK, Irons BM, Zienkiewicz OC (1965) Triangular elements in bending: confirming and non-conforming solutions. *Proc Conf Matrix Meth Struct Mech, Air Force Institute of Technology, WPAFB*, Dayton, OH, AFFDL-TR-66-80:547-576.

Ben-Israel A, Greville TNE (2003) *Generalized Inverses: Theory and Applications*, 2 edition. Springer, New York.

Bogner FK, Fox RL, Schmidt LA Jr (1965) The generation of interelement-compatible stiffness and mass matrices by the use of interpolation formulas. *Proc Conf Matrix Meth Struct Mech, Air Force Institute of Technology, WPAFB*, Dayton, OH, AFFDL-TR-66-80:397-443.

Clough RW, Tocher JL (1965) Finite element stiffness matrices for analysis of plate bending. *Proc Conf Matrix Meth Struct Mech, Air Force Institute of Technology, WPAFB*, Dayton, OH, AFFDL-TR-66-80:515-546.

Fraeijis de Veubeke B (1968) A conforming finite element for plate bending. *Int J Solids Struct* 4(1):95–108.

Gan BS (2018) *An Isogeometric Approach to Beam Structures*. Springer, Switzerland.

Hermann LR (1967) Finite element bending analysis for plates. *Proc ASCE J Engng Mech Div* 93(8):13–26.

Hrabok MM, Hrudey TM (1984) A review and catalog of plate bending finite elements. *Comput Struct* 19(3):479–495.

Irons BM (1969) A conforming quartic triangular element for plate bending. *Int J Numerical Meth Engng* 1:29–45.

Melosh RJ (1963) Basis of derivation of matrices for the direct stiffness method. *AIAA J* 1:1631–1637.

Mindlin RD (1951) Influence of rotatory inertia and shear on flexural motions of isotropic, elastic plates. *J Appl Mech (ASME)* 18:31–38.

Reissner E (1945) The effect of transverse shear deformation on the bending of elastic plates. *J Appl Mech (ASME)* 12(2):69–77.
Reissner E (1947) On bending of elastic plates. *Q Appl Math* 5(1):55–68.
Zienkiewicz OC, Cheung YK (1964) The finite element method for analysis of elastic isotropic and orthotropic slabs. *Proc Inst Civ Engrs, London* 28:471–488.

6

Isogeometric Analysis for Plate and Shell

6.1 Isogeometric Analysis

The isogeometric approach is pioneered by several researchers (Basilevs et al. 2010; Cottrell et al. 2006, 2007, 2009; Gontier and Vollmer 1995; Schramm and Pilkey 1993), where the NURBS curves and surfaces are used to represent the real geometry and direct mesh modeling in the finite element method. The NURBS, with its informative geometrical information which is usually created in the CAD environment, can be conveyed seamlessly into the finite element formulation. In the formulation of the NURBS-based element, the shape function, which is a fundamental component in element formulation, can also be constructed directly.

Table 6.1 shows the matching relationship between the NURBS and finite element concepts. This comparison will help the reader who is familiar with the finite element concept to understand the isogeometric analysis (IGA) concept of modeling.

6.2 NURBS for Plate and Shell Elements

Plate and shell elements are considered as bidimensional structural problems. In IGA, the coordinates of the geometry of a plate can be represented using the NURBS surface (see Section 1.6) as follows:

TABLE 6.1
Interpretation of the Isogeometric Approach in the Finite Element Context

IGA	FEM
Exact geometry	Approximate geometry
Control points	Nodal points
Control variables	Nodal variables
NURBS shape function	Polynomial shape function
Changeable level of C^i—continuity	C^0—continuity

$$x(\xi,\eta)=\sum_{i=0}^{m}\sum_{j=0}^{n}R_{i,j}^{p,q}(\xi,\eta)P_{x(\xi,\eta)}$$

$$y(\xi,\eta)=\sum_{i=0}^{m}\sum_{j=0}^{n}R_{i,j}^{p,q}(\xi,\eta)P_{y(\xi,\eta)},\quad\begin{cases}-1\le\xi\le 1\\-1\le\eta\le 1\end{cases}$$

$$z(\xi,\eta)=\sum_{i=0}^{m}\sum_{j=0}^{n}R_{i,j}^{p,q}(\xi,\eta)P_{z(\xi,\eta)} \tag{6.1}$$

where

$$R_{i,j}^{p,q}(\xi,\eta)=\frac{S_{i,p}(\xi)S_{j,q}(\eta)w_{i,j}}{\sum_{k=0}^{m}\sum_{l=0}^{n}S_{k,p}(\xi)S_{l,q}(\eta)w_{k,l}},\quad w_{i,j}>0$$

$w_{i,j}$ are the weighting parameters of points in the *control mesh.* $R_{i,j}^{p,q}(\xi,\eta)$ are the rationalized $S_{i,p}(\xi)S_{j,q}(\eta)$ basis functions of the rational B-spline curves. Similar to the Bézier surface formulation, the points $P_{x(\xi,\eta)}, P_{y(\xi,\eta)}, P_{z(\xi,\eta)}$ are the coordinates of the corresponding control mesh that maps a set of control points in a rectangular grid (ξ,η) as illustrated in Figure 1.12.

In IGA, the classical shape functions, $\mathbf{N}_{w/\theta_x/\theta_y}(\xi,\eta)$, and their derivatives in Equation (5.33) are substituted by using the rational basis functions of the NURBS surface $R_{i,j}^{p,q}(\xi,\eta)$ which consist of the B-spline curves, $S_{i,p}(\xi)$ and $S_{j,q}(\eta)$, given from Equation (1.15) and the derivatives of the rational basis functions of the NURBS surface which are given in detail in Section 1.8.

In a plate that has three degrees of freedom (DOFs) of generalized displacements (w, θ_x, θ_y) at a node, an arbitrary point on the plate can be interpolated by using the shape functions of the NURBS surface which are given as follows:

$$w(\xi,\eta)=\sum_{i=0}^{m}\sum_{j=0}^{n}R_{i,j}^{p,q}(\xi,\eta)w_{ij}$$

$$\theta_x(\xi,\eta)=\sum_{i=0}^{m}\sum_{j=0}^{n}R_{i,j}^{p,q}(\xi,\eta)\theta_{xij} \tag{6.2}$$

$$\theta_y(\xi,\eta)=\sum_{i=0}^{m}\sum_{j=0}^{n}R_{i,j}^{p,q}(\xi,\eta)\theta_{yij}$$

It is worth noting that in IGA of a plate, the nodal displacements and rotations are treated independently by using NURBS surface functions. Thus, they are not coupled in the governing equations (see Mindlin 1951; Reddy 1993) such as classical plate theory (CPT) and first-order shear deformation theory (FSDT) described in Chapter 5.

Depending on the order of the pth and q^{th} polynomials and the knot vectors Ξ and H being used, we need to have the associated number of control points on a control mesh. The increasing number of control points being used to construct the NURBS surface will increase the DOFs of the plate, as illustrated in Figure 6.1. The figure shows the DOF illustratively at each control point of the plate.

Compared to the plate element in finite element method (FEM) (see Chapter 5) where there are only four nodes having three DOFs at each node, the element developed by using the NURBS surface can have an excessive number of DOF at the control points on the control mesh of the plate which depends on the order of polynomial being considered. For example, in the second-order polynomial of CPT element, there is a 3×3 control mesh with nine control points and a total of 27 DOFs, which means that there are $3 \times 27 = 81$ basis functions to construct the governing equation into the matrix form.

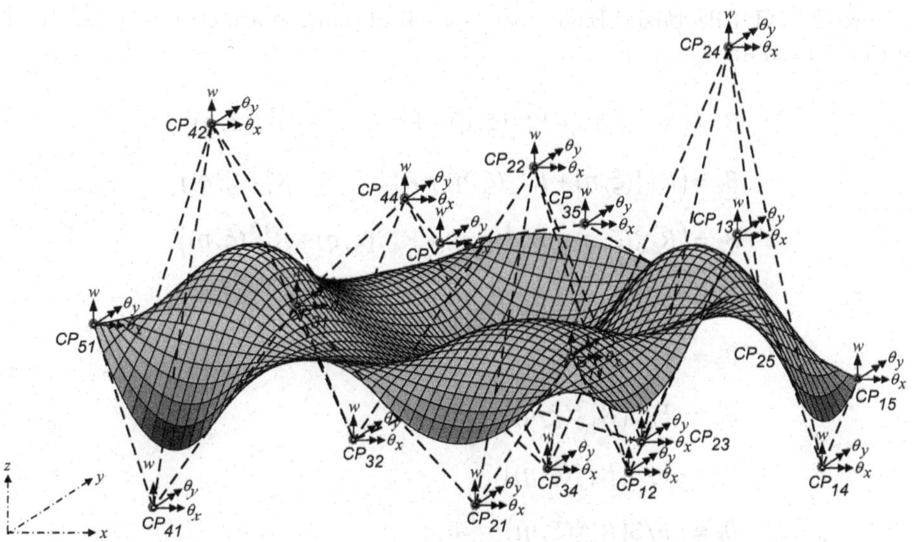

FIGURE 6.1
Generalized displacements (DOF) on a plate element in IGA.

6.3 NURBS for Kirchhoff Thin Plate Element

Similar to FEM, the governing equation of a plate element in IGA can be developed conveniently from the theory of minimum potential energy principle as follows:

$$\begin{aligned} \Pi &= \frac{1}{2}\{\mathbf{d}\}^T[\mathbf{K}]\{\mathbf{d}\} + \frac{1}{2}\{\ddot{\mathbf{d}}\}^T[\mathbf{M}]\{\ddot{\mathbf{d}}\} - \{\mathbf{d}\}^T\{\mathbf{f}\} \\ \frac{d\Pi}{d\{\mathbf{d}\}^T} &= [\mathbf{K}]\{\mathbf{d}\} + [\mathbf{M}]\{\ddot{\mathbf{d}}\} - \{\mathbf{f}\} = \{\mathbf{0}\} \end{aligned} \tag{6.3}$$

where the stiffness matrix $[\mathbf{K}]$, the mass matrix $[\mathbf{M}]$, and the loading vector $\{\mathbf{f}\}$ are described below.

As proven by Gan (2018), the classical Euler–Bernoulli beam shape functions used in IGA do not follow the complete polynomials composed by NURBS. Instead, incomplete polynomials selected from the rational basis B-spline functions are necessary. Referring to Equation (5.41), the stiffness matrix of an isotropic CPT plate based on IGA can be rewritten as

$$[\mathbf{K}] = \int_0^b \int_0^a \left(\mathbf{B}_{,xx}^T D \mathbf{B}_{,xx} + \mathbf{B}_{,yy}^T D \mathbf{B}_{,yy} + \mathbf{B}_{,xx}^T D_{xxyy} \mathbf{B}_{,yy} + \mathbf{B}_{,xy}^T D_{xy} \mathbf{B}_{,xy} \right) dx\,dy \tag{6.4}$$

where the NURBS-based basis functions $\mathbf{B}$ of p and q are equal to the third order, which are given as

$$\begin{aligned}
B_1 &= \left(R_{1,1}^{1,1}(\xi,\eta) + R_{1,2}^{1,2}(\xi,\eta) + R_{2,1}^{2,1}(\xi,\eta) + R_{2,2}^{2,2}(\xi,\eta)\right) \\
B_2 &= \left(R_{3,1}^{3,1}(\xi,\eta) + R_{3,2}^{3,2}(\xi,\eta) + R_{4,1}^{4,1}(\xi,\eta) + R_{4,2}^{4,2}(\xi,\eta)\right) \\
B_3 &= \left(R_{3,3}^{3,3}(\xi,\eta) + R_{3,4}^{3,4}(\xi,\eta) + R_{4,3}^{4,3}(\xi,\eta) + R_{4,4}^{4,4}(\xi,\eta)\right) \\
B_4 &= \left(R_{1,3}^{1,3}(\xi,\eta) + R_{1,4}^{1,4}(\xi,\eta) + R_{2,3}^{2,3}(\xi,\eta) + R_{2,4}^{2,4}(\xi,\eta)\right) \\
B_5 &= +b/3\left(R_{1,2}^{1,2}(\xi,\eta)\right) \\
B_6 &= +b/3\left(R_{4,2}^{4,2}(\xi,\eta)\right) \\
B_7 &= -b/3\left(R_{4,3}^{4,3}(\xi,\eta)\right) \\
B_8 &= -b/3\left(R_{1,3}^{1,3}(\xi,\eta)\right) \\
B_9 &= -a/3\left(R_{2,1}^{2,1}(\xi,\eta)\right) \\
B_{10} &= +a/3\left(R_{3,1}^{3,1}(\xi,\eta)\right)
\end{aligned}$$

$$B_{11} = +a/3\left(R_{3,4}^{3,4}(\xi,\eta)\right)$$
$$B_{12} = -a/3\left(R_{2,4}^{2,4}(\xi,\eta)\right) \tag{6.5}$$

and

$$\mathbf{B}_{,xx} = \frac{d^2\mathbf{B}}{dx^2} = \frac{d^2\mathbf{B}}{d\xi^2} \times \frac{1}{J_x^4} = \frac{1}{J_x^4}\sum_{i=0}^{m}\sum_{j=0}^{n}\left(\frac{d^2R_{i,j}^{p,q}(\xi,\eta)}{d\xi^2}\right)$$

$$\mathbf{B}_{,xy} = \frac{d^2\mathbf{B}}{dxdy} = \frac{d^2\mathbf{B}}{d\xi d\eta} \times \frac{1}{J_x^2 J_y^2} = \frac{1}{J_x^2 J_y^2}\sum_{i=0}^{m}\sum_{j=0}^{n}\left(\frac{d^2R_{i,j}^{p,q}(\xi,\eta)}{d\xi d\eta}\right)$$

$$\mathbf{B}_{,yy} = \frac{d^2\mathbf{B}}{dy^2} = \frac{d^2\mathbf{B}}{d\eta^2} \times \frac{1}{J_y^4} = \frac{1}{J_y^4}\sum_{i=0}^{m}\sum_{j=0}^{n}\left(\frac{d^2R_{i,j}^{p,q}(\xi,\eta)}{d\eta^2}\right) \tag{6.6}$$

where $J_x = d\xi/dx$ and $J_y = d\eta/dy$ are the Jacobian operators in the x- and y-directions, respectively.

Equation (6.4) will be numerically integrated by using the Gaussian quadrature as follows:

$$[\mathbf{K}] = \sum_{i=0}^{m}\sum_{j=0}^{n}\begin{bmatrix}\mathbf{B}_{,xx}\\ \mathbf{B}_{,yy}\\ \mathbf{B}_{,xy}\end{bmatrix}^T\begin{bmatrix} D & \frac{1}{2}D_{xxyy} & 0\\ \frac{1}{2}D_{xxyy} & D & 0\\ 0 & 0 & D_{xy}\end{bmatrix}\begin{bmatrix}\mathbf{B}_{,xx}\\ \mathbf{B}_{,yy}\\ \mathbf{B}_{,xy}\end{bmatrix} Jw_iw_j \tag{6.7}$$

where J is the determinant value of the bidirectional Jacobian operator defined in Equation (2.7).

Referring to Equation (5.42), the mass matrix of a CPT plate based on IGA can be written as

$$[\mathbf{M}] = \int_0^b\int_0^a \begin{bmatrix}\mathbf{B}_w\\ \mathbf{B}_{\theta_x}\\ \mathbf{B}_{\theta_y}\end{bmatrix}^T\begin{bmatrix}\rho I_{xy} & 0 & 0\\ 0 & \rho I & 0\\ 0 & 0 & \rho I\end{bmatrix}\begin{bmatrix}\mathbf{B}_w\\ \mathbf{B}_{\theta_x}\\ \mathbf{B}_{\theta_y}\end{bmatrix} dx\,dy \tag{6.8}$$

Equation (6.9) will be numerically integrated by using the Gaussian quadrature as follows:

$$[\mathbf{M}] = \sum_{i=0}^{m}\sum_{j=0}^{n}\begin{bmatrix}\mathbf{B}_w\\ \mathbf{B}_{\theta_x}\\ \mathbf{B}_{\theta_y}\end{bmatrix}^T\begin{bmatrix}\rho I_{xy} & 0 & 0\\ 0 & \rho I & 0\\ 0 & 0 & \rho I\end{bmatrix}\begin{bmatrix}\mathbf{B}_w\\ \mathbf{B}_{\theta_x}\\ \mathbf{B}_{\theta_y}\end{bmatrix} Jw_jw_i \tag{6.9}$$

where J is the determinant value of Jacobian operator defined in Equation (2.7).

Referring to Equation (5.43), the loading vector of a CPT plate based on IGA is given by

$$\{\mathbf{f}\}=\int_0^b\int_0^a\begin{bmatrix}1\\ \mathbf{B}_{w,x}\\ \mathbf{B}_{w,y}\end{bmatrix}^T\begin{bmatrix}q_z & 0 & 0\\ 0 & N_x & N_{xy}\\ 0 & N_{yx} & N_y\end{bmatrix}\begin{bmatrix}\mathbf{B}_w\\ \mathbf{B}_{w,x}\\ \mathbf{B}_{w,y}\end{bmatrix}dx\,dy \tag{6.10}$$

where the NURBS-based basis functions are obtained from Equation (6.1) by omitting the DOF terms:

$$\begin{aligned}\mathbf{B}_{,x}&=\frac{d\mathbf{B}}{dx}=\frac{d\mathbf{B}}{d\xi}\times\frac{1}{J_x^2}=\frac{1}{J_x^2}\sum_{i=0}^{m}\sum_{j=0}^{n}\left(\frac{dR_{i,j}^{p,q}(\xi,\eta)}{d\xi}\right)\\ \mathbf{B}_{,y}&=\frac{d\mathbf{B}}{dy}=\frac{d\mathbf{B}}{d\eta}\times\frac{1}{J_y^2}=\frac{1}{J_y^2}\sum_{i=0}^{m}\sum_{j=0}^{n}\left(\frac{dR_{i,j}^{p,q}(\xi,\eta)}{d\eta}\right)\end{aligned} \tag{6.11}$$

Equation (6.14) will be numerically integrated by using the Gaussian quadrature as follows:

$$\{\mathbf{f}\}=\sum_{i=0}^{m}\sum_{j=0}^{n}\begin{bmatrix}1\\ \mathbf{B}_{w,x}\\ \mathbf{B}_{w,y}\end{bmatrix}^T\begin{bmatrix}q_z & 0 & 0\\ 0 & N_x & N_{xy}\\ 0 & N_{yx} & N_y\end{bmatrix}\begin{bmatrix}\mathbf{B}_w\\ \mathbf{B}_{w,x}\\ \mathbf{B}_{w,y}\end{bmatrix}Jw_jw_i \tag{6.12}$$

The generalized displacement vectors $\{\mathbf{d}\}^T$ are given as

$$\begin{aligned}&\mathbf{w}=\left[w_{ij}\ldots w_{ij=p\times m+1,q\times n+1}\right]^T;\quad \ddot{\mathbf{w}}=\left[\ddot{w}_{ij}\ldots\ddot{w}_{ij=p\times m+1,q\times n+1}\right]^T\\ &\theta_x=\left[\theta_{xij}\ldots\theta_{xij=p\times m+1,q\times n+1}\right]^T;\quad \theta_x=\left[\ddot{\theta}_{xij}\ldots\ddot{\theta}_{xij=p\times m+1,q\times n+1}\right]^T\\ &\theta_y=\left[\theta_{yij}\ldots\theta_{yij=p\times m+1,q\times n+1}\right]^T;\quad \theta_y=\left[\ddot{\theta}_{yij}\ldots\ddot{\theta}_{yij=p\times m+1,q\times n+1}\right]^T\end{aligned} \tag{6.13}$$

where the notation ($\ddot{\ }$) means the second derivative with respect to time. The subscripts p and q are the orders of the NURBS surface. The subscripts m and n are the number of integration points of the Gaussian quadrature in the x- and y-directions, respectively.

6.4 Making the Stiffness and Mass Matrices of the Kirchhoff Plate Element

MATLAB® code KMmatrixKirchhoffNurbs.m uses the functions NurbsSurface.m and Legendre.m to construct the stiffness and mass matrices of the Kirchhoff plate theory (CPT) of an isotropic plate, as shown in Figure 6.2.

6.4.1 KMmatrixKirchhoffNurbs Program List

```
% KMmatrixKirchhoffNurbs.m (Using NurbsSurface.m)  : R12
% Construct matrices K and M
% Based on the Kirchhoff Plate Theory using NURBS surface
shape functions
clear all; clc;
a      = 1.5;      b     = 1.2;      h     = 0.1;
% isotropic material
E      = 1525;     nu    = 0.3;      rho   = 630;
D      = E*h^3/12/(1-nu^2);
Dxxyy  = nu*E*h^3/6/(1-nu^2);
Dxy    = E*h^3/6/(1+nu);   % Different with the Mindlin's Theory
I      = h^3/12;
Ixy    = h;
% NURBS for w and theta displacement data in x and y
directions
npx    = 3;
npy    = 3;
```

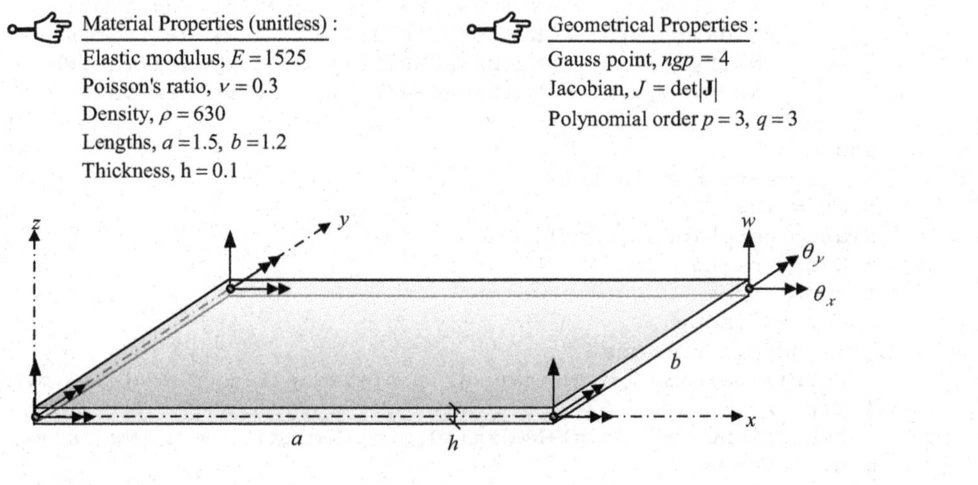

FIGURE 6.2
A rectangular isotropic flat Kirchhoff plate IGA model.

```
knotx = [-ones(1,npx+1) ones(1,npx+1)]; knotx = [knotx
ones(1,npx)];
knoty = [-ones(1,npy+1) ones(1,npy+1)]; knoty = [knoty
ones(1,npy)];
nbx   = size(knotx,2)-npx-1;
nby   = size(knoty,2)-npy-1;
wtxy  = ones(nbx-npx,nby-npy);
% Gauss
ngpx  = 4;                  ngpy  = 4;
Gpwx  = Legendre(ngpx);     Gpwy  = Legendre(ngpy);
% Jac
Jac = a/2*b/2;
Jx = a/2;
Jy = b/2;
% Shape functions of NURBS surface
dim   = (nbx-npx)*(nby-npy);
Nw    = zeros(dim,ngpx,ngpy);
Ntx   = zeros(dim,ngpx,ngpy);
Nty   = zeros(dim,ngpx,ngpy);
Nwdxx = zeros(dim,ngpx,ngpy);
Nwdxy = zeros(dim,ngpx,ngpy);
Nwdyy = zeros(dim,ngpx,ngpy);
% Filling NURBS surface shape functions into the gauss
points
for m=1:ngpx
    s = Gpwx(m,1);
    for n=1:ngpy
        t = Gpwy(n,1);
        NurbsKRCHF     = NurbsSurface(npx,knotx,npy,knoty,s,t);
        Nw(:,m,n)      = NurbsKRCHF(1,:);   % 0th derivative
        Ntx(:,m,n)     = NurbsKRCHF(1,:);   % 0th derivative
        Nty(:,m,n)     = NurbsKRCHF(1,:);   % 0th derivative
        Nwdxx(:,m,n)   = NurbsKRCHF(4,:);   % xx derivative
        Nwdxy(:,m,n)   = NurbsKRCHF(5,:);   % xy derivative
        Nwdyy(:,m,n)   = NurbsKRCHF(6,:);   % yy derivative
    end
end
% Stiffness Matrix Kmat
ndof = 12;
Kmat = zeros(ndof,ndof);
% D-dxx parts
for m=1:ngpx
  for n=1:ngpy
  % Shape functions
  kchf(1)=Nwdxx(1,m,n)+Nwdxx(2,m,n)+Nwdxx(5,m,n)+Nwdxx(6,m,n);
% N1w
  kchf(2)=Nwdxx(9,m,n)+Nwdxx(10,m,n)+Nwdxx(13,m,n)+Nwdxx(14,
m,n);  % N2w
  kchf(3)=Nwdxx(11,m,n)+Nwdxx(12,m,n)+Nwdxx(15,m,n)+Nwdxx(16,
m,n); % N3w
```

```
  kchf(4)=Nwdxx(3,m,n)+Nwdxx(4,m,n)+Nwdxx(7,m,n)+Nwdxx(8,m,n);
% N4w
  kchf(5)=+b^2/9*Nwdxx(2,m,n);              % N1tx
  kchf(6)=+b^2/9*Nwdxx(14,m,n);             % N2tx
  kchf(7)=-b^2/9*Nwdxx(15,m,n);             % N3tx
  kchf(8)=-b^2/9*Nwdxx(3,m,n);              % N4tx
  kchf(9) =-a^2/9*Nwdxx(5,m,n);             % N1ty
  kchf(10)=+a^2/9*Nwdxx(9,m,n);             % N2ty
  kchf(11)=+a^2/9*Nwdxx(12,m,n);            % N3ty
  kchf(12)=-a^2/9*Nwdxx(8,m,n);             % N4ty
  for i=1:ndof
    for j=1:ndof
        Kmat(i,j) = Kmat(i,j) + kchf(j)/Jx^4*D* ...
            kchf(i)/Jx^4*Jac*Gpwx(m,2)*Gpwy(n,2);
    end
  end
  end
end
% D-dyy parts
for m=1:ngpx
  for n=1:ngpy
  % Shape functions
  kchf(1)=Nwdyy(1,m,n)+Nwdyy(2,m,n)+Nwdyy(5,m,n)+Nwdyy(6,m,n);
% N1w
  kchf(2)=Nwdyy(9,m,n)+Nwdyy(10,m,n)+Nwdyy(13,m,n)+Nwdyy(14,
m,n);  % N2w
  kchf(3)=Nwdyy(11,m,n)+Nwdyy(12,m,n)+Nwdyy(15,m,n)+Nwdyy(16,
m,n); % N3w
  kchf(4)=Nwdyy(3,m,n)+Nwdyy(4,m,n)+Nwdyy(7,m,n)+Nwdyy(8,m,n);
% N4w
  kchf(5)=+b^2/9*Nwdyy(2,m,n);              % N1tx
  kchf(6)=+b^2/9*Nwdyy(14,m,n);             % N2tx
  kchf(7)=-b^2/9*Nwdyy(15,m,n);             % N3tx
  kchf(8)=-b^2/9*Nwdyy(3,m,n);              % N4tx
  kchf(9) =-a^2/9*Nwdyy(5,m,n);             % N1ty
  kchf(10)=+a^2/9*Nwdyy(9,m,n);             % N2ty
  kchf(11)=+a^2/9*Nwdyy(12,m,n);            % N3ty
  kchf(12)=-a^2/9*Nwdyy(8,m,n);             % N4ty
  for i=1:ndof
    for j=1:ndof
        Kmat(i,j) = Kmat(i,j) + kchf(j)/Jy^4*D* ...
            kchf(i)/Jy^4*Jac*Gpwx(m,2)*Gpwy(n,2);
    end
  end
  end
end
% Dxy parts
for m=1:ngpx
  for n=1:ngpy
  % Shape functions
```

```
  kchf(1)=Nwdxy(1,m,n)+Nwdxy(2,m,n)+Nwdxy(5,m,n)+Nwdxy(6,m,n);
% N1w
  kchf(2)=Nwdxy(9,m,n)+Nwdxy(10,m,n)+Nwdxy(13,m,n)+Nwdxy(14,
m,n);  % N2w
  kchf(3)=Nwdxy(11,m,n)+Nwdxy(12,m,n)+Nwdxy(15,m,n)+Nwdxy(16,
m,n); % N3w
  kchf(4)=Nwdxy(3,m,n)+Nwdxy(4,m,n)+Nwdxy(7,m,n)+Nwdxy(8,m,n);
% N4w
  kchf(5)=+b^2/9*Nwdxy(2,m,n);                % N1tx
  kchf(6)=+b^2/9*Nwdxy(14,m,n);               % N2tx
  kchf(7)=-b^2/9*Nwdxy(15,m,n);               % N3tx
  kchf(8)=-b^2/9*Nwdxy(3,m,n);                % N4tx
  kchf(9) =-a^2/9*Nwdxy(5,m,n);               % N1ty
  kchf(10)=+a^2/9*Nwdxy(9,m,n);               % N2ty
  kchf(11)=+a^2/9*Nwdxy(12,m,n);              % N3ty
  kchf(12)=-a^2/9*Nwdxy(8,m,n);               % N4ty
  for i=1:ndof
    for j=1:ndof
        Kmat(i,j) = Kmat(i,j) + kchf(j)/Jx^2/Jy^2*Dxy* ...
            kchf(i)/Jx^2/Jy^2*Jac*Gpwx(m,2)*Gpwy(n,2);
    end
  end
  end
end
% Dxxyy parts
for m=1:ngpx
  for n=1:ngpy
  % Shape functions
  kchfx(1)=Nwdxx(1,m,n)+Nwdxx(2,m,n)+Nwdxx(5,m,n)+Nwdxx(6,
m,n);      % N1w
  kchfx(2)=Nwdxx(9,m,n)+Nwdxx(10,m,n)+Nwdxx(13,m,n)+Nwdxx(14,
m,n);  % N2w
  kchfx(3)=Nwdxx(11,m,n)+Nwdxx(12,m,n)+Nwdxx(15,m,n)+Nwdxx(16,
m,n); % N3w
  kchfx(4)=Nwdxx(3,m,n)+Nwdxx(4,m,n)+Nwdxx(7,m,n)+Nwdxx(8,
m,n);      % N4w
  kchfx(5)=+b^2/9*Nwdxx(2,m,n);               % N1tx
  kchfx(6)=+b^2/9*Nwdxx(14,m,n);              % N2tx
  kchfx(7)=-b^2/9*Nwdxx(15,m,n);              % N3tx
  kchfx(8)=-b^2/9*Nwdxx(3,m,n);               % N4tx
  kchfx(9) =-a^2/9*Nwdxx(5,m,n);              % N1ty
  kchfx(10)=+a^2/9*Nwdxx(9,m,n);              % N2ty
  kchfx(11)=+a^2/9*Nwdxx(12,m,n);             % N3ty
  kchfx(12)=-a^2/9*Nwdxx(8,m,n);              % N4ty
  kchfy(1)=Nwdyy(1,m,n)+Nwdyy(2,m,n)+Nwdyy(5,m,n)+Nwdyy(6,
m,n);      % N1w
  kchfy(2)=Nwdyy(9,m,n)+Nwdyy(10,m,n)+Nwdyy(13,m,n)+Nwdyy(14,
m,n);  % N2w
  kchfy(3)=Nwdyy(11,m,n)+Nwdyy(12,m,n)+Nwdyy(15,m,n)+Nwdyy(16,
m,n); % N3w
```

```
  kchfy(4)=Nwdyy(3,m,n)+Nwdyy(4,m,n)+Nwdyy(7,m,n)+Nwdyy(8,
m,n);     % N4w
  kchfy(5)=-a^2/9*Nwdyy(2,m,n);                % N1ty
  kchfy(6)=+a^2/9*Nwdyy(14,m,n);               % N2ty
  kchfy(7)=+a^2/9*Nwdyy(15,m,n);               % N3ty
  kchfy(8)=-a^2/9*Nwdyy(3,m,n);                % N4ty
  kchfy(9) =+b^2/9*Nwdyy(5,m,n);               % N1tx
  kchfy(10)=+b^2/9*Nwdyy(9,m,n);               % N2tx
  kchfy(11)=-b^2/9*Nwdyy(12,m,n);              % N3tx
  kchfy(12)=-b^2/9*Nwdyy(8,m,n);               % N4tx
  for i=1:ndof
    for j=1:ndof
        Kmat(i,j) = Kmat(i,j) + kchfy(j)/Jy^4*Dxxyy* ...
            kchfx(i)/Jx^4*Jac*Gpwx(m,2)*Gpwy(n,2);
    end
  end
  end
end
% Swapping the matrices to the general displacement vector order
Kmat=Kmat([1,4,7,10,2,5,8,11,3,6,9,12],:); % row
Kmat=Kmat(:,[1,4,7,10,2,5,8,11,3,6,9,12]); % column
disp('Stiffness Matrix of Kirchhoff Plate');
disp(Kmat);
%
% Mass matrix Mmat
Mmat = zeros(ndof,ndof);
% Rho-Ixy parts
for m=1:ngpx
  for n=1:ngpy
  % Shape functions
  kchf(1)=Nw(1,m,n)+Nw(2,m,n)+Nw(5,m,n)+Nw(6,m,n);      % N1w
  kchf(2)=Nw(9,m,n)+Nw(10,m,n)+Nw(13,m,n)+Nw(14,m,n);   % N2w
  kchf(3)=Nw(11,m,n)+Nw(12,m,n)+Nw(15,m,n)+Nw(16,m,n);  % N3w
  kchf(4)=Nw(3,m,n)+Nw(4,m,n)+Nw(7,m,n)+Nw(8,m,n);      % N4w
  kchf(5)=+b^2/9*Nw(2,m,n);            % N1tx
  kchf(6)=+b^2/9*Nw(14,m,n);           % N2tx
  kchf(7)=-b^2/9*Nw(15,m,n);           % N3tx
  kchf(8)=-b^2/9*Nw(3,m,n);            % N4tx
  kchf(9) =-a^2/9*Nw(5,m,n);           % N1ty
  kchf(10)=+a^2/9*Nw(9,m,n);           % N2ty
  kchf(11)=+a^2/9*Nw(12,m,n);          % N3ty
  kchf(12)=-a^2/9*Nw(8,m,n);           % N4ty
  for i=1:ndof
    for j=1:ndof
        Mmat(i,j) = Mmat(i,j)+kchf(j)*rho*Ixy* ...
            kchf(i)*Jac*Gpwx(m,2)*Gpwy(n,2);
    end
  end
  end
end
```

```
% Rho-I-x parts
for m=1:ngpx
  for n=1:ngpy
  % Shape functions
  kchf(1)=Ntx(1,m,n)+Ntx(2,m,n)+Ntx(5,m,n)+Ntx(6,m,n);       %
N1w
  kchf(2)=Ntx(9,m,n)+Ntx(10,m,n)+Ntx(13,m,n)+Ntx(14,m,n);    %
N2w
  kchf(3)=Ntx(11,m,n)+Ntx(12,m,n)+Ntx(15,m,n)+Ntx(16,m,n);   %
N3w
  kchf(4)=Ntx(3,m,n)+Ntx(4,m,n)+Ntx(7,m,n)+Ntx(8,m,n);       %
N4w
  kchf(5)=+b^2/9*Ntx(2,m,n);                    % N1tx
  kchf(6)=+b^2/9*Ntx(14,m,n);                   % N2tx
  kchf(7)=-b^2/9*Ntx(15,m,n);                   % N3tx
  kchf(8)=-b^2/9*Ntx(3,m,n);                    % N4tx
  kchf(9) =-a^2/9*Ntx(5,m,n);                   % N1ty
  kchf(10)=+a^2/9*Ntx(9,m,n);                   % N2ty
  kchf(11)=+a^2/9*Ntx(12,m,n);                  % N3ty
  kchf(12)=-a^2/9*Ntx(8,m,n);                   % N4ty
  for i=1:ndof
    for j=1:ndof
        Mmat(i,j) = Mmat(i,j)+kchf(j)*rho*I* ...
            kchf(i)*Jac*Gpwx(m,2)*Gpwy(n,2);
    end
  end
  end
end
% Rho-I-y parts
for m=1:ngpx
  for n=1:ngpy
  % Shape functions
  kchf(1)=Nty(1,m,n)+Nty(2,m,n)+Nty(5,m,n)+Nty(6,m,n);       %
N1w
  kchf(2)=Nty(9,m,n)+Nty(10,m,n)+Nty(13,m,n)+Nty(14,m,n);    %
N2w
  kchf(3)=Nty(11,m,n)+Nty(12,m,n)+Nty(15,m,n)+Nty(16,m,n);   %
N3w
  kchf(4)=Nty(3,m,n)+Nty(4,m,n)+Nty(7,m,n)+Nty(8,m,n);       %
N4w
  kchf(5)=+b^2/9*Nty(2,m,n);                    % N1tx
  kchf(6)=+b^2/9*Nty(14,m,n);                   % N2tx
  kchf(7)=-b^2/9*Nty(15,m,n);                   % N3tx
  kchf(8)=-b^2/9*Nty(3,m,n);                    % N4tx
  kchf(9) =-a^2/9*Nty(5,m,n);                   % N1ty
  kchf(10)=+a^2/9*Nty(9,m,n);                   % N2ty
  kchf(11)=+a^2/9*Nty(12,m,n);                  % N3ty
  kchf(12)=-a^2/9*Nty(8,m,n);                   % N4ty
  for i=1:ndof
    for j=1:ndof
```

```
        Mmat(i,j) = Mmat(i,j)+kchf(j)*rho*I* ...
            kchf(i)*Jac*Gpwx(m,2)*Gpwy(n,2);
    end
  end
  end
end
% Swapping the matrices to the general displacement vector
order
Mmat=Mmat([1,4,7,10,2,5,8,11,3,6,9,12],:); % row
Mmat=Mmat(:,[1,4,7,10,2,5,8,11,3,6,9,12]); % column
disp('Mass Matrix of Kirchhoff Plate');
disp(Mmat);
```

6.4.2 NurbsSurface Function List

```
function NurbsSFC = NurbsSurface(p,knotp,q,knotq,ss,tt)
% NurbsSurface.m using DNurbsLeibnitzR7
% p,q = degree of Bspline curves, knot = knot vector, ss,tt =
parameter
%-------------------------------------------------------------
% Parameter ksi=s and eta=t
ns=size(knotp,2)-p-1;
nt=size(knotq,2)-q-1;
ws=ones(ns-p);
wt=ones(nt-q);
wst=ones(ns-p,nt-q);
NurbsSFC=zeros(6,ns-p,nt-q);
% 1 for NurbsSurface
% 2 for 1st derivative R,x of NurbsSurface
% 3 for 1st derivative R,y of NurbsSurface
% 4 for 2nd derivative R,xx of NurbsSurface
% 5 for 2nd derivative R,yx of NurbsSurface
% 6 for 2nd derivative R,yy of NurbsSurface
Nurbsksi = DNurbsLeibnitz(p,knotp,ws,ss);
SX0(:) = Nurbsksi(:,p+5);
SX1(:) = Nurbsksi(:,p+7);
SX2(:) = Nurbsksi(:,p+8);
Nurbseta = DNurbsLeibnitz(q,knotq,wt,tt);
SY0(:) = Nurbseta(:,q+5);
SY1(:) = Nurbseta(:,q+7);
SY2(:) = Nurbseta(:,q+8);
T = zeros(6,ns-p,nt-q); % Numerator    :T,Tx,Ty,Txx,Txy,Tyy
Z = zeros(6);           % Denominator :Z,Zx,Zy,Zxx,Zxy,Zyy
for i=1:ns-p            % order of NURBS in ksi
  for j=1:nt-q          % order of NURBS in eta
    T(1,i,j) = SX0(i)*SY0(j)*wst(i,j); % NurbsSurface checked
Fig1.18R3.m
    T(2,i,j) = SX1(i)*SY0(j)*wst(i,j);
    T(3,i,j) = SX0(i)*SY1(j)*wst(i,j);
    T(4,i,j) = SX2(i)*SY0(j)*wst(i,j);
```

```
    T(5,i,j)  = SX1(i)*SY1(j)*wst(i,j);
    T(6,i,j)  = SX0(i)*SY2(j)*wst(i,j);
    Z(1)  = Z(1)  + SX0(i)*SY0(j)*wst(i,j); % 0th-der
    Z(2)  = Z(2)  + SX1(i)*SY0(j)*wst(i,j);
    Z(3)  = Z(3)  + SX0(i)*SY1(j)*wst(i,j);
    Z(4)  = Z(4)  + SX2(i)*SY0(j)*wst(i,j);
    Z(5)  = Z(5)  + SX1(i)*SY1(j)*wst(i,j);
    Z(6)  = Z(6)  + SX0(i)*SY2(j)*wst(i,j);
  end
end
% Rationalized NurbsSurface
for i=1:ns-p             % order of NURBS in ksi
    for j=1:nt-q         % order of NURBS in eta
        NurbsSFC(1,i,j) = T(1,i,j)/Z(1);
        % Derivative of NurbsSurface 2:R,x 3:R,y 4:R,xx 5:R,xy
6:R,yy
        NurbsSFC(2,i,j) = T(2,i,j)/Z(1)
- T(1,i,j)*Z(2)/Z(1)/Z(1);
        NurbsSFC(3,i,j) = T(3,i,j)/Z(1)
- T(1,i,j)*Z(3)/Z(1)/Z(1);
        NurbsSFC(4,i,j) = T(4,i,j)/Z(1) -
2*T(2,i,j)*Z(2)/Z(1)^2 ...
          +2*T(1,i,j)*Z(2)^2/Z(1)^3 - T(1,i,j)*Z(4)/Z(1)^2;
        NurbsSFC(5,i,j) = T(5,i,j)/Z(1) - T(2,i,j)*Z(3)/Z(1)^2
...
          -T(3,i,j)*Z(2)/Z(1)^2 + 2*T(1,i,j)*Z(2)*Z(3)/Z(1)^3
...
          -T(1,i,j)*Z(5)/Z(1)^2;
        NurbsSFC(6,i,j) = T(6,i,j)/Z(1) -
2*T(3,i,j)*Z(3)/Z(1)^2 ...
          +2*T(1,i,j)*Z(3)^2/Z(1)^3 - T(1,i,j)*Z(6)/Z(1)^2;
    end
end
return
```

6.5 NURBS for Mindlin Thick Plate Element

Similar to FEM, the governing equation of a plate element in IGA can be developed conveniently from the theory of minimum potential energy principle as follows:

$$\Pi = \frac{1}{2}\{\mathbf{d}\}^T[\mathbf{K}]\{\mathbf{d}\} + \frac{1}{2}\{\ddot{\mathbf{d}}\}^T[\mathbf{M}]\{\ddot{\mathbf{d}}\} - \{\mathbf{d}\}^T\{\mathbf{f}\}$$

$$\frac{d\Pi}{d\{\mathbf{d}\}^T} = [\mathbf{K}]\{\mathbf{d}\} + [\mathbf{M}]\{\ddot{\mathbf{d}}\} - \{\mathbf{f}\} = \{\mathbf{0}\} \tag{6.14}$$

where the stiffness matrix $[\mathbf{K}]$, the mass matrix $[\mathbf{M}]$, and the loading vector $\{\mathbf{f}\}$ are described below. Referring to Equation (5.80), the stiffness matrix of an FSDT plate based on IGA can be rewritten as

$$[\mathbf{K}] = \int_0^b \int_0^a \begin{bmatrix} K_{ww} & K_{w\theta_x} & K_{w\theta_y} \\ K_{\theta_x w} & K_{\theta_x\theta_x} & K_{\theta_x\theta_y} \\ K_{\theta_y w} & K_{\theta_y\theta_x} & K_{\theta_y\theta_y} \end{bmatrix} dxdy \tag{6.15}$$

where

$$\begin{aligned}
K_{ww} &= \mathbf{N}_{w,x}^T D_{xz} \mathbf{N}_{w,x} + \mathbf{N}_{w,y}^T D_{yz} \mathbf{N}_{w,y} \\
K_{w\theta_x} &= -\mathbf{N}_{w,y}^T D_{yz} \mathbf{N}_{\theta_x} \\
K_{w\theta_y} &= -\mathbf{N}_{w,x}^T D_{xz} \mathbf{N}_{\theta_y} \\
K_{\theta_x w} &= -\mathbf{N}_{\theta_x}^T D_{yz} \mathbf{N}_{w,y} \\
K_{\theta_y w} &= -\mathbf{N}_{\theta_y}^T D_{xz} \mathbf{N}_{w,x} \\
K_{\theta_x\theta_x} &= \mathbf{N}_{\theta_{x,y}}^T D \mathbf{N}_{\theta_{x,y}} + \mathbf{N}_{\theta_{x,x}}^T D_{yx} \mathbf{N}_{\theta_{x,x}} + \mathbf{N}_{\theta_x}^T D_{yz} \mathbf{N}_{\theta_x} \\
K_{\theta_x\theta_y} &= \mathbf{N}_{\theta_{x,y}}^T D_{yyxx} \mathbf{N}_{\theta_{y,x}} + \mathbf{N}_{\theta_{x,x}}^T D_{yx} \mathbf{N}_{\theta_{y,y}} \\
K_{\theta_y\theta_x} &= \mathbf{N}_{\theta_{y,x}}^T D_{xxyy} \mathbf{N}_{\theta_{x,y}} + \mathbf{N}_{\theta_{y,y}}^T D_{xy} \mathbf{N}_{\theta_{x,x}} \\
K_{\theta_y\theta_y} &= \mathbf{N}_{\theta_{y,x}}^T D \mathbf{N}_{\theta_{y,x}} + \mathbf{N}_{\theta_{y,y}}^T D_{xy} \mathbf{N}_{\theta_{y,y}} + \mathbf{N}_{\theta_y}^T D_{xz} \mathbf{N}_{\theta_y}
\end{aligned} \tag{6.16}$$

where $D_{xz} = D_{yz}$, $D_{xy} = D_{yx}$, and $D_{xxyy} = D_{yyxx}$.

The NURBS-based basis functions are obtained from Equation (6.1) by omitting the DOF terms so that it becomes

$$\begin{aligned}
\mathbf{N}_w &= \sum_{i=0}^{m} \sum_{j=0}^{n} R_{i,j}^{p,q}(\xi,\eta) \\
\mathbf{N}_{\theta_x} &= \sum_{i=0}^{m} \sum_{j=0}^{n} R_{i,j}^{p,q}(\xi,\eta) \\
\mathbf{N}_{\theta_y} &= \sum_{i=0}^{m} \sum_{j=0}^{n} R_{i,j}^{p,q}(\xi,\eta)
\end{aligned} \tag{6.17}$$

where

$$\mathbf{N}_{w,x}=\frac{d\mathbf{N}_w}{dx}=\frac{d\mathbf{N}_w}{d\xi}\times\frac{1}{J_x}=\frac{1}{J_x}\sum_{i=0}^{m}\sum_{j=0}^{n}\left\{\frac{dR_{i,j}^{p,q}(\xi,\eta)}{d\xi}\right\}$$

$$\mathbf{N}_{w,y}=\frac{d\mathbf{N}_w}{dy}=\frac{d\mathbf{N}_w}{d\eta}\times\frac{1}{J_y}=\frac{1}{J_y}\sum_{i=0}^{m}\sum_{j=0}^{n}\left\{\frac{dR_{i,j}^{p,q}(\xi,\eta)}{d\eta}\right\} \tag{6.18}$$

and

$$\mathbf{N}_{\theta_x,x}=\frac{d\mathbf{N}_{\theta_x}}{dx}=\frac{d\mathbf{N}_{\theta_x}}{d\xi}\times\frac{1}{J_x}=\frac{1}{J_x}\sum_{i=0}^{m}\sum_{j=0}^{n}\left\{\frac{dR_{i,j}^{p,q}(\xi,\eta)}{d\xi}\right\}$$

$$\mathbf{N}_{\theta_x,y}=\frac{d\mathbf{N}_{\theta_x}}{dy}=\frac{d\mathbf{N}_{\theta_x}}{d\eta}\times\frac{1}{J_y}=\frac{1}{J_y}\sum_{i=0}^{m}\sum_{j=0}^{n}\left\{\frac{dR_{i,j}^{p,q}(\xi,\eta)}{d\eta}\right\}$$

$$\mathbf{N}_{\theta_y,x}=\frac{d\mathbf{N}_{\theta_y}}{dx}=\frac{d\mathbf{N}_{\theta_y}}{d\xi}\times\frac{1}{J_x}=\frac{1}{J_x}\sum_{i=0}^{m}\sum_{j=0}^{n}\left\{\frac{dR_{i,j}^{p,q}(\xi,\eta)}{d\xi}\right\}$$

$$\mathbf{N}_{\theta_y,y}=\frac{d\mathbf{N}_{\theta_y}}{dy}=\frac{d\mathbf{N}_{\theta_y}}{d\eta}\times\frac{1}{J_y}=\frac{1}{J_y}\sum_{i=0}^{m}\sum_{j=0}^{n}\left\{\frac{dR_{i,j}^{p,q}(\xi,\eta)}{d\eta}\right\} \tag{6.19}$$

where $J_x = d\xi/dx$ and $J_y = d\eta/dy$ are the Jacobian operators in the x- and y-directions, respectively.

Equation (6.15) will be numerically integrated by using the Gaussian quadrature as follows:

$$[\mathbf{K}]=\sum_{i=0}^{m}\sum_{j=0}^{n}\begin{bmatrix} K_{ww} & K_{w\theta_x} & K_{w\theta_y} \\ K_{\theta_x w} & K_{\theta_x\theta_x} & K_{\theta_x\theta_y} \\ K_{\theta_y w} & K_{\theta_y\theta_x} & K_{\theta_y\theta_y} \end{bmatrix} Jw_jw_i \tag{6.20}$$

where J is the determinant value of the bidirectional Jacobian operator defined in Equation (2.7).

Referring to Equation (5.42), the mass matrix of an FSDT plate based on IGA can be written as

$$[\mathbf{M}]=\int_0^b\int_0^a\begin{bmatrix} \mathbf{N}_w \\ \mathbf{N}_{\theta_x} \\ \mathbf{N}_{\theta_y} \end{bmatrix}^T\begin{bmatrix} \rho I_{xy} & 0 & 0 \\ 0 & \rho I & 0 \\ 0 & 0 & \rho I \end{bmatrix}\begin{bmatrix} \mathbf{N}_w \\ \mathbf{N}_{\theta_x} \\ \mathbf{N}_{\theta_y} \end{bmatrix} dx\,dy \tag{6.21}$$

where the NURBS-based basis functions are obtained from Equation (6.1) by omitting the DOF terms:

$$\mathbf{N}_w = \sum_{i=0}^{m}\sum_{j=0}^{n} R_{i,j}^{p,q}(\xi,\eta)$$

$$\mathbf{N}_{\theta_x} = \sum_{i=0}^{m}\sum_{j=0}^{n} R_{i,j}^{p,q}(\xi,\eta) \tag{6.22}$$

$$\mathbf{N}_{\theta_y} = \sum_{i=0}^{m}\sum_{j=0}^{n} R_{i,j}^{p,q}(\xi,\eta)$$

Equation (6.21) will be numerically integrated by using the Gaussian quadrature as follows:

$$[\mathbf{M}] = \sum_{i=0}^{m}\sum_{j=0}^{n} \begin{bmatrix} \mathbf{N}_w \\ \mathbf{N}_{\theta_x} \\ \mathbf{N}_{\theta_y} \end{bmatrix}^T \begin{bmatrix} \rho I_{xy} & 0 & 0 \\ 0 & \rho I & 0 \\ 0 & 0 & \rho I \end{bmatrix} \begin{bmatrix} \mathbf{N}_w \\ \mathbf{N}_{\theta_x} \\ \mathbf{N}_{\theta_y} \end{bmatrix} J w_j w_i \tag{6.23}$$

Referring to Equation (5.43), the loading vector of an FSDT plate based on IGA is given by

$$\{\mathbf{f}\} = \int_0^b \int_0^a \begin{bmatrix} 1 \\ \mathbf{N}_{w,x} \\ \mathbf{N}_{w,y} \end{bmatrix}^T \begin{bmatrix} q_z & 0 & 0 \\ 0 & N_x & N_{xy} \\ 0 & N_{yx} & N_y \end{bmatrix} \begin{bmatrix} \mathbf{N}_w \\ \mathbf{N}_{w,x} \\ \mathbf{N}_{w,y} \end{bmatrix} dx\,dy \tag{6.24}$$

where the NURBS-based basis functions are obtained from Equation (6.1) by omitting the DOF terms:

$$\mathbf{N}_{w,x} = \frac{d\mathbf{N}_w}{dx} = \frac{d\mathbf{N}_w}{d\xi} \times \frac{1}{J} = \frac{1}{J}\sum_{i=0}^{m}\sum_{j=0}^{n}\left\{\frac{dR_{i,j}^{p,q}(\xi,\eta)}{d\xi}\right\} \tag{6.25}$$

and

$$\mathbf{N}_{w,y} = \frac{d\mathbf{N}_w}{dy} = \frac{d\mathbf{N}_w}{d\eta} \times \frac{1}{J} = \frac{1}{J}\sum_{i=0}^{m}\sum_{j=0}^{n}\left\{\frac{dR_{i,j}^{p,q}(\xi,\eta)}{d\eta}\right\} \tag{6.26}$$

Equation (6.24) will be numerically integrated by using the Gaussian quadrature as follows:

$$\{\mathbf{f}\} = \sum_{i=0}^{m}\sum_{j=0}^{n}\begin{bmatrix} 1 \\ \mathbf{N}_{w,x} \\ \mathbf{N}_{w,y} \end{bmatrix}^T \begin{bmatrix} q_z & 0 & 0 \\ 0 & N_x & N_{xy} \\ 0 & N_{yx} & N_y \end{bmatrix} \begin{bmatrix} \mathbf{N}_w \\ \mathbf{N}_{w,x} \\ \mathbf{N}_{w,y} \end{bmatrix} J w_j w_i \tag{6.27}$$

The generalized displacement vectors $\{\mathbf{d}\}^T$ are given as

$$\begin{aligned} \mathbf{w} &= \left[w_{ij} \dots w_{ij=p\times m+1, q\times n+1}\right]^T; \quad \ddot{\mathbf{w}} = \left[\ddot{w}_{ij} \dots \ddot{w}_{ij=p\times m+1, q\times n+1}\right]^T \\ \boldsymbol{\theta}_x &= \left[\theta_{xij} \dots \theta_{xij=p\times m+1, q\times n+1}\right]^T; \quad \ddot{\boldsymbol{\theta}}_x = \left[\ddot{\theta}_{xij} \dots \ddot{\theta}_{xij=p\times m+1, q\times n+1}\right]^T \\ \boldsymbol{\theta}_y &= \left[\theta_{yij} \dots \theta_{yij=p\times m+1, q\times n+1}\right]^T; \quad \ddot{\boldsymbol{\theta}}_y = \left[\ddot{\theta}_{yij} \dots \ddot{\theta}_{yij=p\times m+1, q\times n+1}\right]^T \end{aligned} \tag{6.28}$$

where the notation ($\ddot{\ }$) means the second derivative with respect to time. The subscripts p and q are the orders of the NURBS surface. The subscripts m and n are the number of integration points of the Gaussian quadrature in the x- and y-directions, respectively.

6.6 Making the Stiffness and Mass Matrices of the Mindlin Plate Element

MATLAB code KMmatrixMindlinNurbs.m uses the functions NurbsSurface.m and Legendre.m to construct the stiffness and mass matrices of the Mindlin plate theory (FSDT) of an isotropic plate, as shown in Figure 6.3.

6.6.1 KMmatrixMindlinIGA Program List

```
% KMmatrixMindlinIGA.m (Using NurbsSurface.m)  : R6
% Construct matrices K and M
% Based on the Mindlin Plate Theory by using NURBS surface
shape functions
clear all; clc;
% isotropic material
a     = 1.5;          b     = 1.2;       h     = 0.1;
E     = 1525;         nu    = 0.3;       rho   = 630;
I     = h^3/12;       Ixy   = h;         kappa = 5/6;
D     = E*h^3/12/(1-nu^2);
```

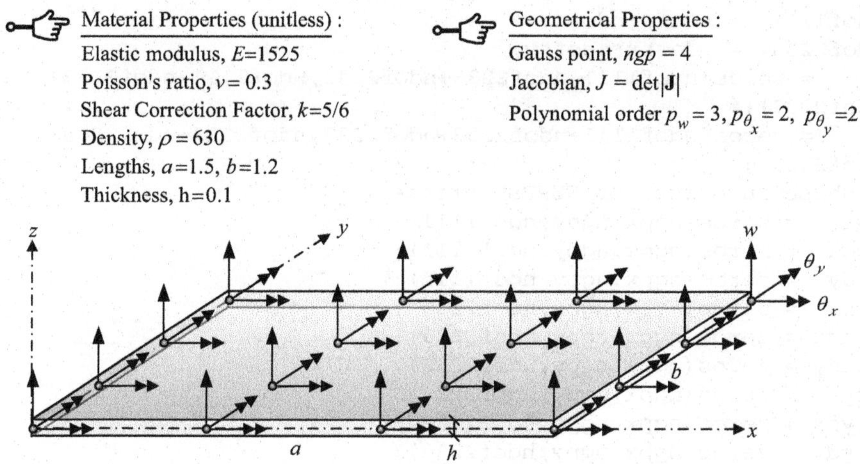

FIGURE 6.3
A rectangular isotropic flat Mindlin plate IGA model.

```
Dxxyy = nu*E*h^3/6/(1-nu^2);
Dxy   = E*h^3/24/(1+nu);
G = E/2/(1+nu);
Dxz = kappa*G*h;
% NURBS for w and theta displacement data in x and y directions
npw     = 3;
nptx    = 2;
npty    = 2;
knotw   = [-ones(1,npw+1) ones(1,npw+1)]; knotw = [knotw ones(1,npw)];
knottx = [-ones(1,nptx+1) ones(1,nptx+1)]; knottx = [knottx ones(1,nptx)];
knotty = [-ones(1,npty+1) ones(1,npty+1)]; knotty = [knotty ones(1,npty)];
nbw     = size(knotw,2)-npw-1;
nbtx    = size(knottx,2)-nptx-1;
nbty    = size(knotty,2)-npty-1;
% Gauss
ngpx  = 4;                    ngpy  = 4;
Gpwx  = Legendre(ngpx);       Gpwy  = Legendre(ngpy);
% Jac
Jac = a/2*b/2;
Jx = a/2;
Jy = b/2;
% Initialization
ndofw       = (nbw-npw);
ndoftx      = (nbtx-nptx);
ndofty      = (nbty-npty);
```

```
ndof1111  = ndofw*ndofw;
ndof2233  = ndoftx*ndofty;
K    = zeros(ndof1111+ndof2233+ndof2233,ndof1111+ndof2233+
ndof2233);
M    = zeros(ndof1111+ndof2233+ndof2233,ndof1111+ndof2233+
ndof2233);
% Shape functions of NURBS surface
Nw    = zeros(ngpx,ngpy,ndof1111);
Nwdx  = zeros(ngpx,ngpy,ndof1111);
Nwdy  = zeros(ngpx,ngpy,ndof1111);
Ntx   = zeros(ngpx,ngpy,ndof2233);
Ntxdx = zeros(ngpx,ngpy,ndof2233);
Ntxdy = zeros(ngpx,ngpy,ndof2233);
Nty   = zeros(ngpx,ngpy,ndof2233);
Ntydx = zeros(ngpx,ngpy,ndof2233);
Ntydy = zeros(ngpx,ngpy,ndof2233);
% Stiffness and Mass Matrices K & M
% K11 and M11 of w displacements
% Filling NURBS surface shape functions into the gauss points
for m=1:ngpx
    s = Gpwx(m,1);
    for n=1:ngpy
      t = Gpwy(n,1);
      NurbsMNDLNw = NurbsSurface(npw,knotw,npw,knotw,s,t);
      for i=1:ndofw
        for j=1:ndofw
          ij=(i-1)*ndofw+j; % Columnwise arrangement of DOF
w-w
          Nw(m,n,ij)      = NurbsMNDLNw(1,i,j);    % 0th
derivative
          Nwdx(m,n,ij)    = NurbsMNDLNw(2,i,j);    % x
derivative
          Nwdy(m,n,ij)    = NurbsMNDLNw(3,i,j);    % y
derivative
        end
      end
      NurbsMNDLNt = NurbsSurface(nptx,knottx,npty,knotty,s,t);
      for i=1:ndoftx
        for j=1:ndofty
          ij=(i-1)*ndoftx+j; % Columnwise arrangement of DOF w-w
          Ntx(m,n,ij)     = NurbsMNDLNt(1,i,j);    % 0th
derivative
          Ntxdx(m,n,ij)   = NurbsMNDLNt(2,i,j);    % x
derivative
          Ntxdy(m,n,ij)   = NurbsMNDLNt(3,i,j);    % y
derivative
          Nty(m,n,ij)     = NurbsMNDLNt(1,i,j);    % 0th
derivative
          Ntydx(m,n,ij)   = NurbsMNDLNt(2,i,j);    % x
derivative
```

```
          Ntydy(m,n,ij)  = NurbsMNDLNt(3,i,j);    % y
derivative
        end
      end
    end
end
kmat11 = zeros(ndof1111,ndof1111);
mmat11 = zeros(ndof1111,ndof1111);
for i=1:ndof1111
  for j=1:ndof1111
    for m=1:ngpx
      for n=1:ngpy
        % Stiffness
        kmat11(i,j) = kmat11(i,j)+Nwdx(m,n,j)/Jx*Dxz* ...
            Nwdx(m,n,i)/Jx*Jac*Gpwx(m,2)*Gpwy(n,2);
        kmat11(i,j) = kmat11(i,j)+Nwdy(m,n,j)/Jy*Dxz* ...
            Nwdy(m,n,i)/Jy*Jac*Gpwx(m,2)*Gpwy(n,2);
        % Mass
        mmat11(i,j) = mmat11(i,j)+Nw(m,n,j)*rho*Ixy* ...
                    Nw(m,n,i)*Jac*Gpwx(m,2)*Gpwy(n,2);
      end
    end
  end
end
K(1:ndof1111,1:ndof1111)=kmat11; % K11 inclusion
M(1:ndof1111,1:ndof1111)=mmat11; % M11 inclusion
% Stiffness and Mass Matrices K
% K12,K21,K13,K31 of w-txy and w-ty displacements and
rotations
kmat12 = zeros(ndof1111,ndof2233);
kmat21 = zeros(ndof2233,ndof1111);
kmat13 = zeros(ndof1111,ndof2233);
kmat31 = zeros(ndof2233,ndof1111);
for i=1:ndof1111
  for j=1:ndof2233
    for m=1:ngpx
      for n=1:ngpy
        % Stiffness
        kmat12(i,j) = kmat12(i,j)-Nwdy(m,n,i)/Jy*Dxz* ...
            Ntx(m,n,j)*Jac*Gpwx(m,2)*Gpwy(n,2);
        kmat21(j,i) = kmat21(j,i)-Nwdy(m,n,i)/Jy*Dxz* ...
            Ntx(m,n,j)*Jac*Gpwx(m,2)*Gpwy(n,2);
        kmat13(i,j) = kmat13(i,j)-Nwdx(m,n,i)/Jx*Dxz* ...
            Nty(m,n,j)*Jac*Gpwx(m,2)*Gpwy(n,2);
        kmat31(j,i) = kmat31(j,i)-Nwdx(m,n,i)/Jx*Dxz* ...
            Nty(m,n,j)*Jac*Gpwx(m,2)*Gpwy(n,2);
      end
    end
  end
end
```

```
K(1:ndof1111,ndof1111+1:ndof1111+ndof2233)=kmat12;
% K12
K(ndof1111+1:ndof1111+ndof2233,1:ndof1111)=kmat21;
% K21
K(1:ndof1111,ndof1111+ndof2233+1:ndof1111+ndof2233+ndof2233)=k
mat13;  % K13
K(ndof1111+ndof2233+1:ndof1111+ndof2233+ndof2233,1:ndof1111)=k
mat31;  % K31
% Stiffness and Mass Matrices K & M
% K22,K33 and M22,M33 of theta-x,y rotations
kmat22 = zeros(ndof2233,ndof2233);
kmat23 = zeros(ndof2233,ndof2233);
kmat32 = zeros(ndof2233,ndof2233);
kmat33 = zeros(ndof2233,ndof2233);
mmat22 = zeros(ndof2233,ndof2233);
mmat33 = zeros(ndof2233,ndof2233);
for i=1:ndof2233
  for j=1:ndof2233
    for m=1:ngpx
      for n=1:ngpy
        % Stiffness
        kmat22(i,j) = kmat22(i,j)+ ...
          Ntxdy(m,n,j)/Jy*D*Ntxdy(m,n,i)/Jy*Jac*Gpwx(m,2)*
Gpwy(n,2)+ ...
          Ntxdx(m,n,j)/Jx*Dxy*Ntxdx(m,n,i)/Jx*Jac*Gpwx(m,2)*
Gpwy(n,2)+...
          Ntx(m,n,j)*Dxz*Ntx(m,n,i)*Jac*Gpwx(m,2)*Gpwy(n,2);
        kmat23(i,j) = kmat23(i,j)+ ...
          Ntxdy(m,n,j)/Jy*Dxxyy*Ntydx(m,n,i)/Jx*Jac*Gpwx(m,2)*
Gpwy(n,2)+ ...
          Ntxdx(m,n,j)/Jx*Dxy*Ntydy(m,n,i)/Jy*Jac*Gpwx(m,2)*
Gpwy(n,2);
        kmat32(j,i) = kmat32(j,i)+ ...
          Ntydx(m,n,i)/Jx*Dxxyy*Ntxdy(m,n,j)/Jy*Jac*Gpwx(m,2)*
Gpwy(n,2)+ ...
          Ntydy(m,n,i)/Jy*Dxy*Ntxdx(m,n,j)/Jx*Jac*Gpwx(m,2)*
Gpwy(n,2);
        kmat33(i,j) = kmat33(i,j)+ ...
          Ntydx(m,n,j)/Jx*D*Ntydx(m,n,i)/Jx*Jac*Gpwx(m,2)*
Gpwy(n,2)+ ...
          Ntydy(m,n,j)/Jy*Dxy*Ntydy(m,n,i)/Jy*Jac*Gpwx(m,2)*
Gpwy(n,2)+...
          Nty(m,n,j)*Dxz*Nty(m,n,i)*Jac*Gpwx(m,2)*Gpwy(n,2);
        % Mass
        mmat22(i,j) = mmat22(i,j)+Ntx(m,n,j)*rho*I* ...
            Ntx(m,n,i)*Jac*Gpwx(m,2)*Gpwy(n,2);
        mmat33(i,j) = mmat33(i,j)+Nty(m,n,j)*rho*I* ...
            Nty(m,n,i)*Jac*Gpwx(m,2)*Gpwy(n,2);
      end
    end
```

```
  end
end
K(ndof1111+1:ndof1111+ndof2233,ndof1111+1:ndof1111+ndof2233)=k
mat22;  % K22
K(ndof1111+1:ndof1111+ndof2233,ndof1111+ndof2233+1: ...
    ndof1111+ndof2233+ndof2233)=kmat23;
% K23
K(ndof1111+ndof2233+1:ndof1111+ndof2233+ndof2233,ndof1111+1:
...
    ndof1111+ndof2233)=kmat32;
% K32
K(ndof1111+ndof2233+1:ndof1111+ndof2233+ndof2233,ndof1111+n
dof2233+1: ...
    ndof1111+ndof2233+ndof2233)=kmat33;
% K33
M(ndof1111+1:ndof1111+ndof2233,ndof1111+1:ndof1111+ndof2233)=m
mat22;  % K22
M(ndof1111+ndof2233+1:ndof1111+ndof2233+ndof2233,ndof1111+n
dof2233+1: ...
    ndof1111+ndof2233+ndof2233)=mmat33;
% M33
disp('Stiffness Matrix of Mindlin Plate');
disp(K);
disp('Mass Matrix of Mindlin Plate');
disp(M);
```

References

Basilevs Y, Calo VM, Cottrell JA, Evans J, Hughes TJR, Lipton S, Scott MA, Sederberg TW (2010) Isogeometric analysis using T-splines. *Comput Meth Appl Mech Eng* 199:229–263.

Cottrell JA, Reali A, Bazilevs Y, Hughes TJR (2006) Isogeometric analysis of structural vibrations. *Comput Meth Appl Mech Eng* 195:5257–5296.

Cottrell JA, Hughes TJR, Reali A (2007) Studies of refinement and continuity in isogeometric analysis. *Comput Meth Appl Mech Eng* 196:4160–4183.

Cottrell JA, Huges TJR, Bazilevs Y (2009) *Isogeometric Analysis: Toward Integration of CAD and FEA*. John Wiley & Sons, Ltd, Chichester.

Gan BS (2018) *An Isogeometric Approach to Beam Structures*. Springer, Switzerland.

Gontier C, Vollmer C (1995) A large displacement analysis of a beam using a CAD geometric definition. *Comp & Struct* 57(6):981–989.

Mindlin RD (1951) Influence of rotatory inertia and shear on flexural motions of isotropic, elastic plates. *J Appl Mech (ASME)* 18:31–38.

Reddy JN (1993) *An Introduction to the Finite Element Method*, 2nd edition. McGraw-Hill, New York.

Schramm U, Pilkey WD (1993) The coupling of geometric descriptions and finite elements using NURBs—A study in shape optimization. *Finite Elem in Anal and Des* 15:11–34.

7

Recoverable Sandwich Condensation

7.1 Condensation in Finite Element Method

The conventional static and dynamic condensations are known to be effective in reducing the number of degrees of freedom (DOFs) in a structure that has plenty of nodes. The DOFs of the unnecessary nodes are "hidden" so that only the DOFs at particular locations are still retained. DOF reduction is mostly found in the structural dynamic problems, which helps decrease the cost of computation in time-dependent solutions (Bathe 1982; Cook 1981; Hart and Wong 2000) considerably.

Although the importance of condensation in the finite element method (FEM) is not significant for plate and shell problems due to the number of DOFs that are not substantial, condensation of the unnecessary DOFs to the standard four-node 12-DOF plate and shell element has many benefits in real applications. First, the introduction of nonuniform rational basis spline (NURBS), which significantly increases the number of DOFs, into the design practice will hinder some design practitioners that are not familiar with the new NURBS geometry concept. Second, the condensed NURBS element codes can be integrated seamlessly into the existing plate and shell finite element codes.

In the old-style static and dynamic condensation methods, the plate and shell element matrix is usually partitioned into two sub-matrices which consist of retained and condensed parts. Therefore, there are only two simultaneous equations to be solved, that is a linear relationship between two variables. The condensed part will be eliminated from the governing equations, whereas the modified retained part is used to solve the plate and shell problems. This concept is very suitable for the NURBS functions where the DOFs of the unnecessary nodes are increased while using high order degree of the polynomial.

7.2 Sandwich Condensation for Isogeometric Analysis

In the *sandwich condensation* method (Gan 2018), the element matrices are divided into three sub-matrix simultaneous equations, where the unnecessary DOF matrix is subjected to the condensation. We have found that rather dividing the matrix into two sub-matrices, the method gives better results for beam elements.

By breaking down the matrix into three simultaneous sub-matrix equations, we can obtain a second-order relationship between the general displacements, because three unknowns have to be solved. The conventional dynamic condensation is then modified by taking twice derivatives to the free vibration equilibrium equations with respect to time t. Finally, we can obtain a relationship between the elements of the stiffness matrix $\mathbf{k}$ and the acceleration vector $\ddot{\mathbf{d}}$. The following sections will give the detail derivations of the sandwich condensation method.

7.3 Static Condensation

The sandwich condensation method starts from reducing the static equilibrium of three simultaneous equations which can be represented in the matrix form as

$$\begin{bmatrix} \mathbf{k}_{11} & \mathbf{k}_{12} & \mathbf{k}_{13} \\ \mathbf{k}_{21} & \mathbf{k}_{22} & \mathbf{k}_{23} \\ \mathbf{k}_{31} & \mathbf{k}_{32} & \mathbf{k}_{33} \end{bmatrix} \cdot \begin{Bmatrix} \mathbf{d}_1 \\ \mathbf{d}_2 \\ \mathbf{d}_3 \end{Bmatrix} = \begin{Bmatrix} \mathbf{f}_1 \\ \mathbf{f}_2 \\ \mathbf{f}_3 \end{Bmatrix} \tag{7.1}$$

From the above matrix equation, we can have three simultaneous equations as follows:

$$\text{(a) } \mathbf{k}_{11}\mathbf{d}_1 + \mathbf{k}_{12}\mathbf{d}_2 + \mathbf{k}_{13}\mathbf{d}_3 = \mathbf{f}_1$$

$$\text{(b) } \mathbf{k}_{21}\mathbf{d}_1 + \mathbf{k}_{22}\mathbf{d}_2 + \mathbf{k}_{23}\mathbf{d}_3 = \mathbf{f}_2$$

$$\text{(c) } \mathbf{k}_{31}\mathbf{d}_1 + \mathbf{k}_{32}\mathbf{d}_2 + \mathbf{k}_{33}\mathbf{d}_3 = \mathbf{f}_3$$

Equation (b) consists of the unnecessary DOFs that will be condensed. By extracting the generalized displacement vector $\mathbf{d}_2$ from Equation (b), we can have

$$\mathbf{k}_{22}\mathbf{d}_2 = \mathbf{f}_2 - \mathbf{k}_{21}\mathbf{d}_1 - \mathbf{k}_{23}\mathbf{d}_3$$

$$\mathbf{d}_2 = \mathbf{k}_{22}^{-1}\left(\mathbf{f}_2 - \mathbf{k}_{21}\mathbf{d}_1 - \mathbf{k}_{23}\mathbf{d}_3\right)$$

Next, substituting $\mathbf{d}_2$ into Equation (a) results in

$$\mathbf{k}_{11}\mathbf{d}_1 + \mathbf{k}_{12}\mathbf{k}_{22}^{-1}(\mathbf{f}_2 - \mathbf{k}_{21}\mathbf{d}_1 - \mathbf{k}_{23}\mathbf{d}_3) + \mathbf{k}_{13}\mathbf{d}_3 = \mathbf{f}_1$$

$$\left(\mathbf{k}_{11} - \mathbf{k}_{12}\mathbf{k}_{22}^{-1}\mathbf{k}_{21}\right)\mathbf{d}_1 + \left(\mathbf{k}_{13} - \mathbf{k}_{12}\mathbf{k}_{22}^{-1}\mathbf{k}_{23}\right)\mathbf{d}_3 = \mathbf{f}_1 - \mathbf{k}_{12}\mathbf{k}_{22}^{-1}\mathbf{f}_2$$

where

$$\bar{\mathbf{k}}_{11}\mathbf{d}_1 + \bar{\mathbf{k}}_{13}\mathbf{d}_3 = \bar{\mathbf{f}}_1 \tag{7.2}$$

where

$$\begin{aligned} \bar{\mathbf{k}}_{11} &= \mathbf{k}_{11} - \mathbf{k}_{12}\mathbf{k}_{22}^{-1}\mathbf{k}_{21} \\ \bar{\mathbf{k}}_{13} &= \mathbf{k}_{13} - \mathbf{k}_{12}\mathbf{k}_{22}^{-1}\mathbf{k}_{23} \\ \bar{\mathbf{f}}_1 &= \mathbf{f}_1 - \mathbf{k}_{12}\mathbf{k}_{22}^{-1}\mathbf{f}_2 \end{aligned} \tag{7.3}$$

Again, substituting $\mathbf{d}_2$ into Equation (c) results in

$$\mathbf{k}_{31}\mathbf{d}_1 + \mathbf{k}_{32}\mathbf{k}_{22}^{-1}(\mathbf{f}_2 - \mathbf{k}_{21}\mathbf{d}_1 - \mathbf{k}_{23}\mathbf{d}_3) + \mathbf{k}_{33}\mathbf{d}_3 = \mathbf{f}_3$$

$$\left(\mathbf{k}_{31} - \mathbf{k}_{32}\mathbf{k}_{22}^{-1}\mathbf{k}_{21}\right)\mathbf{d}_1 + \left(\mathbf{k}_{33} - \mathbf{k}_{32}\mathbf{k}_{22}^{-1}\mathbf{k}_{23}\right)\mathbf{d}_3 = \mathbf{f}_3 - \mathbf{k}_{32}\mathbf{k}_{22}^{-1}\mathbf{f}_2$$

The previous equation can be written as

$$\bar{\mathbf{k}}_{31}\mathbf{d}_1 + \bar{\mathbf{k}}_{33}\mathbf{d}_3 = \bar{\mathbf{f}}_3 \tag{7.4}$$

where

$$\begin{aligned} \bar{\mathbf{k}}_{31} &= \mathbf{k}_{31} - \mathbf{k}_{32}\mathbf{k}_{22}^{-1}\mathbf{k}_{21} \\ \bar{\mathbf{k}}_{33} &= \mathbf{k}_{33} - \mathbf{k}_{32}\mathbf{k}_{22}^{-1}\mathbf{k}_{23} \\ \bar{\mathbf{f}}_3 &= \mathbf{f}_3 - \mathbf{k}_{32}\mathbf{k}_{22}^{-1}\mathbf{f}_2 \end{aligned} \tag{7.5}$$

Combining Equations (7.2) and (7.4) with the associated Equations (7.3) and (7.5), we can have the result of static condensation by using sandwich condensation, which is given by

$$\begin{bmatrix} \bar{\mathbf{k}}_{11} & \bar{\mathbf{k}}_{13} \\ \bar{\mathbf{k}}_{31} & \bar{\mathbf{k}}_{33} \end{bmatrix} \cdot \begin{Bmatrix} \mathbf{d}_1 \\ \mathbf{d}_3 \end{Bmatrix} = \begin{Bmatrix} \bar{\mathbf{f}}_1 \\ \bar{\mathbf{f}}_3 \end{Bmatrix} \tag{7.6}$$

7.4 Dynamic Condensation

Further, we continue the condensation to the free vibration governing equation, where the equilibrium equation of free vibration in the matrix form can be expressed by

$$\begin{bmatrix} \mathbf{k}_{11} & \mathbf{k}_{12} & \mathbf{k}_{13} \\ \mathbf{k}_{21} & \mathbf{k}_{22} & \mathbf{k}_{23} \\ \mathbf{k}_{31} & \mathbf{k}_{32} & \mathbf{k}_{33} \end{bmatrix} \cdot \begin{Bmatrix} \mathbf{d}_1 \\ \mathbf{d}_2 \\ \mathbf{d}_3 \end{Bmatrix} + \begin{bmatrix} \mathbf{m}_{11} & \mathbf{m}_{12} & \mathbf{m}_{13} \\ \mathbf{m}_{21} & \mathbf{m}_{22} & \mathbf{m}_{23} \\ \mathbf{m}_{31} & \mathbf{m}_{32} & \mathbf{m}_{33} \end{bmatrix} \cdot \begin{Bmatrix} \ddot{\mathbf{d}}_1 \\ \ddot{\mathbf{d}}_2 \\ \ddot{\mathbf{d}}_3 \end{Bmatrix} = \begin{Bmatrix} 0 \\ 0 \\ 0 \end{Bmatrix} \tag{7.7}$$

To relate the generalized displacement vector $\mathbf{d}_2$ and the generalized acceleration vector $\ddot{\mathbf{d}}_2$ in the governing equations, let us derive Equation (7.7) two times with respect to time t, which results in

$$\begin{bmatrix} \mathbf{k}_{11} & \mathbf{k}_{12} & \mathbf{k}_{13} \\ \mathbf{k}_{21} & \mathbf{k}_{22} & \mathbf{k}_{23} \\ \mathbf{k}_{31} & \mathbf{k}_{32} & \mathbf{k}_{33} \end{bmatrix} \cdot \begin{Bmatrix} \ddot{\mathbf{d}}_1 \\ \ddot{\mathbf{d}}_2 \\ \ddot{\mathbf{d}}_3 \end{Bmatrix} + \begin{bmatrix} \mathbf{m}_{11} & \mathbf{m}_{12} & \mathbf{m}_{13} \\ \mathbf{m}_{21} & \mathbf{m}_{22} & \mathbf{m}_{23} \\ \mathbf{m}_{31} & \mathbf{m}_{32} & \mathbf{m}_{33} \end{bmatrix} \cdot \begin{Bmatrix} \ddddot{\mathbf{d}}_1 \\ \ddddot{\mathbf{d}}_2 \\ \ddddot{\mathbf{d}}_3 \end{Bmatrix} = \begin{Bmatrix} 0 \\ 0 \\ 0 \end{Bmatrix} \tag{7.8}$$

Neglecting the fourth-order generalized displacements vectors, we can have

$$\begin{bmatrix} \mathbf{k}_{11} & \mathbf{k}_{12} & \mathbf{k}_{13} \\ \mathbf{k}_{21} & \mathbf{k}_{22} & \mathbf{k}_{23} \\ \mathbf{k}_{31} & \mathbf{k}_{32} & \mathbf{k}_{33} \end{bmatrix} \cdot \begin{Bmatrix} \ddot{\mathbf{d}}_1 \\ \ddot{\mathbf{d}}_2 \\ \ddot{\mathbf{d}}_3 \end{Bmatrix} = \begin{Bmatrix} 0 \\ 0 \\ 0 \end{Bmatrix} \tag{7.9}$$

The above matrix equation can be expressed by three simultaneous equations as follows:

$$\text{(a) } \mathbf{k}_{11}\ddot{\mathbf{d}}_1 + \mathbf{k}_{12}\ddot{\mathbf{d}}_2 + \mathbf{k}_{13}\ddot{\mathbf{d}}_3 = 0$$

$$\text{(b) } \mathbf{k}_{21}\ddot{\mathbf{d}}_1 + \mathbf{k}_{22}\ddot{\mathbf{d}}_2 + \mathbf{k}_{23}\ddot{\mathbf{d}}_3 = 0$$

$$\text{(c) } \mathbf{k}_{31}\ddot{\mathbf{d}}_1 + \mathbf{k}_{32}\ddot{\mathbf{d}}_2 + \mathbf{k}_{33}\ddot{\mathbf{d}}_3 = 0$$

Equation (b) consists of the unnecessary DOFs that will be condensed. By extracting the generalized acceleration vector $\ddot{\mathbf{d}}_2$ from Equation (b), we can have

$$\ddot{\mathbf{d}}_2 = \mathbf{k}_{22}^{-1}\left(-\mathbf{k}_{21}\ddot{\mathbf{d}}_1 - \mathbf{k}_{23}\ddot{\mathbf{d}}_3\right) \tag{7.10}$$

Substituting the vector $\ddot{\mathbf{d}}_2$ into Equation (7.7) and solving for $\mathbf{d}_2$ from the second simultaneous equation of Equation (7.7),

$$\mathbf{k}_{21}\mathbf{d}_1 + \mathbf{k}_{22}\mathbf{d}_2 + \mathbf{k}_{23}\mathbf{d}_3 + \mathbf{m}_{21}\ddot{\mathbf{d}}_1 + \mathbf{m}_{22}\mathbf{k}_{22}^{-1}\left(-\mathbf{k}_{21}\ddot{\mathbf{d}}_1 - \mathbf{k}_{23}\ddot{\mathbf{d}}_3\right) + \mathbf{m}_{23}\ddot{\mathbf{d}}_3 = 0$$

$$\mathbf{d}_2 = -\mathbf{k}_{22}^{-1}\mathbf{k}_{21}\mathbf{d}_1 - \mathbf{k}_{22}^{-1}\mathbf{k}_{23}\mathbf{d}_3 - \mathbf{k}_{22}^{-1}\mathbf{m}_{21}\ddot{\mathbf{d}}_1 + \mathbf{k}_{22}^{-1}\mathbf{m}_{22}\mathbf{k}_{22}^{-1}\left(\mathbf{k}_{21}\ddot{\mathbf{d}}_1 + \mathbf{k}_{23}\ddot{\mathbf{d}}_3\right) - \mathbf{k}_{22}^{-1}\mathbf{m}_{23}\ddot{\mathbf{d}}_3$$

After the substitution, the vector $\ddot{\mathbf{d}}_2$ is eliminated from the above equation; hence, it results in

$$\mathbf{d}_2 = -\mathbf{k}_{22}^{-1}\mathbf{k}_{21}\mathbf{d}_1 - \mathbf{k}_{22}^{-1}\mathbf{k}_{23}\mathbf{d}_3 - \mathbf{k}_{22}^{-1}\bar{\mathbf{m}}_{21}\ddot{\mathbf{d}}_1 - \mathbf{k}_{22}^{-1}\bar{\mathbf{m}}_{23}\ddot{\mathbf{d}}_3 \tag{7.11}$$

where

$$\begin{aligned} \bar{\mathbf{m}}_{21} &= \mathbf{m}_{21} - \mathbf{m}_{22}\mathbf{k}_{22}^{-1}\mathbf{k}_{21} \\ \bar{\mathbf{m}}_{23} &= \mathbf{m}_{23} - \mathbf{m}_{22}\mathbf{k}_{22}^{-1}\mathbf{k}_{23} \end{aligned} \tag{7.12}$$

Substituting the vector $\ddot{\mathbf{d}}_2$ from Equation (7.10) and $\mathbf{d}_2$ from Equation (7.11) into the first simultaneous equation in Equation (7.7) results in

$$\begin{aligned} &\mathbf{k}_{11}\mathbf{d}_1 + \mathbf{k}_{12}\left(-\mathbf{k}_{22}^{-1}\mathbf{k}_{21}\mathbf{d}_1 - \mathbf{k}_{22}^{-1}\mathbf{k}_{23}\mathbf{d}_3 - \mathbf{k}_{22}^{-1}\bar{\mathbf{m}}_{21}\ddot{\mathbf{d}}_1 - \mathbf{k}_{22}^{-1}\bar{\mathbf{m}}_{23}\ddot{\mathbf{d}}_3\right) \\ &+ \mathbf{k}_{13}\mathbf{d}_3 + \mathbf{m}_{11}\ddot{\mathbf{d}}_1 + \mathbf{m}_{12}\mathbf{k}_{22}^{-1}\left(-\mathbf{k}_{21}\ddot{\mathbf{d}}_1 - \mathbf{k}_{23}\ddot{\mathbf{d}}_3\right) + \mathbf{m}_{13}\ddot{\mathbf{d}}_3 = 0 \end{aligned}$$

which leads to an equation:

$$\bar{\mathbf{k}}_{11}\mathbf{d}_1 + \bar{\mathbf{k}}_{13}\mathbf{d}_3 + \bar{\mathbf{m}}_{11}\ddot{\mathbf{d}}_1 + \bar{\mathbf{m}}_{13}\ddot{\mathbf{d}}_3 = 0 \tag{7.13}$$

where

$$\begin{aligned} \bar{\mathbf{m}}_{11} &= \mathbf{m}_{11} - \mathbf{m}_{12}\mathbf{k}_{22}^{-1}\mathbf{k}_{21} - \mathbf{k}_{12}\mathbf{k}_{22}^{-1}\bar{\mathbf{m}}_{21} \\ \bar{\mathbf{m}}_{13} &= \mathbf{m}_{13} - \mathbf{m}_{12}\mathbf{k}_{22}^{-1}\mathbf{k}_{23} - \mathbf{k}_{12}\mathbf{k}_{22}^{-1}\bar{\mathbf{m}}_{23} \end{aligned} \tag{7.14}$$

Finally, substituting the vector $\ddot{\mathbf{d}}_2$ from Equation (7.10) and $\mathbf{d}_2$ from Equation (7.11) into the third simultaneous equation in Equation (7.7) results in

$$\begin{aligned} &\mathbf{k}_{31}\mathbf{d}_1 + \mathbf{k}_{32}\left(-\mathbf{k}_{22}^{-1}\mathbf{k}_{21}\mathbf{d}_1 - \mathbf{k}_{22}^{-1}\mathbf{k}_{23}\mathbf{d}_3 - \mathbf{k}_{22}^{-1}\bar{\mathbf{m}}_{21}\ddot{\mathbf{d}}_1 - \mathbf{k}_{22}^{-1}\bar{\mathbf{m}}_{23}\ddot{\mathbf{d}}_3\right) \\ &+ \mathbf{k}_{33}\mathbf{d}_3 + \mathbf{m}_{31}\ddot{\mathbf{d}}_1 + \mathbf{m}_{32}\mathbf{k}_{22}^{-1}\left(-\mathbf{k}_{21}\ddot{\mathbf{d}}_1 - \mathbf{k}_{23}\ddot{\mathbf{d}}_3\right) + \mathbf{m}_{33}\ddot{\mathbf{d}}_3 = 0 \end{aligned}$$

which leads to an equation:

$$\bar{\mathbf{k}}_{31}\mathbf{d}_1 + \bar{\mathbf{k}}_{33}\mathbf{d}_3 + \bar{\mathbf{m}}_{31}\ddot{\mathbf{d}}_1 + \bar{\mathbf{m}}_{33}\ddot{\mathbf{d}}_3 = \mathbf{0} \tag{7.15}$$

where

$$\begin{aligned} \bar{\mathbf{m}}_{31} &= \mathbf{m}_{31} - \mathbf{m}_{32}\mathbf{k}_{22}^{-1}\mathbf{k}_{21} - \mathbf{k}_{32}\mathbf{k}_{22}^{-1}\bar{\mathbf{m}}_{21} \\ \bar{\mathbf{m}}_{33} &= \mathbf{m}_{33} - \mathbf{m}_{32}\mathbf{k}_{22}^{-1}\mathbf{k}_{23} - \mathbf{k}_{32}\mathbf{k}_{22}^{-1}\bar{\mathbf{m}}_{23} \end{aligned} \tag{7.16}$$

Finally, summarizing Equations (7.13) and (7.15), we can have condensed two governing equations that can be used for solving the free vibration problems with only four nodes at the edges of the plate and shell element as

$$\begin{bmatrix} \bar{\mathbf{k}}_{11} & \bar{\mathbf{k}}_{13} \\ \bar{\mathbf{k}}_{31} & \bar{\mathbf{k}}_{33} \end{bmatrix} \cdot \begin{Bmatrix} \mathbf{d}_1 \\ \mathbf{d}_3 \end{Bmatrix} + \begin{bmatrix} \bar{\mathbf{m}}_{11} & \bar{\mathbf{m}}_{13} \\ \bar{\mathbf{m}}_{31} & \bar{\mathbf{m}}_{33} \end{bmatrix} \cdot \begin{Bmatrix} \ddot{\mathbf{d}}_1 \\ \ddot{\mathbf{d}}_3 \end{Bmatrix} = \begin{Bmatrix} 0 \\ 0 \end{Bmatrix} \tag{7.17}$$

Following the same procedure described previously, we can establish the sandwich condensation method for free vibration with damping and obtain the dynamic condensation equation to include the external dynamic loading without any difficulties.

7.5 Procedure of Condensation

The element matrix of the plate and shell constructed by using the NURBS functions was initially arranged in each node with its generalized displacement (DOF) order as

$$\mathbf{d} = \begin{bmatrix} \overbrace{w_1 w_2 \ldots}^{\text{retained DOF}} \underbrace{w_i \ldots w_j}_{\text{condensed DOF}} \overbrace{\ldots w_{k-1} w_k}^{\text{retained DOF}} \quad \overbrace{\theta_{x1}\theta_{x2}\ldots}^{\text{retained DOF}} \underbrace{\theta_{xi}\ldots\theta_{xj}}_{\text{condensed DOF}} \overbrace{\ldots\theta_{xk-1}\theta_{xk}}^{\text{retained DOF}} \quad \ldots \\ \overbrace{\theta_{y1}\theta_{y2}\ldots}^{\text{retained DOF}} \underbrace{\theta_{yi}\ldots\theta_{yj}}_{\text{condensed DOF}} \overbrace{\ldots\theta_{yk-1}\theta_{yk}}^{\text{retained DOF}} \end{bmatrix}^T \tag{7.18}$$

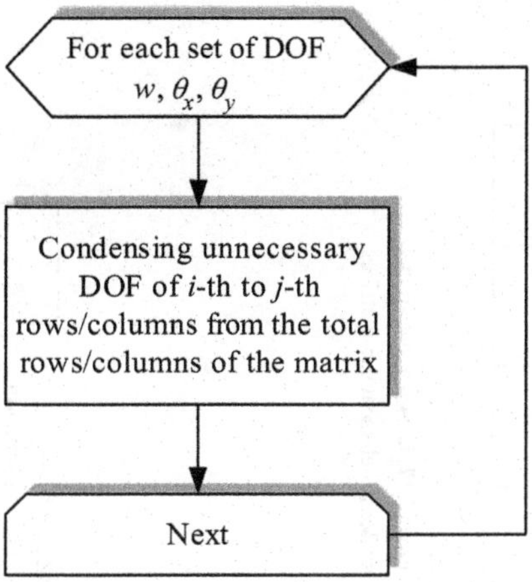

FIGURE 7.1
Flowchart of the sandwich condensation method.

The procedure of the sandwich condensation is shown in Figure 7.1. In the procedure, the condensation is done to all the unnecessary nodes' DOF matrix equations.

Illustratively, the processes of three stages condensations are shown in Figures 7.2–7.4. The sub-matrices are constructed in every stage by using Equation (7.17), repeatedly eliminating the unnecessary DOFs of w, θ_x, and θ_y rows/columns in the matrix equations. The final condensed matrix will be the desired number of nodes and its associated number of DOFs in the plate and shell element matrix (Figure 7.5).

7.6 Condensing the Stiffness and Mass Matrices of the Mindlin Plate Element

MATLAB® code KMmatrixMindlinIGACondensed.m uses the functions NurbsSurface.m and Legendre.m to construct the stiffness and mass matrices of the Mindlin plate theory (FSDT) of an isotropic plate and condensed the 34 DOFs to a four-node 12-DOF thick plate element. The same example given in Section 6.5 is considered for the condensation example. The material and geometric properties are shown in Figure 7.6.

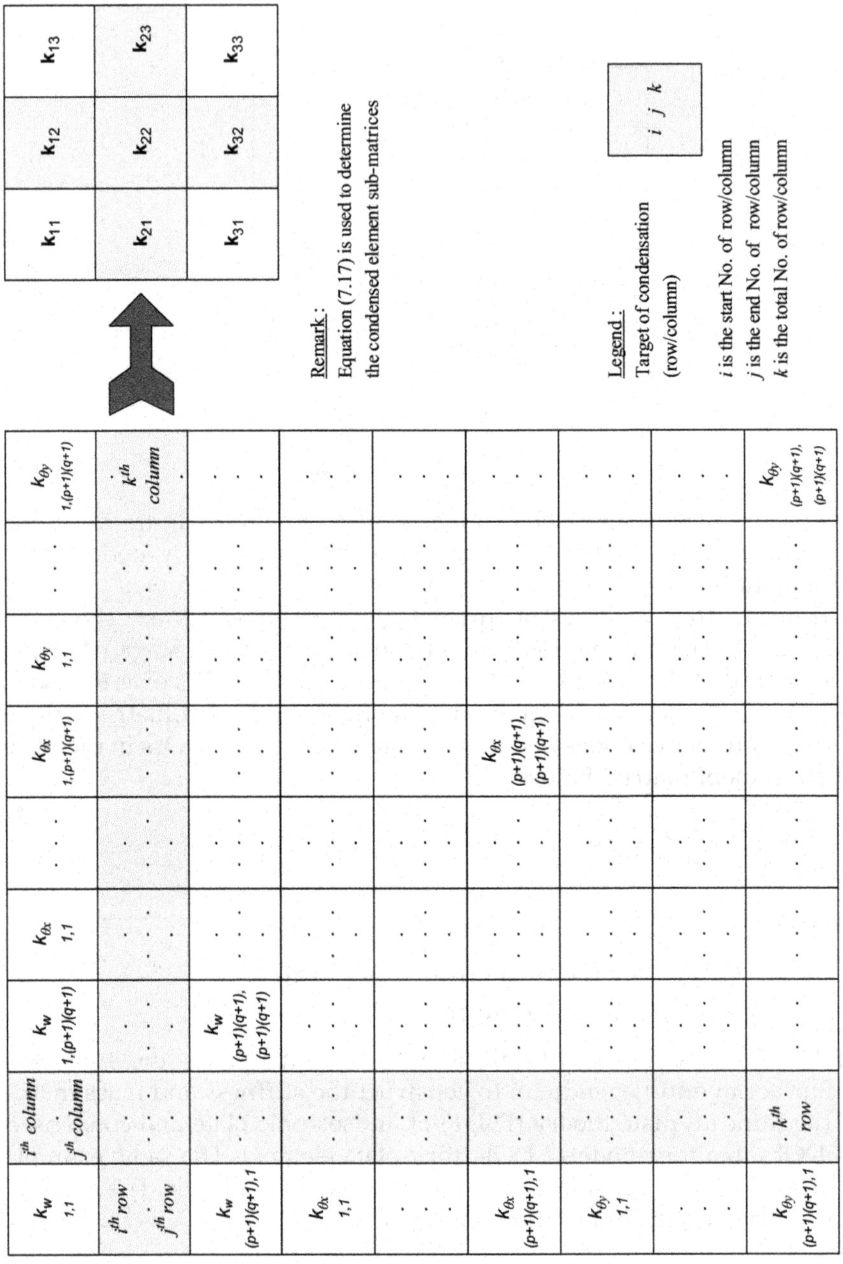

FIGURE 7.2
Illustration of the sandwich condensation procedure: First stage.

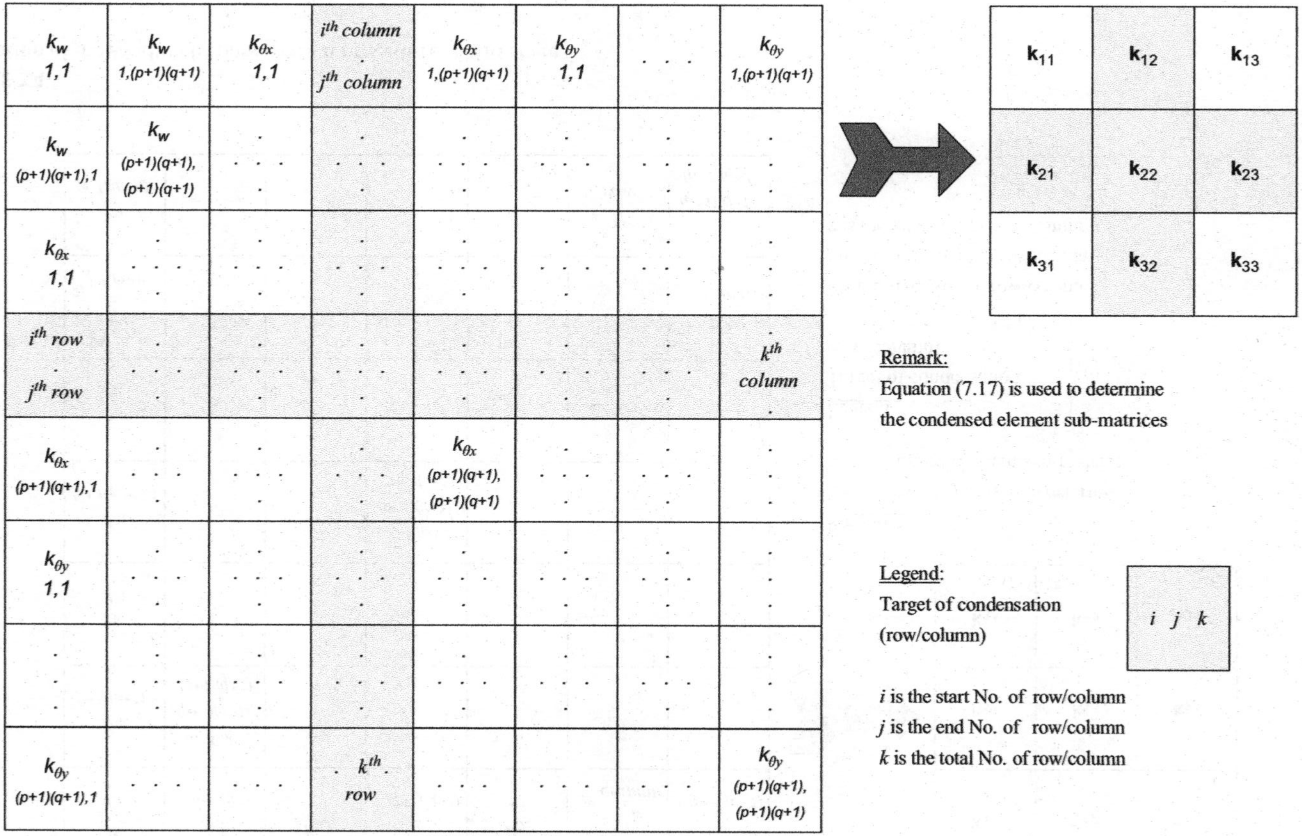

FIGURE 7.3
Illustration of the sandwich condensation procedure: Second stage.

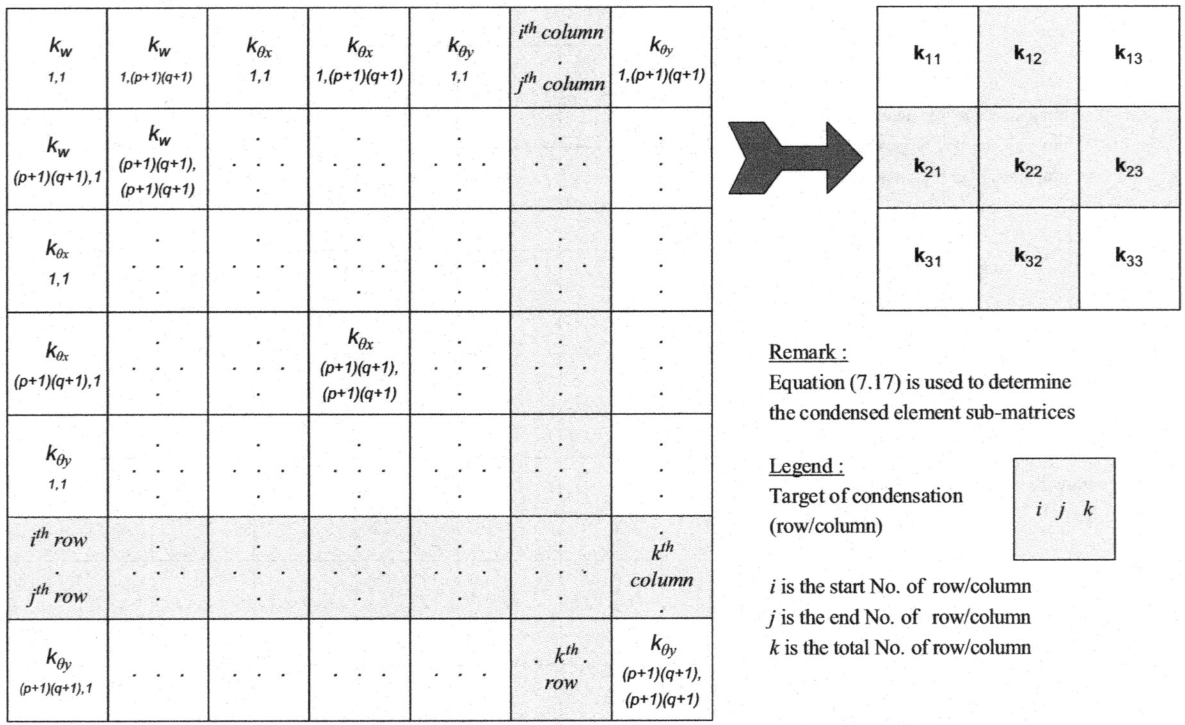

FIGURE 7.4
Illustration of the sandwich condensation procedure: Third stage.

k_w 1,1	k_w 1,(p+1)(q+1)	$k_{\theta x}$ 1,1	$k_{\theta x}$ 1,(p+1)(q+1)	$k_{\theta y}$ 1,1	$k_{\theta y}$ 1,(p+1)(q+1)
k_w (p+1)(q+1),1	k_w (p+1)(q+1), (p+1)(q+1)	⋯	⋯	⋯	⋯
$k_{\theta x}$ 1,1	⋯	⋯	⋯	⋯	⋯
$k_{\theta x}$ (p+1)(q+1),1	⋯	⋯	$k_{\theta x}$ (p+1)(q+1), (p+1)(q+1)	⋯	⋯
$k_{\theta y}$ 1,1	⋯	⋯	⋯	⋯	⋯
$k_{\theta y}$ (p+1)(q+1),1	⋯	⋯	⋯	⋯	$k_{\theta y}$ (p+1)(q+1), (p+1)(q+1)

FIGURE 7.5
Illustration of the sandwich condensation procedure: Result.

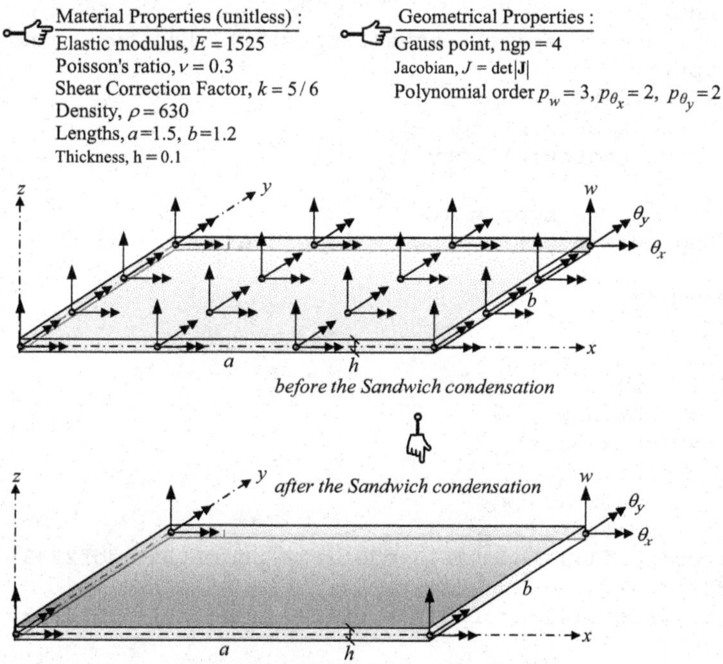

FIGURE 7.6 A condensed rectangular isotropic flat Kirchhoff plate model.

7.6.1 KMmatrixMindlinIGACondensed Program List

```
% KMmatrixMindlinIGACondensed.m (Using NurbsSurface.m) : R6
% Construct matrices K and M
% Based on the Mindlin Plate Theory by using NURBS surface
shape functions
% Condensing the matrix from 34 DOF to 12 DOF Mindlin plate
element
% isotropic material
clear all; clc;
a  = 1.5;     b  = 1.2;    h  = 0.1;
E  = 1525;     nu  = 0.3;    rho = 630;
I  = h^3/12;    Ixy = h;     kappa = 5/6;
D  = E*h^3/12/(1-nu^2);
Dxxyy = nu*E*h^3/6/(1-nu^2);
Dxy  = E*h^3/24/(1+nu);
G = E/2/(1+nu);
Dxz = kappa*G*h;
% NURBS for w and theta displacement data in x and y
directions
npw  = 3;
nptx  = 2;
npty  = 2;
knotw = [-ones(1,npw+1) ones(1,npw+1)]; knotw = [knotw
ones(1,npw)];
knottx = [-ones(1,nptx+1) ones(1,nptx+1)]; knottx = [knottx
ones(1,nptx)];
knotty = [-ones(1,npty+1) ones(1,npty+1)]; knotty = [knotty
ones(1,npty)];
nbw  = size(knotw,2)-npw-1;
nbtx  = size(knottx,2)-nptx-1;
nbty  = size(knotty,2)-npty-1;
% Gauss
ngpx = 4;          ngpy = 4;
Gpwx = Legendre(ngpx);   Gpwy = Legendre(ngpy);
% Jac
Jac = a/2*b/2;
Jx = a/2;
Jy = b/2;
% Initialization
ndofw   = (nbw-npw);
ndoftx  = (nbtx-nptx);
ndofty  = (nbty-npty);
ndof1111 = ndofw*ndofw;
ndof2233 = ndoftx*ndofty;
K  = zeros(ndof1111+ndof2233+ndof2233,ndof1111+ndof2233
+ndof2233);
M  = zeros(ndof1111+ndof2233+ndof2233,ndof1111+ndof2233
+ndof2233);
% Shape functions of NURBS surface
```

```
Nw   = zeros(ngpx,ngpy,ndof1111);
Nwdx = zeros(ngpx,ngpy,ndof1111);
Nwdy = zeros(ngpx,ngpy,ndof1111);
Ntx  = zeros(ngpx,ngpy,ndof2233);
Ntxdx = zeros(ngpx,ngpy,ndof2233);
Ntxdy = zeros(ngpx,ngpy,ndof2233);
Nty  = zeros(ngpx,ngpy,ndof2233);
Ntydx = zeros(ngpx,ngpy,ndof2233);
Ntydy = zeros(ngpx,ngpy,ndof2233);
% Stiffness and Mass Matrices K & M
% K11 and M11 of w displacements
% Filling NURBS surface shape functions into the gauss points
for m=1:ngpx
  s = Gpwx(m,1);
  for n=1:ngpy
   t = Gpwy(n,1);
   NurbsMNDLNw = NurbsSurface(npw,knotw,npw,knotw,s,t);
   for i=1:ndofw
    for j=1:ndofw
     ij=(i-1)*ndofw+j; % Columnwise arrangement of DOF w-w
     Nw(m,n,ij)    = NurbsMNDLNw(1,i,j);  % 0th derivative
     Nwdx(m,n,ij)   = NurbsMNDLNw(2,i,j);  % x derivative
     Nwdy(m,n,ij)   = NurbsMNDLNw(3,i,j);  % y derivative
    end
   end
   NurbsMNDLNt = NurbsSurface(nptx,knottx,npty,knotty,s,t);
   for i=1:ndoftx
    for j=1:ndofty
     ij=(i-1)*ndoftx+j; % Columnwise arrangement of DOF w-w
     Ntx(m,n,ij)   = NurbsMNDLNt(1,i,j);  % 0th derivative
     Ntxdx(m,n,ij) = NurbsMNDLNt(2,i,j);  % x derivative
     Ntxdy(m,n,ij) = NurbsMNDLNt(3,i,j);  % y derivative
     Nty(m,n,ij)   = NurbsMNDLNt(1,i,j);  % 0th derivative
     Ntydx(m,n,ij) = NurbsMNDLNt(2,i,j);  % x derivative
     Ntydy(m,n,ij) = NurbsMNDLNt(3,i,j);  % y derivative
    end
   end
  end
end
kmat11 = zeros(ndof1111,ndof1111);
mmat11 = zeros(ndof1111,ndof1111);
for i=1:ndof1111
 for j=1:ndof1111
  for m=1:ngpx
   for n=1:ngpy
    % Stiffness
    kmat11(i,j) = kmat11(i,j)+Nwdx(m,n,j)/Jx^2*Dxz* ...
      Nwdx(m,n,i)/Jx^2*Jac*Gpwx(m,2)*Gpwy(n,2);
    kmat11(i,j) = kmat11(i,j)+Nwdy(m,n,j)/Jy^2*Dxz* ...
      Nwdy(m,n,i)/Jy^2*Jac*Gpwx(m,2)*Gpwy(n,2);
```

```
    % Mass
    mmat11(i,j) = mmat11(i,j)+Nw(m,n,j)*rho*Ixy* ...
          Nw(m,n,i)*Jac*Gpwx(m,2)*Gpwy(n,2);
   end
  end
 end
end
K(1:ndof1111,1:ndof1111)=kmat11; % K11 inclusion
M(1:ndof1111,1:ndof1111)=mmat11; % M11 inclusion
% Stiffness and Mass Matrices K
% K12,K21,K13,K31 of w-txy and w-ty displacements and
rotations
kmat12 = zeros(ndof1111,ndof2233);
kmat21 = zeros(ndof2233,ndof1111);
kmat13 = zeros(ndof1111,ndof2233);
kmat31 = zeros(ndof2233,ndof1111);
for i=1:ndof1111
 for j=1:ndof2233
  for m=1:ngpx
   for n=1:ngpy
    %. Stiffness
    kmat12(i,j) = kmat12(i,j)-Nwdy(m,n,i)/Jy^2*Dxz* ...
      Ntx(m,n,j)*Jac*Gpwx(m,2)*Gpwy(n,2);
    kmat21(j,i) = kmat21(j,i)-Nwdy(m,n,i)/Jy^2*Dxz* ...
      Ntx(m,n,j)*Jac*Gpwx(m,2)*Gpwy(n,2);
    kmat13(i,j) = kmat13(i,j)-Nwdx(m,n,i)/Jx^2*Dxz* ...
      Nty(m,n,j)*Jac*Gpwx(m,2)*Gpwy(n,2);
    kmat31(j,i) = kmat31(j,i)-Nwdx(m,n,i)/Jx^2*Dxz* ...
      Nty(m,n,j)*Jac*Gpwx(m,2)*Gpwy(n,2);
   end
  end
 end
end
K(1:ndof1111,ndof1111+1:ndof1111+ndof2233)=kmat12;        % K12
K(ndof1111+1:ndof1111+ndof2233,1:ndof1111)=kmat21;        % K21
K(1:ndof1111,ndof1111+ndof2233+1:ndof1111+ndof2233+ndof2233)=
kmat13; % K13
K(ndof1111+ndof2233+1:ndof1111+ndof2233+ndof2233,1:ndof1111)=
kmat31; % K31
% Stiffness and Mass Matrices K & M
% K22,K33 and M22,M33 of theta-x,y rotations
kmat22 = zeros(ndof2233,ndof2233);
kmat23 = zeros(ndof2233,ndof2233);
kmat32 = zeros(ndof2233,ndof2233);
kmat33 = zeros(ndof2233,ndof2233);
mmat22 = zeros(ndof2233,ndof2233);
mmat33 = zeros(ndof2233,ndof2233);
for i=1:ndof2233
 for j=1:ndof2233
  for m=1:ngpx
```

```
    for n=1:ngpy
     % Stiffness
     kmat22(i,j) = kmat22(i,j)+ ...
      Ntxdy(m,n,j)/Jy^2*D*Ntxdy(m,n,i)/Jy^2*Jac*Gpwx(m,2)*
Gpwy(n,2)+ ...
      Ntxdx(m,n,j)/Jx^2*Dxy*Ntxdx(m,n,i)/Jx^2*Jac*Gpwx(m,2)*
Gpwy(n,2)+...
      Ntx(m,n,j)*Dxz*Ntx(m,n,i)*Jac*Gpwx(m,2)*Gpwy(n,2);
     kmat23(i,j) = kmat23(i,j)+ ...
      Ntxdy(m,n,j)/Jy^2*Dxxyy*Ntydx(m,n,i)/Jx^2*Jac*Gpwx(m,2)*
Gpwy(n,2)+ ...
      Ntxdx(m,n,j)/Jx^2*Dxy*Ntydy(m,n,i)/Jy^2*Jac*Gpwx(m,2)*
Gpwy(n,2);
     kmat32(j,i) = kmat32(j,i)+ ...
      Ntydx(m,n,i)/Jx^2*Dxxyy*Ntxdy(m,n,j)/Jy^2*Jac*Gpwx(m,2)*
Gpwy(n,2)+ ...
      Ntydy(m,n,i)/Jy^2*Dxy*Ntxdx(m,n,j)/Jx^2*Jac*Gpwx(m,2)*
Gpwy(n,2);
     kmat33(i,j) = kmat33(i,j)+ ...
      Ntydx(m,n,j)/Jx^2*D*Ntydx(m,n,i)/Jx^2*Jac*Gpwx(m,2)*
Gpwy(n,2)+ ...
      Ntydy(m,n,j)/Jy^2*Dxy*Ntydy(m,n,i)/Jy^2*Jac*Gpwx(m,2)*
Gpwy(n,2)+...
      Nty(m,n,j)*Dxz*Nty(m,n,i)*Jac*Gpwx(m,2)*Gpwy(n,2);
     % Mass
     mmat22(i,j) = mmat22(i,j)+Ntx(m,n,j)*rho*I* ...
       Ntx(m,n,i)*Jac*Gpwx(m,2)*Gpwy(n,2);
     mmat33(i,j) = mmat33(i,j)+Nty(m,n,j)*rho*I* ...
       Nty(m,n,i)*Jac*Gpwx(m,2)*Gpwy(n,2);
    end
   end
  end
end
K(ndof1111+1:ndof1111+ndof2233,ndof1111+1:ndof1111+ndof2233)=
kmat22; % K22
K(ndof1111+1:ndof1111+ndof2233,ndof1111+ndof2233+1: ...
  ndof1111+ndof2233+ndof2233)=kmat23;                  % K23
K(ndof1111+ndof2233+1:ndof1111+ndof2233+ndof2233,ndof1111+1:
...
  ndof1111+ndof2233)=kmat32;                      % K32
K(ndof1111+ndof2233+1:ndof1111+ndof2233+ndof2233,ndof1111+
ndof2233+1: ...
  ndof1111+ndof2233+ndof2233)=kmat33;                  % K33
M(ndof1111+1:ndof1111+ndof2233,ndof1111+1:ndof1111+ndof2233)=
mmat22; % K22
M(ndof1111+ndof2233+1:ndof1111+ndof2233+ndof2233,ndof1111+
ndof2233+1: ...
  ndof1111+ndof2233+ndof2233)=mmat33;                  % M33
disp('Stiffness Matrix of Mindlin Plate');
disp(K);
```

```
disp('Mass Matrix of Mindlin Plate');
disp(M);
% Loading Vector
Lvec=zeros(ndof1111+ndof2233+ndof2233,1);
% Condensation to four-node plate element
i=2; j=3; k=34;
[KMC,MMC,FVC] = Condensation(K,M,Lvec,i,j,k); % Condensing
K11a
i=3; j=10; k=32;
[KMC,MMC,FVC] = Condensation(KMC,MMC,FVC,i,j,k); % Condensing
K11b
i=4; j=5; k=24;
[KMC,MMC,FVC] = Condensation(KMC,MMC,FVC,i,j,k); % Condensing
K11c
i=6; j=6; k=22;
[KMC,MMC,FVC] = Condensation(KMC,MMC,FVC,i,j,k); % Condensing
K22a
i=7; j=9; k=21;
[KMC,MMC,FVC] = Condensation(KMC,MMC,FVC,i,j,k); % Condensing
K22b
i=8; j=8; k=18;
[KMC,MMC,FVC] = Condensation(KMC,MMC,FVC,i,j,k); % Condensing
K22c
i=10; j=10; k=17;
[KMC,MMC,FVC] = Condensation(KMC,MMC,FVC,i,j,k); % Condensing
K33a
i=11; j=13; k=16;
[KMC,MMC,FVC] = Condensation(KMC,MMC,FVC,i,j,k); % Condensing
K33b
i=12; j=12; k=13;
[KMC,MMC,FVC] = Condensation(KMC,MMC,FVC,i,j,k); % Condensing
K33c
% Swapping the matrices
KMC=KMC([1,4,7,10,2,5,8,11,3,6,9,12],:); % row
KMC=KMC(:,[1,4,7,10,2,5,8,11,3,6,9,12]); % column
MMC=MMC([1,4,7,10,2,5,8,11,3,6,9,12],:); % row
MMC=MMC(:,[1,4,7,10,2,5,8,11,3,6,9,12]); % column
disp('The Stiffness Matrix after condensation');
disp(KMC);
disp('The Mass Matrix after condensation');
disp(MMC);
```

7.6.2 Condensation Function List

```
function [KMC,MMC,FVC] = Condensation(KM,MM,FV,nA,nB,nC)
n1 = nA - 1;
n2 = nB + 1;
% Extracting KM
```

```
K11 = KM(1:n1,1:n1);  K12 = KM(1:n1,nA:nB);  K13 =
KM(1:n1,n2:nC);
K21 = KM(nA:nB,1:n1); K22 = KM(nA:nB,nA:nB); K23 =
KM(nA:nB,n2:nC);
K31 = KM(n2:nC,1:n1); K32 = KM(n2:nC,nA:nB); K33 =
KM(n2:nC,n2:nC);
% Extracting MM
M11 = MM(1:n1,1:n1);  M12 = MM(1:n1,nA:nB);  M13 =
MM(1:n1,n2:nC);
M21 = MM(nA:nB,1:n1); M22 = MM(nA:nB,nA:nB); M23 =
MM(nA:nB,n2:nC);
M31 = MM(n2:nC,1:n1); M32 = MM(n2:nC,nA:nB); M33 =
MM(n2:nC,n2:nC);
% Extracting FV
F1 = FV(1:n1);      F2 = FV(nA:nB);      F3 = FV(n2:nC);
% Condensation
K22I =inv(K22);
%
M21B = M21 - M22 * K22I * K21;
M23B = M23 - M22 * K22I * K23;
K11B = K11 - K12 * K22I * K21;
M11BB = M11 - M12 * K22I * K21 - K12 * K22I * M21B;
F1B  = F1 - K12 * K22I * F2;
K13B = K13 - K12 * K22I * K23;
M13BB = M13 - M12 * K22I * K23 - K12 * K22I * M23B;
K31B = K31 - K32 * K22I * K21;
M31BB = M31 - M32 * K22I * K21 - K32 * K22I * M21B;
K33B = K33 - K32 * K22I * K23;
M33BB = M33 - M32 * K22I * K23 - K32 * K22I * M23B;
F3B  = F3 - K32 * K22I * F2;
%
KMC = [K11B  K13B; K31B K33B];
MMC = [M11BB M13BB; M31BB M33BB];
FVC = [    F1B;      F3B];
return
```

References

Bathe KJ (1982) *Finite Element Procedures in Engineering Analysis*. Prentice Hall, Englewood Cliffs.

Cook RD (1981) *Concepts and Applications of Finite Element Analysis*, 2nd edition. John Wiley & Sons, Canada.

Gan BS (2018) *An Isogeometric Approach to Beam Structures*. Springer, Switzerland.

Hart GC, Wong K (2000) *Structural Dynamics for Structural Engineers*. John Wiley & Sons, New York.

8

Square Flat Plate Example

8.1 Square Thick Flat Plate Analyzed Using Condensed Isogeometric Analysis

In the Mindlin flat plate theory, the plane sections of the edges do not remain straight to the midplane during the deformation. A square thick flat plate problem is adopted here for solving the static and free vibration problems.

To integrate numerically the stiffness, mass matrices, and loading vector, the Jacobian operators are necessary. Because the plate has the square dimension, the Jacobian operators are constants, which can be determined by using the following equations:

$$
\begin{aligned}
J &= \frac{ab}{4} = \frac{b^2}{4} \\
J_x &= \frac{a}{2} = \frac{b}{2} \\
J_y &= \frac{b}{2} = J_x
\end{aligned}
\tag{8.1}
$$

It should be noted that all the program lists given in the following chapters are solved by using the least degrees of freedom (DOFs) after the sandwich condensation scheme. The standard procedure such as assembling technique, coordinate transformation, and sub-element division method in general finite element method (FEM) program will not be discussed here.

8.2 Static Problem

Consider a square flat plate that has simply supported edges as shown in Figure 8.1. The plate has geometry and property depicted inside the figure. The degrees of the polynomial used in the nonuniform rational basis spline

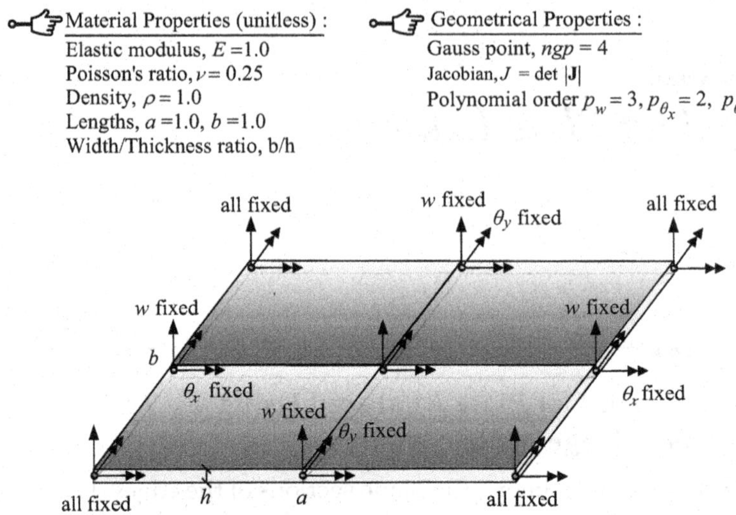

FIGURE 8.1
A square isotropic flat Mindlin plate example.

(NURBS) surface functions are $p_w = 3, p_{\theta_x} = 2, p_{\theta_x} = 2$. The plate is subjected to the distributed loading q_0 on the top surface. Four elements with four Gauss integration points are used in the analysis. The distributed loading is computed by the following equation:

$$\{\mathbf{f}_w\} = \sum_{i=0}^{m} \sum_{j=0}^{n} \mathbf{N}_w^T q_z J w_j w_i \tag{8.2}$$

The plate is analyzed by using the sandwich condensed isogeometric analysis (IGA) discussed in Chapter 7. Hence, there are a total of 12 DOFs, with 3 DOFs condensed at each of the four nodes.

The plate is considered simply supported on the four boundary edges, and the rotational DOFs at the middle edges are not constrained to simulate the pin/hinge support conditions.

Because we need to observe the maximum vertical displacement of the middle point on the plate, the element has to be discretized into four element. The element numbering, nodal numbering, and DOF numbering are shown in Figure 8.2.

Table 8.1 presents the static analysis results of the maximum nondimensional vertical displacements at the middle point on the top of the plate.

The following nondimensional formulas ($\nu = 0.25$) are used:

$$\overline{w}_b = w_0 \left(\frac{Eh^3}{b^4 q_0} \right) \tag{8.3}$$

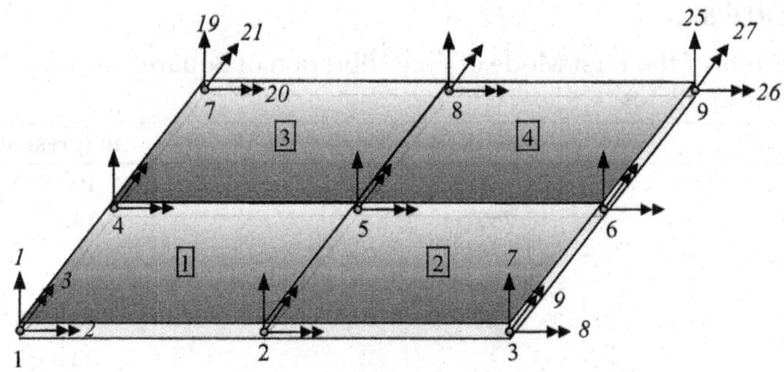

FIGURE 8.2
The numbering system of the square isotropic flat Mindlin plate example.

TABLE 8.1

Results of the Square Isotropic Mindlin Plates Subjected to Distributed Loading

bb/h/h	$\bar{w}$ **(Reddy 2002)**	$\bar{w}$ **(present)**
5	–	0.1208
10	0.4259	0.0354
20	0.4111	0.0130
50	0.4070	0.0075
100	0.4060	0.0033

8.3 Free Vibration Problem

Consider the same plate, as shown in Figure 8.1, as a free vibration problem. In this example, the first natural frequencies of various width-to-thickness ratios of the square plate are analyzed.

The nondimensional natural frequency ($\nu = 0.30$) of the plate is given as

$$\bar{\omega} = \omega_0 \left(\frac{a^2}{h} \right) \sqrt{\frac{\rho}{E}} \tag{8.4}$$

Table 8.2 presents the free vibration analysis results of the plate.

8.4 Flat Mindlin Plate Problem

MATLAB® code FlatMindlinPlateProblem.m calls the functions NurbsSurface.m, Condensation.m, Legendre.m, and DNurbsLeibnitz.m to integrate the stiffness and mass matrices of a square isotropic Mindlin plate

TABLE 8.2

Results of the First Mode of Free Vibration of Square Isotropic Mindlin Plates

bb/h/h	$\bar{w}$ (Reddy 2002)	$\bar{w}$ (present)
5	–	10.5842
10	5.769 ($m = 1, n = 1$) 13.764 ($m = 2, n = 1$) 21.121 ($m = 2, n = 2$)	14.4152
20	–	17.4000
50	–	23.0533
100	–	28.8873

and solve the static response displacements and the first free vibration mode of the plate numerically by using NURBS functions as the shape functions.

8.4.1 FlatMindlinPlateProblem Program List

```
% FlatMindlinPlateProblem.m (Using NurbsSurface.m and
Condensation.m)
% R6
% Solving Static and Free Vibration of a rectangular Mindlin
plate
% Based on the Mindlin Plate Theory using NURBS surface shape
functions
clear variables; clc;
elediv = 4; % Element divisions
a    = 1.0/elediv;           b   = a;
h    = (b*elediv/5);
E    = 1.0;      %nu   = 0.25;   % Static problem
rho   = 1.00;     nu   = 0.30;   % Free Vibration
I    = h^3/12;    Ixy = h;     kappa = 5/6;
D    = E*h^3/12/(1-nu^2);
Dxxyy = nu*E*h^3/6/(1-nu^2);
Dxy   = E*h^3/24/(1+nu);
G    = E/2/(1+nu);
Dxz   = kappa*G*h;
% Distributed loading
q0 = -1.0;
% NURBS for w and theta displacement data in x and y
directions
npw   = 3;
nptx  = 2;
npty  = 2;
knotw = [-ones(1,npw+1) ones(1,npw+1)]; knotw = [knotw
ones(1,npw)];
knottx = [-ones(1,nptx+1) ones(1,nptx+1)]; knottx = [knottx
ones(1,nptx)];
```

```
knotty = [-ones(1,npty+1) ones(1,npty+1)]; knotty = [knotty
ones(1,npty)];
nbw  = size(knotw,2)-npw-1;
nbtx  = size(knottx,2)-nptx-1;
nbty  = size(knotty,2)-npty-1;
% Gauss
ngpx = 4;          ngpy = 4;
Gpwx = Legendre(ngpx);    Gpwy = Legendre(ngpy);
% Jac
Jac = a/2*b/2;
Jx = a/2;
Jy = b/2;
% Initialization
ndofw   = (nbw-npw);
ndoftx  = (nbtx-nptx);
ndofty  = (nbty-npty);
ndof1111 = ndofw*ndofw;
ndof2233 = ndoftx*ndofty;
K  = zeros(ndof1111+ndof2233+ndof2233,ndof1111+ndof2233+
ndof2233);
M  = zeros(ndof1111+ndof2233+ndof2233,ndof1111+ndof2233+
ndof2233);
% Loading Vector
Lvec=zeros(ndof1111+ndof2233+ndof2233,1);
% Shape functions of NURBS surface
Nw  = zeros(ngpx,ngpy,ndof1111);
Nwdx = zeros(ngpx,ngpy,ndof1111);
Nwdy = zeros(ngpx,ngpy,ndof1111);
Ntx  = zeros(ngpx,ngpy,ndof2233);
Ntxdx = zeros(ngpx,ngpy,ndof2233);
Ntxdy = zeros(ngpx,ngpy,ndof2233);
Nty  = zeros(ngpx,ngpy,ndof2233);
Ntydx = zeros(ngpx,ngpy,ndof2233);
Ntydy = zeros(ngpx,ngpy,ndof2233);
% Stiffness and Mass Matrices K & M of 1/elediv element
% K11 and M11 of w displacements
% Filling NURBS surface shape functions into the gauss points
for m=1:ngpx
  s = Gpwx(m,1);
  for n=1:ngpy
   t = Gpwy(n,1);
   NurbsMNDLNw = NurbsSurface(npw,knotw,npw,knotw,s,t);
   for i=1:ndofw
    for j=1:ndofw
     ij=(i-1)*ndofw+j; % Columnwise arrangement of DOF w-w
     Nw(m,n,ij)   = NurbsMNDLNw(1,i,j);  % 0th derivative
     Nwdx(m,n,ij)  = NurbsMNDLNw(2,i,j);  % x derivative
     Nwdy(m,n,ij)  = NurbsMNDLNw(3,i,j);  % y derivative
    end
   end
```

```
   NurbsMNDLNt = NurbsSurface(nptx,knottx,npty,knotty,s,t);
   for i=1:ndoftx
    for j=1:ndofty
     ij=(i-1)*ndoftx+j; % Columnwise arrangement of DOF w-w
     Ntx(m,n,ij)   = NurbsMNDLNt(1,i,j);  % 0th derivative
     Ntxdx(m,n,ij) = NurbsMNDLNt(2,i,j);   % x derivative
     Ntxdy(m,n,ij) = NurbsMNDLNt(3,i,j);   % y derivative
     Nty(m,n,ij)   = NurbsMNDLNt(1,i,j);  % 0th derivative
     Ntydx(m,n,ij) = NurbsMNDLNt(2,i,j);   % x derivative
     Ntydy(m,n,ij) = NurbsMNDLNt(3,i,j);   % y derivative
    end
   end
  end
end
kmat11 = zeros(ndof1111,ndof1111);
mmat11 = zeros(ndof1111,ndof1111);
lvecww = zeros(ndof1111,1);
for i=1:ndof1111
 for j=1:ndof1111
  for m=1:ngpx
   for n=1:ngpy
    % Stiffness
    kmat11(i,j) = kmat11(i,j)+Nwdx(m,n,j)/Jx*Dxz* ...
      Nwdx(m,n,i)/Jx*Jac*Gpwx(m,2)*Gpwy(n,2);
    kmat11(i,j) = kmat11(i,j)+Nwdy(m,n,j)/Jy*Dxz* ...
      Nwdy(m,n,i)/Jy*Jac*Gpwx(m,2)*Gpwy(n,2);
    % Mass
    mmat11(i,j) = mmat11(i,j)+Nw(m,n,j)*rho*Ixy* ...
          Nw(m,n,i)*Jac*Gpwx(m,2)*Gpwy(n,2);
    % Distributed Loading
    lvecww(j,1) = lvecww(j,1) + q0*Nw(m,n,j)*Jac*Gpwx(m,2)*
Gpwy(n,2);
   end
  end
 end
end
K(1:ndof1111,1:ndof1111)=kmat11; % K11 inclusion
M(1:ndof1111,1:ndof1111)=mmat11; % M11 inclusion
Lvec(1:ndof1111,1)=lvecww;     % Loading inclusion
% Stiffness and Mass Matrices K
% K12,K21,K13,K31 of w-txy and w-ty displacements and
rotations
kmat12 = zeros(ndof1111,ndof2233);
kmat21 = zeros(ndof2233,ndof1111);
kmat13 = zeros(ndof1111,ndof2233);
kmat31 = zeros(ndof2233,ndof1111);
for i=1:ndof1111
 for j=1:ndof2233
  for m=1:ngpx
```

```
    for n=1:ngpy
     % Stiffness
     kmat12(i,j) = kmat12(i,j)-Nwdy(m,n,i)/Jy*Dxz* ...
       Ntx(m,n,j)*Jac*Gpwx(m,2)*Gpwy(n,2);
     kmat21(j,i) = kmat21(j,i)-Nwdy(m,n,i)/Jy*Dxz* ...
       Ntx(m,n,j)*Jac*Gpwx(m,2)*Gpwy(n,2);
     kmat13(i,j) = kmat13(i,j)-Nwdx(m,n,i)/Jx*Dxz* ...
       Nty(m,n,j)*Jac*Gpwx(m,2)*Gpwy(n,2);
     kmat31(j,i) = kmat31(j,i)-Nwdx(m,n,i)/Jx*Dxz* ...
       Nty(m,n,j)*Jac*Gpwx(m,2)*Gpwy(n,2);
    end
   end
  end
end
K(1:ndof1111,ndof1111+1:ndof1111+ndof2233)=kmat12;        % K12
K(ndof1111+1:ndof1111+ndof2233,1:ndof1111)=kmat21;        % K21
K(1:ndof1111,ndof1111+ndof2233+1:ndof1111+ndof2233+ndof2233)=
kmat13; % K13
K(ndof1111+ndof2233+1:ndof1111+ndof2233+ndof2233,1:ndof1111)=
kmat31; % K31
% Stiffness and Mass Matrices K & M
% K22,K33 and M22,M33 of theta-x,y rotations
kmat22 = zeros(ndof2233,ndof2233);
kmat23 = zeros(ndof2233,ndof2233);
kmat32 = zeros(ndof2233,ndof2233);
kmat33 = zeros(ndof2233,ndof2233);
mmat22 = zeros(ndof2233,ndof2233);
mmat33 = zeros(ndof2233,ndof2233);
for i=1:ndof2233
 for j=1:ndof2233
  for m=1:ngpx
   for n=1:ngpy
    % Stiffness
    kmat22(i,j) = kmat22(i,j)+ ...
     Ntxdy(m,n,j)/Jy*D*Ntxdy(m,n,i)/Jy*Jac*Gpwx(m,2)*
Gpwy(n,2)+ ...
     Ntxdx(m,n,j)/Jx*Dxy*Ntxdx(m,n,i)/Jx*Jac*Gpwx(m,2)*
Gpwy(n,2)+...
     Ntx(m,n,j)*Dxz*Ntx(m,n,i)*Jac*Gpwx(m,2)*Gpwy(n,2);
    kmat23(i,j) = kmat23(i,j)+ ...
     Ntxdy(m,n,j)/Jy*Dxxyy*Ntydx(m,n,i)/Jx*Jac*Gpwx(m,2)*
Gpwy(n,2)+ ...
     Ntxdx(m,n,j)/Jx*Dxy*Ntydy(m,n,i)/Jy*Jac*Gpwx(m,2)*
Gpwy(n,2);
    kmat32(j,i) = kmat32(j,i)+ ...
     Ntydx(m,n,i)/Jx*Dxxyy*Ntxdy(m,n,j)/Jy*Jac*Gpwx(m,2)*
Gpwy(n,2)+ ...
     Ntydy(m,n,i)/Jy*Dxy*Ntxdx(m,n,j)/Jx*Jac*Gpwx(m,2)*
Gpwy(n,2);
```

```
    kmat33(i,j) = kmat33(i,j)+ ...
     Ntydx(m,n,j)/Jx*D*Ntydx(m,n,i)/Jx*Jac*Gpwx(m,2)*
Gpwy(n,2)+ ...
     Ntydy(m,n,j)/Jy*Dxy*Ntydy(m,n,i)/Jy*Jac*Gpwx(m,2)*
Gpwy(n,2)+...
     Nty(m,n,j)*Dxz*Nty(m,n,i)*Jac*Gpwx(m,2)*Gpwy(n,2);
    % Mass
    mmat22(i,j) = mmat22(i,j)+Ntx(m,n,j)*rho*I* ...
      Ntx(m,n,i)*Jac*Gpwx(m,2)*Gpwy(n,2);
    mmat33(i,j) = mmat33(i,j)+Nty(m,n,j)*rho*I* ...
      Nty(m,n,i)*Jac*Gpwx(m,2)*Gpwy(n,2);
   end
  end
 end
end
K(ndof1111+1:ndof1111+ndof2233,ndof1111+1:ndof1111+ndof2233)=
kmat22; % K22
K(ndof1111+1:ndof1111+ndof2233,ndof1111+ndof2233+1: ...
  ndof1111+ndof2233+ndof2233)=kmat23;                    % K23
K(ndof1111+ndof2233+1:ndof1111+ndof2233+ndof2233,ndof1111+1:
...
  ndof1111+ndof2233)=kmat32;                        % K32
K(ndof1111+ndof2233+1:ndof1111+ndof2233+ndof2233,ndof1111+
ndof2233+1: ...
  ndof1111+ndof2233+ndof2233)=kmat33;                    % K33
M(ndof1111+1:ndof1111+ndof2233,ndof1111+1:ndof1111+ndof2233)=
mmat22; % K22
M(ndof1111+ndof2233+1:ndof1111+ndof2233+ndof2233,ndof1111+
ndof2233+1: ...
  ndof1111+ndof2233+ndof2233)=mmat33;                    % M33
disp('Stiffness Matrix of Mindlin Plate');
disp(K);
disp('Mass Matrix of Mindlin Plate');
disp(M);
% Condensation to four-node plate element
i=2; j=3; k=34;
[KMC,MMC,FVC] = Condensation(K,M,Lvec,i,j,k); % Condensing
K11a
i=3; j=10; k=32;
[KMC,MMC,FVC] = Condensation(KMC,MMC,FVC,i,j,k); % Condensing
K11b
i=4; j=5; k=24;
[KMC,MMC,FVC] = Condensation(KMC,MMC,FVC,i,j,k); % Condensing
K11c
i=6; j=6; k=22;
[KMC,MMC,FVC] = Condensation(KMC,MMC,FVC,i,j,k); % Condensing
K22a
i=7; j=9; k=21;
[KMC,MMC,FVC] = Condensation(KMC,MMC,FVC,i,j,k); % Condensing
K22b
```

```
i=8; j=8; k=18;
[KMC,MMC,FVC] = Condensation(KMC,MMC,FVC,i,j,k); % Condensing
K22c
i=10; j=10; k=17;
[KMC,MMC,FVC] = Condensation(KMC,MMC,FVC,i,j,k); % Condensing
K33a
i=11; j=13; k=16;
[KMC,MMC,FVC] = Condensation(KMC,MMC,FVC,i,j,k); % Condensing
K33b
i=12; j=12; k=13;
[KMC,MMC,FVC] = Condensation(KMC,MMC,FVC,i,j,k); % Condensing
K33c
% Swapping the matrices
KMC=KMC([1,4,7,10,2,5,8,11,3,6,9,12],:); % row
KMC=KMC(:,[1,4,7,10,2,5,8,11,3,6,9,12]); % column
MMC=MMC([1,4,7,10,2,5,8,11,3,6,9,12],:); % row
MMC=MMC(:,[1,4,7,10,2,5,8,11,3,6,9,12]); % column
FVC=FVC([1,4,7,10,2,5,8,11,3,6,9,12],1); % row
% disp('The Stiffness Matrix after condensation');
% disp(KMC);
% disp('The Mass Matrix after condensation');
% disp(MMC);
% disp('The Loading Vector after condensation');
% disp(FVC);
% Assembling the four elements of the plate into global
coordinates
KGL=zeros(27,27);
MGL=zeros(27,27);
FGL=zeros(27,1);
% Element 1
L=[ 1 0 0 0 0 0 0 0 0 0 0 0 0 0 0 0 0 0 0 0 0 0 0 0 0 0 0;
  0 1 0 0 0 0 0 0 0 0 0 0 0 0 0 0 0 0 0 0 0 0 0 0 0 0 0;
  0 0 1 0 0 0 0 0 0 0 0 0 0 0 0 0 0 0 0 0 0 0 0 0 0 0 0;
  0 0 0 1 0 0 0 0 0 0 0 0 0 0 0 0 0 0 0 0 0 0 0 0 0 0 0;
  0 0 0 0 1 0 0 0 0 0 0 0 0 0 0 0 0 0 0 0 0 0 0 0 0 0 0;
  0 0 0 0 0 1 0 0 0 0 0 0 0 0 0 0 0 0 0 0 0 0 0 0 0 0 0;
  0 0 0 0 0 0 0 0 0 0 0 0 1 0 0 0 0 0 0 0 0 0 0 0 0 0 0;
  0 0 0 0 0 0 0 0 0 0 0 0 0 1 0 0 0 0 0 0 0 0 0 0 0 0 0;
  0 0 0 0 0 0 0 0 0 0 0 0 0 0 1 0 0 0 0 0 0 0 0 0 0 0 0;
  0 0 0 0 0 0 0 0 0 1 0 0 0 0 0 0 0 0 0 0 0 0 0 0 0 0 0;
  0 0 0 0 0 0 0 0 0 0 1 0 0 0 0 0 0 0 0 0 0 0 0 0 0 0 0;
  0 0 0 0 0 0 0 0 0 0 0 1 0 0 0 0 0 0 0 0 0 0 0 0 0 0 0;];
KGL=KGL+L'*KMC*L;
MGL=MGL+L'*MMC*L;
FGL=FGL+L'*FVC;
% ELement 2
L=[ 0 0 0 1 0 0 0 0 0 0 0 0 0 0 0 0 0 0 0 0 0 0 0 0 0 0 0;
  0 0 0 0 1 0 0 0 0 0 0 0 0 0 0 0 0 0 0 0 0 0 0 0 0 0 0;
  0 0 0 0 0 1 0 0 0 0 0 0 0 0 0 0 0 0 0 0 0 0 0 0 0 0 0;
  0 0 0 0 0 0 1 0 0 0 0 0 0 0 0 0 0 0 0 0 0 0 0 0 0 0 0;
```

```
  0 0 0 0 0 0 0 1 0 0 0 0 0 0 0 0 0 0 0 0 0 0 0 0 0 0 0;
  0 0 0 0 0 0 0 0 1 0 0 0 0 0 0 0 0 0 0 0 0 0 0 0 0 0 0;
  0 0 0 0 0 0 0 0 0 0 0 0 0 0 0 1 0 0 0 0 0 0 0 0 0 0 0;
  0 0 0 0 0 0 0 0 0 0 0 0 0 0 0 0 1 0 0 0 0 0 0 0 0 0 0;
  0 0 0 0 0 0 0 0 0 0 0 0 0 0 0 0 0 1 0 0 0 0 0 0 0 0 0;
  0 0 0 0 0 0 0 0 0 0 0 0 1 0 0 0 0 0 0 0 0 0 0 0 0 0 0;
  0 0 0 0 0 0 0 0 0 0 0 0 0 1 0 0 0 0 0 0 0 0 0 0 0 0 0;
  0 0 0 0 0 0 0 0 0 0 0 0 0 0 1 0 0 0 0 0 0 0 0 0 0 0 0;];
KGL=KGL+L'*KMC*L;
MGL=MGL+L'*MMC*L;
FGL=FGL+L'*FVC;
% Element 3
L=[ 0 0 0 0 0 0 0 0 0 1 0 0 0 0 0 0 0 0 0 0 0 0 0 0 0 0 0;
  0 0 0 0 0 0 0 0 0 0 1 0 0 0 0 0 0 0 0 0 0 0 0 0 0 0 0;
  0 0 0 0 0 0 0 0 0 0 0 1 0 0 0 0 0 0 0 0 0 0 0 0 0 0 0;
  0 0 0 0 0 0 0 0 0 0 0 0 1 0 0 0 0 0 0 0 0 0 0 0 0 0 0;
  0 0 0 0 0 0 0 0 0 0 0 0 0 1 0 0 0 0 0 0 0 0 0 0 0 0 0;
  0 0 0 0 0 0 0 0 0 0 0 0 0 0 1 0 0 0 0 0 0 0 0 0 0 0 0;
  0 0 0 0 0 0 0 0 0 0 0 0 0 0 0 0 0 0 0 0 0 1 0 0 0 0 0;
  0 0 0 0 0 0 0 0 0 0 0 0 0 0 0 0 0 0 0 0 0 0 1 0 0 0 0;
  0 0 0 0 0 0 0 0 0 0 0 0 0 0 0 0 0 0 0 0 0 0 0 1 0 0 0;
  0 0 0 0 0 0 0 0 0 0 0 0 0 0 0 0 0 0 1 0 0 0 0 0 0 0 0;
  0 0 0 0 0 0 0 0 0 0 0 0 0 0 0 0 0 0 0 1 0 0 0 0 0 0 0;
  0 0 0 0 0 0 0 0 0 0 0 0 0 0 0 0 0 0 0 0 1 0 0 0 0 0 0;];
KGL=KGL+L'*KMC*L;
MGL=MGL+L'*MMC*L;
FGL=FGL+L'*FVC;
% ELement 4
L=[ 0 0 0 0 0 0 0 0 0 0 0 0 1 0 0 0 0 0 0 0 0 0 0 0 0 0 0;
  0 0 0 0 0 0 0 0 0 0 0 0 0 1 0 0 0 0 0 0 0 0 0 0 0 0 0;
  0 0 0 0 0 0 0 0 0 0 0 0 0 0 1 0 0 0 0 0 0 0 0 0 0 0 0;
  0 0 0 0 0 0 0 0 0 0 0 0 0 0 0 1 0 0 0 0 0 0 0 0 0 0 0;
  0 0 0 0 0 0 0 0 0 0 0 0 0 0 0 0 1 0 0 0 0 0 0 0 0 0 0;
  0 0 0 0 0 0 0 0 0 0 0 0 0 0 0 0 0 1 0 0 0 0 0 0 0 0 0;
  0 0 0 0 0 0 0 0 0 0 0 0 0 0 0 0 0 0 0 0 0 0 0 0 1 0 0;
  0 0 0 0 0 0 0 0 0 0 0 0 0 0 0 0 0 0 0 0 0 0 0 0 0 1 0;
  0 0 0 0 0 0 0 0 0 0 0 0 0 0 0 0 0 0 0 0 0 0 0 0 0 0 1;
  0 0 0 0 0 0 0 0 0 0 0 0 0 0 0 0 0 0 0 0 0 1 0 0 0 0 0;
  0 0 0 0 0 0 0 0 0 0 0 0 0 0 0 0 0 0 0 0 0 0 1 0 0 0 0;
  0 0 0 0 0 0 0 0 0 0 0 0 0 0 0 0 0 0 0 0 0 0 0 1 0 0 0;];
KGL=KGL+L'*KMC*L;
MGL=MGL+L'*MMC*L;
FGL=FGL+L'*FVC;
% Eliminating fixed DOFs of the plate
KGL=KGL([5,12:15,18,23],:); % extracting free DOF rows
KGL=KGL(:,[5,12:15,18,23]); % extracting free DOF columns
MGL=MGL([5,12:15,18,23],:); % extracting free DOF rows
MGL=MGL(:,[5,12:15,18,23]); % extracting free DOF columns
FGL=FGL([5,12:15,18,23],1); % extracting free DOF rows
disp('Displacement Vector - Static Analysis');
```

```
wvec=KGL\FGL;
bw=b*elediv;
wbar=wvec(3)*E*h^3/bw^4/q0; % displacement at middle point of
plate
disp(wbar)
%
% Free Vibration Analysis (Eigenvalue Analysis)
[vec, lam2] = eig(KGL,MGL);
% Normalized Natural Frequency
lambda=sort(diag(sqrt(lam2)));
omegaN=lambda(1)*(a*elediv)^2/h*sqrt(rho/E);
disp('The First Normalized Natural Frequencies');
disp(omegaN)
```

Reference

Reddy JN (2002) *Energy Principles and Variational Methods in Applied Mechanic*. John Wiley & Sons, New Jersey.

9

Free Surface Plate Example

9.1 Free Surface Plate Analyzed by Using Condensed Isogeometric Analysis

In the Mindlin plate theory, the plane sections of the edges do not remain straight to the midplane during the deformation. The nonuniform rational basis spline (NURBS) surface shown in Figure 1.18 is used to model the free surface plate and solve the static and free vibration problems.

The second order of NURBS basis functions for the free surface, $p = 2$, with the given knot vectors of $\Xi = [-1 \ \ -1 \ \ -1 \ \ -0.2 \ \ 0.5 \ \ 1 \ \ 1 \ \ 1]$ and $H = [-1 \ \ -1 \ \ -1 \ \ -0.2 \ \ 0.5 \ \ 1 \ \ 1 \ \ 1]$ and uniform unit weight parameters $\mathbf{w}_{i,j} = \mathbf{1}$ is considered. The sampling points are created from the 5×5 control mesh given in Table 1.2. The basis functions that are used to create the NURBS surface are shown in Figure 1.17.

The degree of freedom (DOF) of the free surface plate before the condensation is shown in Figure 9.1.

In the isogeometric analysis (IGA) of the flat plate element in Chapter 8, the Jacobian operators are constant values. However, in the free surface plate element, the Jacobian operators depend on the terms of the derivative at a point on the plate. The Jacobian operators can be determined by the following equations (see Equations 1.19 and 2.7):

$$J_x = \frac{dx(\xi,\eta)}{d\xi} + \frac{dx(\xi,\eta)}{d\eta} = \sum_{i=0}^{m}\sum_{j=0}^{n}\left(\frac{dR_{i,j}^{p,q}(\xi,\eta)}{d\xi} + \frac{dR_{i,j}^{p,q}(\xi,\eta)}{d\eta}\right)P_{x(\xi,\eta)}$$

$$J_y = \frac{dy(\xi,\eta)}{d\xi} + \frac{dy(\xi,\eta)}{d\eta} = \sum_{i=0}^{m}\sum_{j=0}^{n}\left(\frac{dR_{i,j}^{p,q}(\xi,\eta)}{d\xi} + \frac{dR_{i,j}^{p,q}(\xi,\eta)}{d\eta}\right)P_{y(\xi,\eta)} \tag{9.1}$$

$$J = \det|J| = \det\begin{vmatrix} \frac{dx(\xi,\eta)}{d\xi} & \frac{dy(\xi,\eta)}{d\xi} \\ \frac{dx(\xi,\eta)}{d\eta} & \frac{dy(\xi,\eta)}{d\eta} \end{vmatrix} = \det\begin{vmatrix} \frac{dR_{i,j}^{p,q}(\xi,\eta)}{d\xi}P_{x(\xi,\eta)} & \frac{dR_{i,j}^{p,q}(\xi,\eta)}{d\xi}P_{y(\xi,\eta)} \\ \frac{dR_{i,j}^{p,q}(\xi,\eta)}{d\eta}P_{x(\xi,\eta)} & \frac{dR_{i,j}^{p,q}(\xi,\eta)}{d\eta}P_{y(\xi,\eta)} \end{vmatrix}$$

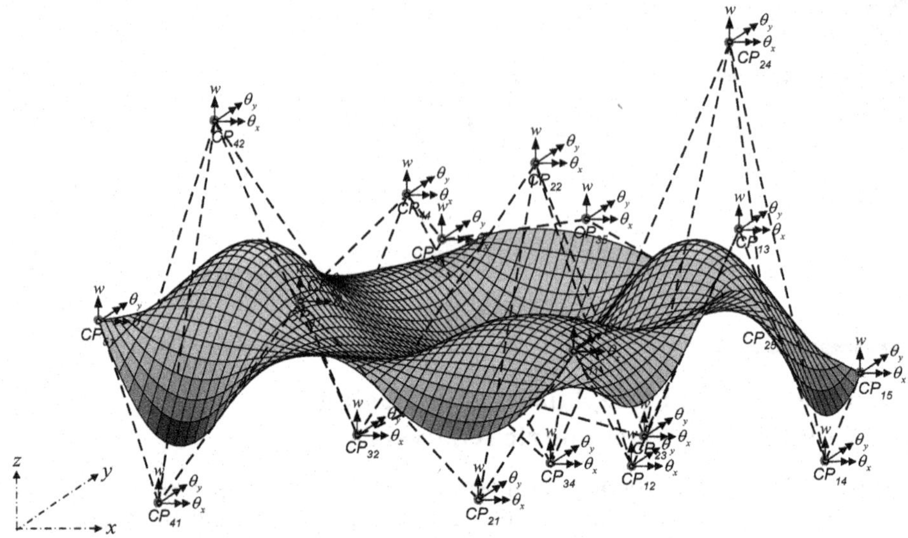

FIGURE 9.1
The DOF of the free surface Mindlin plate example before the condensation.

where $P_{x(\xi,\eta)}$ and $P_{y(\xi,\eta)}$ are the points on the control mesh which are used as the locations of the gauss points on the NURBS surface, which can be determined by solving the vector $\boldsymbol{\alpha}$ in Equation (1.20). It should be noted that all the program lists given in the following chapters are solved by using the least DOFs after following the sandwich condensation process. The standard procedure such as assembling technique and coordinate transformation method in general finite element method program will not be discussed here.

9.2 Static Problem

Consider the free surface plate, which is fixed on both edges of the plate, as shown in Figure 9.2. The geometry and property of the plate are depicted inside the figure. The degrees of the polynomial used in the NURBS surface functions are $p_w = 3, p_{\theta_x} = 2, p_{\theta_x} = 2$. The plate is subjected to the concentrated loadings $P = -10$, $M_x = 10$, and $M_y = -10$ at the front corner of the plate. One element with four Gauss integration points is used in the analysis.

The plate is analyzed by using the sandwich condensed IGA discussed in Chapter 7. Hence, there are a total of 12 DOFs, with 3 DOFs condensed at each of the four nodes. The free DOFs of vertical displacement and rotations are at the front corner of the plate.

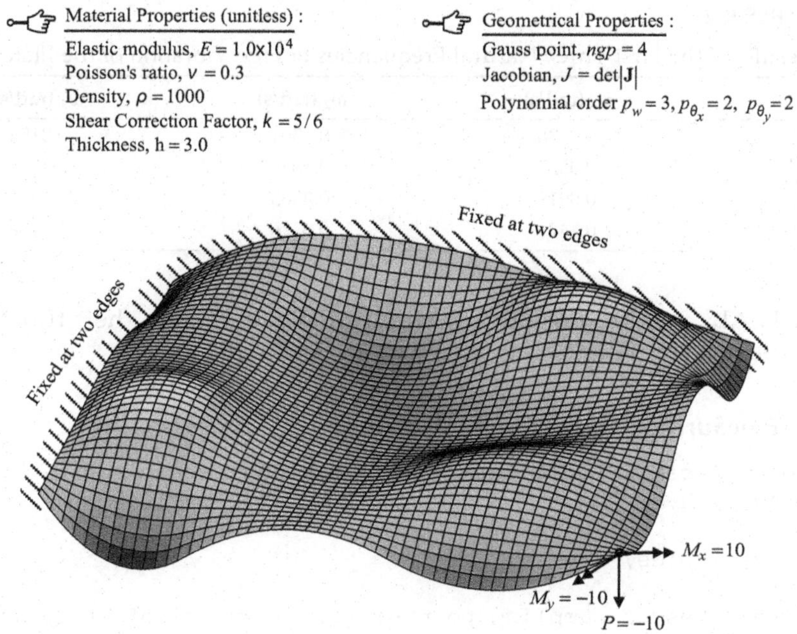

FIGURE 9.2
An isotropic free surface Mindlin plate example.

TABLE 9.1

Results of the Free Surface Mindlin Plates Subjected to Front Corner Loadings

h	w	θ_x (rad)	θ_y (rad)
0.3	−0.6370	3.6228	0.8263
0.5	−0.1652	0.8396	0.1717
1.0	−0.0271	0.1228	0.0217
2.0	−0.0049	0.0229	0.0031

Table 9.1 presents the static analysis results of the vertical displacements and rotations at the front corner of the free surface plate with varying thicknesses. The dimension of the plate is 14×5.

9.3 Free Vibration Problem

Consider the same free surface plate, as shown in Figure 9.2, as a free vibration problem. In this example, the natural frequencies of the plate are analyzed. Table 9.2 presents the first natural frequencies of the free vibration results of the plate with varying thicknesses.

TABLE 9.2

Results of the First Three Natural Frequencies of Free Vibration of the Plates

h	ω_1 (rad/s)	ω_2 (rad/s)	ω_3 (rad/s)
0.3	0.0338	0.1504	0.3154
0.5	0.0527	0.1854	0.4959
1.0	0.0917	0.3003	0.9245
2.0	0.1479	0.5376	1.5286

MATLAB® code FreeSurfaceMindlinPlate.m calls the functions NurbsSurface.m, Condensation.m, and Legendre.m.

9.3.1 FreeSurfaceMindlinPlate Program List

```
% FreeSurfaceMindlinPlate.m (Using NurbsSurface.m and
Condensation.m)
% R5
clear variables; clc;
%%
% Coordinates of sampling points: xi, yi and zi of NURBS free
surface
xi = [linspace(-5,5,5) linspace(-5,5,5) linspace(-5,5,5) ...
      linspace(-5,5,5) linspace(-5,5,5)];
yi = [linspace(-5,-5,5) linspace(-2.5,-2.5,5) linspace(0,0,5)
...
      linspace(2.5,2.5,5) linspace(5,5,5)];
rng(400); zi=randi([-20 25],1,25);
p=2; m=0;
q=2; n=0;
xyzi=transpose([xi' yi' zi'])';
% Parameter ksi=s and eta=t
knotksi=[-1 -1 -1 -0.2 0.5 1 1 1 ones(1,p)]; % ones p-dummy
knoteta=[-1 -1 -1 -0.2 0.5 1 1 1 ones(1,q)]; % ones q-dummy
ns=size(knotksi,2)-p-1;
nt=size(knoteta,2)-q-1;
% Gauss
ngpx  = 4;                    ngpy  = 4;
Gpwx  = Legendre(ngpx);       Gpwy  = Legendre(ngpy);
stshape=zeros(ns-p,nt-q,ngpx,ngpy);
for m=1:ngpx
    s = Gpwx(m,1);
    for n=1:ngpy
      t = Gpwy(n,1);
      NurbsSFC = NurbsSurface(p,knotksi,q,knoteta,s,t);
      for i=1:ns-p          % order of NURBS in ksi
        for j=1:nt-q        % order of NURBS in eta
          stshape(i,j,m,n) = NurbsSFC(1,i,j); % Rij at gauss
points
```

```
        end
      end
    end
end
% [T] matrix
st=reshape(stshape,[ngpx*ngpy,(ns-p)*(nt-q)])';
% Control Mesh Px's, Py's and Pz's calculation
pxyz=st\xyzi;
%%
% Solving Static and Free Vibration of a free surface Mindlin
plate
% Based on the Mindlin Plate Theory using NURBS surface shape
functions
h     = 2.0;
E     = 1.0e4;       nu    = 0.3;      rho   = 1000.0;
I     = h^3/12;      Ixy   = h;        kappa = 5/6;
D     = E*h^3/12/(1-nu^2);
Dxxyy = nu*E*h^3/6/(1-nu^2);
Dxy   = E*h^3/24/(1+nu);
G = E/2/(1+nu);
Dxz = kappa*G*h;
% Concentrated Loadings
P2  = -10.0;
M2x =  10.0;
M2y = -10.0;
% NURBS for w and theta displacement data in x and y
directions
npw    = 3;
nptx   = 2;
npty   = 2;
knotw  = [-ones(1,npw+1) ones(1,npw+1)]; knotw = [knotw
ones(1,npw)];
knottx = [-ones(1,nptx+1) ones(1,nptx+1)]; knottx = [knottx
ones(1,nptx)];
knotty = [-ones(1,npty+1) ones(1,npty+1)]; knotty = [knotty
ones(1,npty)];
nbw    = size(knotw,2)-npw-1;
nbtx   = size(knottx,2)-nptx-1;
nbty   = size(knotty,2)-npty-1;
% Initialization
ndofw      = (nbw-npw);
ndoftx     = (nbtx-nptx);
ndofty     = (nbty-npty);
ndof1111   = ndofw*ndofw;
ndof2233   = ndoftx*ndofty;
K     = zeros(ndof1111+ndof2233+ndof2233,ndof1111+ndof2233+n
dof2233);
M     = zeros(ndof1111+ndof2233+ndof2233,ndof1111+ndof2233+n
dof2233);
% Loading Vector
```

```
Lvec=zeros(ndof1111+ndof2233+ndof2233,1);
% Shape functions of NURBS surface
Nw    = zeros(ngpx,ngpy,ndof1111);
Nwdx  = zeros(ngpx,ngpy,ndof1111);
Nwdy  = zeros(ngpx,ngpy,ndof1111);
Ntx   = zeros(ngpx,ngpy,ndof2233);
Ntxdx = zeros(ngpx,ngpy,ndof2233);
Ntxdy = zeros(ngpx,ngpy,ndof2233);
Nty   = zeros(ngpx,ngpy,ndof2233);
Ntydx = zeros(ngpx,ngpy,ndof2233);
Ntydy = zeros(ngpx,ngpy,ndof2233);
% Jacobian operators
Jw    = zeros(ngpx,ngpy);
Jwx   = zeros(ngpx,ngpy);
Jwy   = zeros(ngpx,ngpy);
Jt    = zeros(ngpx,ngpy);
Jtx   = zeros(ngpx,ngpy);
Jty   = zeros(ngpx,ngpy);
% Stiffness and Mass Matrices K & M
% K11 and M11 of w displacements
% Filling NURBS surface shape functions into the gauss points
for m=1:ngpx
    s = Gpwx(m,1);
    for n=1:ngpy
      t = Gpwy(n,1);
      mn=(m-1)*ngpx+n; % Columnwise arrangement of Gauss
points
      % Shape functions for w
      NurbsMNDLNw = NurbsSurface(npw,knotw,npw,knotw,s,t);
      dxdxi = 0; dxdeta = 0; dydxi = 0; dydeta = 0;
      for i=1:ndofw
        for j=1:ndofw
          ij=(i-1)*ndofw+j; % Columnwise arrangement of DOF
w-w
          Nw(m,n,ij)      = NurbsMNDLNw(1,i,j);   % 0th
derivative
          Nwdx(m,n,ij)    = NurbsMNDLNw(2,i,j);   % x
derivative
          Nwdy(m,n,ij)    = NurbsMNDLNw(3,i,j);   % y
derivative
          % Jt(m,n)obian Operators
          dxdxi           = dxdxi+Nwdx(m,n,ij)*pxyz(mn,1); % dx
          dxdeta          = dxdeta+Nwdy(m,n,ij)*pxyz(mn,1); %
dx
          dydxi           = dydxi+Nwdx(m,n,ij)*pxyz(mn,2); % dx
          dydeta          = dydeta+Nwdy(m,n,ij)*pxyz(mn,2); %
dx
        end
      end
      Jwx(m,n) = dxdxi+dxdeta; Jwx(m,n) = 14.0/2;
```

```
      Jwy(m,n) = dydxi+dydeta; Jwy(m,n) =  5.0/2;
      Jw(m,n)  = abs(det(dxdxi*dydeta-dxdeta*dydxi));
      Jw(m,n)  = Jwx(m,n)*Jwy(m,n);
      % Shape functions for theta-x and theta-y
      NurbsMNDLNt = NurbsSurface(nptx,knottx,npty,knotty,s,t);
      dxdxi = 0; dxdeta = 0; dydxi = 0; dydeta = 0;
      for i=1:ndoftx
        for j=1:ndofty
          ij=(i-1)*ndoftx+j; % Columnwise arrangement of DOF
w-w
          Ntx(m,n,ij)    = NurbsMNDLNt(1,i,j);    % 0th
derivative
          Ntxdx(m,n,ij)  = NurbsMNDLNt(2,i,j);    % x
derivative
          Ntxdy(m,n,ij)  = NurbsMNDLNt(3,i,j);    % y
derivative
          Nty(m,n,ij)    = NurbsMNDLNt(1,i,j);    % 0th
derivative
          Ntydx(m,n,ij)  = NurbsMNDLNt(2,i,j);    % x
derivative
          Ntydy(m,n,ij)  = NurbsMNDLNt(3,i,j);    % y
derivative
          % Jt(m,n)obian Operators
          dxdxi          = dxdxi+Ntxdx(m,n,ij)*pxyz(mn,1); % dx
          dxdeta         = dxdeta+Ntxdy(m,n,ij)*pxyz(mn,1); %
dx
          dydxi          = dydxi+Ntydx(m,n,ij)*pxyz(mn,2); % dx
          dydeta         = dydeta+Ntydy(m,n,ij)*pxyz(mn,2); %
dx
        end
      end
      Jtx(m,n) = dxdxi+dxdeta;  Jtx(m,n) = 14.0/2;
      Jty(m,n) = dydxi+dydeta;  Jty(m,n) =  5.0/2;
      Jt(m,n)  = det(dxdxi*dydeta-dxdeta*dydxi);
      Jt(m,n)  = Jtx(m,n)*Jty(m,n);
    end
end
kmat11 = zeros(ndof1111,ndof1111);
mmat11 = zeros(ndof1111,ndof1111);
for i=1:ndof1111
  for j=1:ndof1111
    for m=1:ngpx
      for n=1:ngpy
        % Stiffness
        kmat11(i,j) = kmat11(i,j)+Nwdx(m,n,j)/Jwx(m,n)*Dxz*
...
          Nwdx(m,n,i)/Jwx(m,n)*Jw(m,n)*Gpwx(m,2)*Gpwy(n,2);
        kmat11(i,j) = kmat11(i,j)+Nwdy(m,n,j)/Jwy(m,n)*Dxz*
...
          Nwdy(m,n,i)/Jwy(m,n)*Jw(m,n)*Gpwx(m,2)*Gpwy(n,2);
```

```
        % Mass
        mmat11(i,j) = mmat11(i,j)+Nw(m,n,j)*rho*Ixy* ...
                      Nw(m,n,i)*Jw(m,n)*Gpwx(m,2)*Gpwy(n,2);
      end
    end
  end
end
K(1:ndof1111,1:ndof1111)=kmat11; % K11 inclusion
M(1:ndof1111,1:ndof1111)=mmat11; % M11 inclusion
% Stiffness and Mass Matrices K
% K12,K21,K13,K31 of w-txy and w-ty displacements and
rotations
kmat12 = zeros(ndof1111,ndof2233);
kmat21 = zeros(ndof2233,ndof1111);
kmat13 = zeros(ndof1111,ndof2233);
kmat31 = zeros(ndof2233,ndof1111);
for i=1:ndof1111
  for j=1:ndof2233
    for m=1:ngpx
      for n=1:ngpy
        % Stiffness
        kmat12(i,j) = kmat12(i,j)-Nwdy(m,n,i)/Jwy(m,n)*Dxz*
...
            Ntx(m,n,j)*Jw(m,n)*Gpwx(m,2)*Gpwy(n,2);
        kmat21(j,i) = kmat21(j,i)-Nwdy(m,n,i)/Jwy(m,n)*Dxz*
...
            Ntx(m,n,j)*Jw(m,n)*Gpwx(m,2)*Gpwy(n,2);
        kmat13(i,j) = kmat13(i,j)-Nwdx(m,n,i)/Jwx(m,n)*Dxz*
...
            Nty(m,n,j)*Jw(m,n)*Gpwx(m,2)*Gpwy(n,2);
        kmat31(j,i) = kmat31(j,i)-Nwdx(m,n,i)/Jwx(m,n)*Dxz*
...
            Nty(m,n,j)*Jw(m,n)*Gpwx(m,2)*Gpwy(n,2);
      end
    end
  end
end
K(1:ndof1111,ndof1111+1:ndof1111+ndof2233)=kmat12;
% K12
K(ndof1111+1:ndof1111+ndof2233,1:ndof1111)=kmat21;
% K21
K(1:ndof1111,ndof1111+ndof2233+1:ndof1111+ndof2233+ndof2233)=k
mat13;  % K13
K(ndof1111+ndof2233+1:ndof1111+ndof2233+ndof2233,1:ndof1111)=k
mat31;  % K31
% Stiffness and Mass Matrices K & M
% K22,K33 and M22,M33 of theta-x,y rotations
kmat22 = zeros(ndof2233,ndof2233);
kmat23 = zeros(ndof2233,ndof2233);
kmat32 = zeros(ndof2233,ndof2233);
```

```
kmat33 = zeros(ndof2233,ndof2233);
mmat22 = zeros(ndof2233,ndof2233);
mmat33 = zeros(ndof2233,ndof2233);
for i=1:ndof2233
  for j=1:ndof2233
    for m=1:ngpx
      for n=1:ngpy
        % Stiffness
        kmat22(i,j) = kmat22(i,j)+ ...
          Ntxdy(m,n,j)/Jty(m,n)*D*Ntxdy(m,n,i)/Jty(m,n)*...
              Jt(m,n)*Gpwx(m,2)*Gpwy(n,2)+ ...
          Ntxdx(m,n,j)/Jtx(m,n)*Dxy*Ntxdx(m,n,i)/Jtx(m,n)*...
              Jt(m,n)*Gpwx(m,2)*Gpwy(n,2)+...
          Ntx(m,n,j)*Dxz*Ntx(m,n,i)*Jt(m,n)*Gpwx(m,2)*Gpw
y(n,2);
        kmat23(i,j) = kmat23(i,j)+ ...
          Ntxdy(m,n,j)/Jty(m,n)*Dxxyy*Ntydx(m,n,i)/
Jtx(m,n)*...
              Jt(m,n)*Gpwx(m,2)*Gpwy(n,2)+ ...
          Ntxdx(m,n,j)/Jtx(m,n)*Dxy*Ntydy(m,n,i)/Jty(m,n)*...
              Jt(m,n)*Gpwx(m,2)*Gpwy(n,2);
        kmat32(j,i) = kmat32(j,i)+ ...
          Ntydx(m,n,i)/Jtx(m,n)*Dxxyy*Ntxdy(m,n,j)/
Jty(m,n)*...
              Jt(m,n)*Gpwx(m,2)*Gpwy(n,2)+ ...
          Ntydy(m,n,i)/Jty(m,n)*Dxy*Ntxdx(m,n,j)/Jtx(m,n)*...
              Jt(m,n)*Gpwx(m,2)*Gpwy(n,2);
        kmat33(i,j) = kmat33(i,j)+ ...
          Ntydx(m,n,j)/Jtx(m,n)*D*Ntydx(m,n,i)/Jtx(m,n)*...
              Jt(m,n)*Gpwx(m,2)*Gpwy(n,2)+ ...
          Ntydy(m,n,j)/Jty(m,n)*Dxy*Ntydy(m,n,i)/Jty(m,n)*...
              Jt(m,n)*Gpwx(m,2)*Gpwy(n,2)+...
          Nty(m,n,j)*Dxz*Nty(m,n,i)*Jt(m,n)*Gpwx(m,2)*Gpw
y(n,2);
        % Mass
        mmat22(i,j) = mmat22(i,j)+Ntx(m,n,j)*rho*I* ...
            Ntx(m,n,i)*Jt(m,n)*Gpwx(m,2)*Gpwy(n,2);
        mmat33(i,j) = mmat33(i,j)+Nty(m,n,j)*rho*I* ...
            Nty(m,n,i)*Jt(m,n)*Gpwx(m,2)*Gpwy(n,2);
      end
    end
  end
end
K(ndof1111+1:ndof1111+ndof2233,ndof1111+1:ndof1111+ndof2233)=k
mat22;  % K22
K(ndof1111+1:ndof1111+ndof2233,ndof1111+ndof2233+1: ...
    ndof1111+ndof2233+ndof2233)=kmat23;
% K23
K(ndof1111+ndof2233+1:ndof1111+ndof2233+ndof2233,ndof1111+1:
...
```

```
    ndof1111+ndof2233)=kmat32;
% K32
K(ndof1111+ndof2233+1:ndof1111+ndof2233+ndof2233,ndof1111+n
dof2233+1: ...
    ndof1111+ndof2233+ndof2233)=kmat33;
% K33
M(ndof1111+1:ndof1111+ndof2233,ndof1111+1:ndof1111+ndof2233)=m
mat22;  % K22
M(ndof1111+ndof2233+1:ndof1111+ndof2233+ndof2233,ndof1111+n
dof2233+1: ...
    ndof1111+ndof2233+ndof2233)=mmat33;
% M33
% disp('Stiffness Matrix of Mindlin Plate');
% disp(K);
% disp('Mass Matrix of Mindlin Plate');
% disp(M);
% Distributed Loading : None
% Condensation to four-node plate element
i=2; j=3; k=34;
[KMC,MMC,FVC] = Condensation(K,M,Lvec,i,j,k); % Condensing
K11a
i=3; j=10; k=32;
[KMC,MMC,FVC] = Condensation(KMC,MMC,FVC,i,j,k); % Condensing
K11b
i=4; j=5; k=24;
[KMC,MMC,FVC] = Condensation(KMC,MMC,FVC,i,j,k); % Condensing
K11c
i=6; j=6; k=22;
[KMC,MMC,FVC] = Condensation(KMC,MMC,FVC,i,j,k); % Condensing
K22a
i=7; j=9; k=21;
[KMC,MMC,FVC] = Condensation(KMC,MMC,FVC,i,j,k); % Condensing
K22b
i=8; j=8; k=18;
[KMC,MMC,FVC] = Condensation(KMC,MMC,FVC,i,j,k); % Condensing
K22c
i=10; j=10; k=17;
[KMC,MMC,FVC] = Condensation(KMC,MMC,FVC,i,j,k); % Condensing
K33a
i=11; j=13; k=16;
[KMC,MMC,FVC] = Condensation(KMC,MMC,FVC,i,j,k); % Condensing
K33b
i=12; j=12; k=13;
[KMC,MMC,FVC] = Condensation(KMC,MMC,FVC,i,j,k); % Condensing
K33c
% Swapping the matrices
KMC=KMC([1,4,7,10,2,5,8,11,3,6,9,12],:); % row
KMC=KMC(:,[1,4,7,10,2,5,8,11,3,6,9,12]); % column
MMC=MMC([1,4,7,10,2,5,8,11,3,6,9,12],:); % row
MMC=MMC(:,[1,4,7,10,2,5,8,11,3,6,9,12]); % column
```

```
FVC=FVC([1,4,7,10,2,5,8,11,3,6,9,12],1); % row
% disp('The Stiffness Matrix after condensation');
% disp(KMC);
% disp('The Mass Matrix after condensation');
% disp(MMC);
% disp('The Loading Vector after condensation');
% disp(FVC);
% Eliminating fixed DOFs of the plate
KMC=KMC([4:6],:); % extracting free DOF rows
KMC=KMC(:,[4:6]); % extracting free DOF columns
MMC=MMC([4:6],:);  % extracting free DOF rows
MMC=MMC(:,[4:6]); % extracting free DOF columns
FVC=FVC([4:6],1); % extracting free DOF rows
% Concentrated Loadings
FVC(1)=P2;
FVC(2)=M2x;
FVC(3)=M2y;
disp('Displacements Vector - Static Analysis');
wvec=KMC\FVC;
% displacement at the front corner of the plate
disp(wvec)
%
% Free Vibration Analysis (Eigenvalue Analysis)
[vec, lam2] = eig(KMC,MMC);
% Normalized Natural Frequency
lambda=sort(diag(sqrt(lam2)));
disp('The First Three Natural Frequencies');
disp(lambda(1:3)).
```

Index

J

K

L

M

N

P